NETWORK DESIGN

Management and Technical Perspectives

Teresa C. Mann-Rubinson
Kornel Terplan

CRC Press

Boca Raton   London   New York   Washington, D.C.

Library of Congress Cataloging-in-Publication Data

Rubinson, Teresa.
    Network design : management and technical perspectives / Teresa
Rubinson, Kornel Terplan.
        p. cm.
    Includes bibliographical references and index.
    ISBN 0-8493-3404-7
    1. Telecommunication. 2. Computer networks. I. Terplan, Kornel.
II. Title.
TK5101.R73 1998
004.6--dc21   98-6583
                                                                         CIP

No claim to original U.S. Government works
International Standard Book Number 0-8493-3404-7
Library of Congress Card Number 98-6583
Printed in the United States of America 1 2 3 4 5 6 7 8 9 0
Printed on acid-free paper

Preface

The objective of this book is to introduce managers and technical professionals in the telecommunications field to the fundamental principles and analytical techniques used in designing advanced communication networks. It is also designed to help managers and technical professionals better understand and appreciate each other's perspective in the network design process. Thus, this book is intended to help bridge the communication gap that often exists between managers and technical staff involved in the design and implementation of networks.

Network managers need a sound grounding in basic design principles to effectively manage, plan, and assess the plethora of new technologies and equipment available for designing networks. They must also understand how requirements should be formulated and specified for design engineers. In turn, network designers and engineers need a sound grounding in basic management principles to fully understand how organizational requirements are best reflected in design recommendations.

In this book, we emphasize basic network design principles and analytical approaches that will survive the test of time, independent of a specific implementation of technology. However, to illustrate how basic principles can be applied in the context of realistic network design problems, we also present specific implementations of technology (such as ATM, etc.) through the use of case studies. In the case studies, both management and technical considerations are discussed. This approach is used to demystify the design process, which traditionally — on anything other than an ad hoc basis — has been limited to the purview of highly trained and specialized professionals. By describing the lingua franca of both managers and design engineers, it is hoped that each will gain a better understanding of the total network design process.

The material for this book is based on our experiences both as industry consultants and as professors teaching graduate network design and management courses. In each of these roles, we have to deal with network managers and with hands-on network designers. We frequently observe suboptimal communication between these two groups of professionals. For example, network managers frequently express a lack of confidence in evaluating whether or not organizational objectives will be satisfied by the recommendations proposed by design engineers. Design engineers, on the other hand, voice concerns that "management does not have any idea of what I'm doing and what is involved," and thus lack confidence that the goals established by management are realistic. Network managers have told us that they need to understand the technical jargon and basic approaches that network designers use so they can evaluate vendor offerings and staff recommendations. Managers want this perspective without getting too immersed in technical details. Network implementers have told us that they need and want practical advice on how to apply sound design principles in the context of a realistic design scenario, when organizational, budgetary, political, and other considerations must enter into the design process. Thus, one of our objectives is to help management and design professionals work together towards achieving their respective goals in the network design process.

Teresa C. Mann-Rubinson
Kornel Terplan

About the authors

Teresa C. Mann-Rubinson is Vice President of TCR, Inc., a management consulting and software development company specializing in advanced technologies.

Prior to TCR, Inc., Dr. Rubinson held a variety of senior management and technical positions at Pepsico, IT&T, Boehringer Ingelheim, and Pitney Bowes, working in both MIS and R&D environments to design state-of-the-art systems incorporating database, networking, and artificial intelligence technologies. In addition, she has provided consulting services to numerous government agencies and Fortune 500 companies.

Dr. Rubinson has been a pioneer of emerging technologies throughout her career, introducing artificial intelligence and information management techniques to many of the companies with which she has worked. She has published and lectured extensively, particularly in the areas of network design, fuzzy logic, and advanced decision support techniques.

Dr. Rubinson holds a B.S. degree from the University of Illinois, an M.B.A. in management information systems from Iona College, and a Ph.D. in industrial engineering and operations research from Polytechnic University. She completed a postdoctoral fellowship at the Man-Machine Intelligence Institute in New Rochelle, New York.

She is a Professor (Adjunct) at Polytechnic University in the Telecommunications and Information Management Executive Degree Program and teaches courses relating to telecommunications and network design, the Internet, and emerging technologies. She is a member of IEEE, INFORMS, and the National Academy of Sciences.

Kornel Terplan is a telecommunications expert with more than 30 years of highly successful multinational consulting experience. His books, *Communication Networks Management* (Prentice Hall) and *Effective Management of Local Area Networks* (McGraw-Hill), both in second editions, are viewed as state-of-the-art compendiums in the international community of corporate users.

Dr. Terplan has provided consulting, training, and product development services to over 75 national and international corporations on 4 continents, following a scholarly career that combined the writing of some 140 articles, 16 books, and 115 papers while serving on editorial boards.

Dr. Terplan received his doctoral degree at the University of Dresden and completed advanced studies, researched, and lectured at Berkeley, Stanford University, the University of California at Los Angeles, and Rensselaer Polytechnic Institute. Over the last 20 years, he has designed 5 network management related seminars and made some 80 seminar presentations in 15 countries.

Dr. Terplan has consulted on network management products and services, outsourcing, central administration of LANs, network management centers, strategy of network management integration, implementation of network design and planning guidelines, product comparisons, and benchmarking network management solutions. His clients include AT&T, GTE, Walt Disney World, Boole and Babbage, BMW, Kaiser Permanente, Siemens, France Telecom, German Telekom, Commerzbank, Union Bank of Switzerland, Creditanstalt Austria, Swiss Credit, State of Washington, Georgia Pacific Corporation, Objective Systems Integrators, Unisource, and the Hungarian Telecommunication Company.

Dr. Terplan is currently Industry Professor at Brooklyn Polytechnic University and at Stevens Institute of Technology.

Acknowledgments

Any undertaking such as this gives one pause to reflect on how much one owes to others. In particular, we would like to acknowledge the following people, without whose contributions the completion of this textbook would not have been possible.

We owe a substantial debt of gratitude to Byron Piliouras, Senior Engineer and Group Leader for MCI Telecommunications' Private Network Engineering Department. Mr. Piliouras, a specialist in global frame relay and ATM network designs, made significant contributions to Chapters 3 and 4, including the case studies and their associated illustrations, the section on SONET, and the closing remarks. In addition, he provided ongoing feedback and technical assistance throughout the preparation of this manuscript. Mr. Piliouras holds a bachelor's degree in electrical engineering from SUNY College of Technology and a masters' degree in telecommunications and computing management from Polytechnic University. He is a member of the IEEE society.

We also wish to thank Mr. Andrew Resnick, a Project Engineer working in communications engineering/planning and development for the Securities Industry Automation Corporation (SIAC). Chapter 6 on Intranets was adapted from the thesis he prepared as part of the requirement for his completion of a graduate degree in telecommunications and computing management at Polytechnic University.

We wish to acknowledge the assistance of David Rubin, President of NAC. Mr. Rubin generously supplied us with realistic network design case studies and use of NAC's premier MIND™ design tool. NAC has provided communications networking pricing and analysis software and services since 1969. In addition to network consulting services, NAC develops and maintains a suite of software products used by over 200 communication service providers, major corporations, and government agencies. Mr. Rubin has conducted many courses on data communications for McGraw Hill, DataPro, and numerous customers, and has authored publications in *Data Communications Magazine, Communications Week,* and many others. He provided many invaluable insights during the preparation of this book.

We also want to thank Mr. Paul Serrano, President and CEO of Network Tools, for providing graphical screen displays and information on his company's Caliper™ and Chisel™ products. Network Tools offers products that are designed for both vendors and end users to help them evaluate competing network designs and product options.

We also want to acknowledgment the many useful comments, suggestions, and proofreading help provided by the following individuals: Audrey Gatling, Brian Miller, Bob Hobson, Kurt Brown, Delia Marino, Ed Kamp, Erick Blanc, Howard Phillip, James Allocco, Simon Tsang, Stephen Martin, Thomas Masotto, Ted Wilk, Anthony Maisano, Anthony Ferraro, Carl Shapiro, Richard Stubits, John Benjamin, Joji Joseph, L. D. Washington, Hormazd Mistry, Cliffton Bogues, Coram Rimes, Dennis Brugger, Debra James-Phillip, Jay Schwartzberg, Martha Harper, Dorothy Wu, and Kim McClain.

Teresa Rubinson also wishes to thank her doctoral advisor and mentor, Dr. Aaron Kershenbaum, Professor of Computer Science at Polytechnic University, and Dr. Robert Cahn, of the IBM T.J. Watson Research Laboratory. Many of the techniques used to design networks were pioneered and developed by these individuals, and the field as a whole has been profoundly impacted by their ongoing accomplishments and contributions. She is especially grateful for their continual support and assistance in keeping current with the latest trends in network design.

Finally, she wishes to thank Barry Rubinson for his encouragement and support during this endeavor.

Contents

chapter one

Making the business case for the network

Contents

1.1   Management overview of network design

The basic goal of network design is to assign links between various hardware devices so that resources can be shared and distributed. Despite the apparent simplicity of this goal, network design is a very complex task that involves balancing a multitude of managerial and technical considerations. In this chapter, we focus on the managerial considerations involved in planning and designing a network.

Business concerns and philosophy have a profound impact on the network planning process. In some organizations, expenses associated with the network are viewed as "overhead." Taken to an extreme, this view can lead to the conclusion that anything but the most basic expenditures on the network are superfluous to the primary business. In this type of environment, it is not uncommon to observe a lack of formal commitment and managerial sponsorship of the network. End users, acting on their own initiative, may buy and install network components and software without working with any central budgeting and planning authority. Although formal budget allocations for the network may not exist, the network is not "free." An informal, reactive network planning process is often associated with frequent downtime, since the network can be dismantled and/or changed at whim, without regard to how the changes might impact the people using it. Thus, there is an opportunity cost associated with the lost staff productivity resulting from network downtime. There is also an opportunity cost associated with lost productivity resulting from the diversion of staff from their primary job function to support the network. Furthermore, when changes to the network are not carefully planned and implemented, the difficulty of maintaining the network in a cost-effective way is compounded. In organizations of any substantial size lacking a formal network planning process, it is not uncommon for audits to reveal that millions of dollars have been invested in technology that is not effectively used and that cannot evolve to support ongoing organizational needs. Although this is network design at its worst, it is not that uncommon. The moral of this is that ignoring or avoiding direct consideration of the true network costs does not make problems go away. A see-no-evil/hear-no-evil/speak-no-evil strategy "works" only when decision-makers are not accountable for their actions. In a cost-conscious, competitive business climate that focuses increasing scrutiny on inefficient processes ripe for reengineering, this is a risky approach.

In contrast, there are organizations that view the network as the corporate lifeblood. In this environment there is considerable management accountability for and scrutiny of network planning and implementation. Increased recognition of the network's importance to the bottom line improves the chances that the network(s) will be well planned and executed. For example, it is vital to the New York Stock Exchange that its networks perform reliably even under conditions of extreme stress. High-profile, high-performance networks require thorough planning to ensure that they can meet the demands that are placed on them. A systems approach is essential to ensuring a comprehensive assessment of critical network requirements.

A systems approach means that the requirements are considered from a global perspective that encompasses both top-down and bottom-up views. In the discussion that follows, we outline a general methodology for performing a systems analysis of the network requirements from a business perspective. The business perspective is a top-down, big-picture view of how the design will impact the organization. We continue this discussion, from a technical perspective, in the chapters that follow. The technical perspective is a bottom-up, narrow focus view concentrating on essential design details.

1.1.1   Define the business objectives

A logical start to the top-down analysis is to define the business objectives to be served by the network. The business objectives should relate to the strategic focus of the organization. There may be many motivations for building and implementing a network. After the business objectives have been made explicit, they can be prioritized, and objective criteria can be developed for measuring the success of network implementation. The objectives will also help to determine the type of network needed and the level of expenditure and support that is appropriate. The business objectives have many impacts on technical decisions regarding the selection of technology, performance requirements, and required resource commitments. When the objectives are poorly defined, there are often many subsequent complications. Thus, defining the business objectives is a vital first step in the network planning process.

The findings of this analysis should be formalized in a written document. We recommend that the following sections be considered for inclusion in this report:

- Management summary
- Business objectives of the network
- Major functional units and staff affecting and affected by the proposed network
- Project infrastructure
  - Project sponsor
  - Project manager
  - Team members
  - Support services and facilities

1.1.2   Determine major potential risks and dependencies

After the business objectives of the design project are well understood, their feasibility should be evaluated in light of the proposed network project. The feasibility of the project should be considered relative to the availability of technology. The use of new and emerging technologies is inherently more risky than the use of proven, well-established technologies and may cause unforeseen delays and/or expense. Another important feasibility concern relates to the organization's experience with similar projects. Often, the network implementation will run into difficulty because the organization is not well positioned to support network maintenance. The ability to manage and maintain the network once it is in place is an important determinant of whether or not the project will succeed. Other risks that should be considered include the possible failure of the project and the resulting impacts on the business. Impacts can be tangible (e.g., financial impacts due to lost profits, budget changes, etc.) or intangible (e.g., impacts due to loss of customer goodwill, industry perceptions, etc.). Strategies to manage these risks should be developed as early on in the project as possible.

A preliminary cost-benefit analysis should be performed to determine the expenditure levels and payback time periods required for a viable project. A recent *Datapro* report shows that about 60% of technology related projects, including network projects, are significantly over budget [Data97, p.1]. In part, this reflects the difficulty of properly estimating costs, particularly when new technologies are being used or a new system is being implemented. It also reflects the fact that in many organizations it is difficult to develop balanced and objective cost estimates. Typically, upper management exerts pressure to keep network spending low, while line management exerts pressure to keep network spending high(er) so that the productivity benefits from automation for which they are held accountable will be achieved [Data97, p.1]. For details on the comparisons and financial calculations needed to produce a comprehensive cost-benefit analysis of the network plan, the reader is referred to the Appendix and to [Mino93] and [Data95].

The findings of the risk and dependency analysis should be formalized in a written document. We recommend that the following sections be considered for inclusion in this report:

- Management summary
- Refined statement of project scope
- Business issues requiring further study
- Major dependencies
    - Other corporate initiatives that may affect project
    - Skill requirements
    - Technology issues
    - Time frame requirements
- Assessment of project risks
    - Potential impacts on the business as a whole
    - Potential security and control risks
    - Potential financial risks
- Preliminary cost/benefit analysis
    - Financial impacts on profitability
    - Financial impacts of hidden costs
    - Financial impacts of tangible and intangible risks
    - Financial analysis of procurement options (e.g., lease or purchase)

1.1.3 Determine project requirements

At this stage of the network planning process, information is collected on what the proposed network is supposed to do. Since many technical decisions will be based on this information, it is important that the requirements analysis be as accurate as possible. This step will likely involve interviewing decision-makers and staff who have a significant role in planning or using the network. It will also involve collecting information from various sources on the following:

- Estimated traffic patterns and flow
- Application programs and services to be supported by the network
- Destination and number of proposed system users
- Estimated equipment and line costs
- Network reliability requirements
- Network security requirements
- Network delay requirements

In general, it is not easy to collect the information needed to perform a complete requirements analysis. We discuss why this is so at length in Chapter 2 and how the requirements affect the network design. However, the main reason the data collection is time consuming and complex is the fact that the data needed are often not available or in the form needed for analysis. Considerable effort is usually required to estimate and derive the real parameters needed to design the network.

The project requirements should be documented in a written report. We recommend the following sections:

- Management summary
- Project requirements
  - Functional requirements
  - Quality assurance and control requirements
  - Security and control requirements
  - Test plan requirements
  - Resource requirements
- Supporting documentation
  - Estimated traffic patterns and flow
  - Application programs and services to be supported by the network
  - Destination and number of proposed system users
  - Estimated equipment and line costs

1.1.4 Develop project implementation approach

Once the business objectives, project feasibility, and major requirements have been determined, a strategy for moving forward on the network design and implementation should be developed. A number of factors need to be considered when formulating the project strategy. Among the most important factors shaping the project strategy are the business and technical constraints and risks. Constraints and risks heavily influence decisions on whether or not to proceed with the design using outside consulting help, or with in-house staff, or with some combination of the two.

As stated earlier, tactics to manage the potential project risks should be developed early on in the project, particularly if the risks are great. One method of dealing with risk is to create an implementation plan that is evolutionary, as opposed to revolutionary. This suggests a phased implementation approach that might, for example, specify how existing systems are to coexist or be transformed to operate in the new network environment. Pilot projects may be helpful in evaluating various network options and can be planned at this stage.

Business processes affected by the network implementation should also be examined, and those needing refinement should be identified. Strategies for making the workflow adjustments should be addressed in the project plan. To promote awareness and support of the network, it may be necessary to incorporate educational and internal communications programs into the project plan.

Given the pace at which technology is evolving, it is important to accept change as a fact of life and to plan accordingly. This involves planning reviews at critical points throughout the network planning and implementation process to identify unforeseen problems or changes that must be accommodated by midcourse adjustments. The urge to resist "scope-creep" must remain strong while making these midcourse adjustments, to prevent the project requirements from changing and burgeoning out of control.

The implementation strategy should be formalized in a written document. We recommend the following sections for this report:

- Management summary
- Assumptions and major constraints
- Work flow and organizational impacts
- Project approach
  - Risk management strategies
  - Project staffing
  - Project schedule
  - Change control procedures

After completing a top-down business analysis of the network plan, the organization is well positioned to begin a detailed analysis of the technical network requirements and to develop network alternatives for consideration. This process is described in detail in the chapters that follow.

1.2   Strategic positioning using networks

There are many ways a network can enhance an organization's competitiveness. Networks can be used to offer new services and capabilities, attract new customers, enhance image and visibility, increase productivity and reduce costs, and improve customer service. Organizations with creativity and insight have produced networks that offer dramatic advantages over competitors who are not as effective in their use of technology.

Automated teller machine networks provide a familiar example of how networking can revolutionize an industry. Automated teller machine networks provide a means for banks to offer a variety of services, often for revenue producing fees. Large financial institutions — such as Citibank and American Express — have international networks that help them maintain their dominance as global leaders in the banking and financial industry.

Federal Express provides another example of how networks can be employed to strategic advantage. The core of Federal Express's service is designed around sophisticated network and tracking systems. Federal Express's success in implementing fast, effective tracking has been a major factor in the erosion of the United States Postal Service's (USPS) share of the very profitable overnight mail delivery market. As part of a continual process to improve its tracking services, Federal Express now offers its customers the ability to track letters and packages using the Internet. It provides the software needed to do this free of charge. Within a short time after Federal Express began to offer this service, Pitney Bowes saw a precipitous decline in their shipping and weighing systems business. Pitney Bowes's shipping products sold for thousands of dollars, and its customers were unwilling to continue to pay for something they could easily get for free. The Pitney Bowes shipping and weighing systems provided mailing rates for a variety of mail services, including those for the USPS, UPS, Federal Express, and others. The demise of a product line facilitating the use of competitor services further solidified Federal Express's position as a leader in the mailing industry.

The Pentagon provides yet another case study on the strategic use of networks. Prior to 1994, the Pentagon relied on a delivery van — which was frequently caught in downtown Washington traffic — to bring secret reports from its headquarters to the White House. In 1994, the Pentagon began operation of Intelink, an international network linking 35 intelligence organizations with over three thousand users with secret or top-secret security clearance. The Intelink allows White House aides, State Department analysts, Pentagon generals, and soldiers in the field to access classified information on a wide range of topics [Wall95].

> During the Gulf War, for example, ground commanders lacked timely satellite photos to prepare for combat because the four computer systems handling the pictures could not talk to one another. Today Intelink users can punch up on their computers the most recent satellite photos as well as thousands of pages of classified reports...
> [Wall95]

There are numerous other examples illustrating the vital contribution networks can make to an organization's success. For instance:

> If General Electric switched off the multimillion-dollar network it uses to design jet engines, it would become obsolete overnight. If Mobil unplugged its 30,000-unit Lotus Notes system, executives would hardly know where to ship the next day's petroleum. Technology and competitiveness have become the same... [Coop97]

In short, the race to adopt new network technology continues to quicken, as more and more organizations extend their reach and influence through the use of global networks. Organizations, large or small, cannot remain complacent in their use of technology if they are to remain competitive in today's environment.

1.3   Major challenges in network design

There are many aspects to network design, all of which present unique challenges. From a business perspective, planning and executing an effective network strategy requires a substantial ongoing organizational commitment. The organization must be committed to developing an infrastructure that facilitates communication of the business objectives to the network planning team. The organization must also develop internal standards, methods, and procedures to promote effective planning. A commitment to do things the "right" way means adhering to the standardized processes and procedures even when there are substantial pressures to take risky shortcuts.

The organization should strive to hire, train, and retain skilled managers and staff who understand technology and how it can be used to satisfy organizational objectives. This is not easy, given the highly competitive job market for network specialists, and the rapid proliferation of new networking technologies. During the planning process, potentially serious political and organizational issues should be identified. For instance, people may feel threatened if they believe that the proposed network will compromise their power or influence. Consequently, they may attempt to hinder the project's progress. The organization must confront these fears and develop strategies for dealing with them.

In addition to organizational challenges, numerous technical challenges must be faced when designing a network. Perhaps the foremost challenge is the sheer multiplicity of options that must be considered. Added to this is the fact that current networks continue to grow in size, scope, and complexity. On top of this, the networking options available are in a constant state of flux. Keeping abreast of new developments and relating them to organizational requirements is a formidable task, and it is rare that an organization will have all the in-house expertise that it needs to do this well.

Often consultants and outside vendors are needed to help plan and implement the network. It is much easier to manage the activities of the consultants if the organization has a firm grip on the business objectives and requirements. However, sometimes consultants are needed to help develop and specify the business objectives and requirements. Although outside consultants offer benefits such as expertise and objectivity, they also present their own set of challenges. For instance, it is important to develop a "technology transfer" plan when working with outside consultants, to make sure that in-house staff can carry on as needed after the consultant leaves.

Through the 1970s and 1980s, if you wanted a network, you could call IBM and they would design your network. It was a common adage that "the only risk is not buying IBM." However, for the foreseeable future, there will be increasing numbers of network vendors in the marketplace and a decreasing likelihood that any one vendor will satisfy all of the organization's network requirements. While often unavoidable, using multiple vendors can pose problems, particularly when there are problems with the network implementation and each vendor is pointing a finger at the other. Since it is increasingly likely that a particular network vendor will provide only a part of the network solution, it is incumbent on the network design team to make sure that the *global* network requirements are addressed.

In short, the sheer volume, complexity, and pace of change in technology complicate the already formidable task of network design. Strategies for meeting these challenges are dictated by common sense and good management principles. We briefly summarize some of these strategies below:

- Develop methods for hiring and retaining good staff.
- Where necessary, augment existing staff with consultants and vendor support.
- Use training and internal communication to reduce the fears of those affected by the network.
- Encourage and offer ongoing education to help staff remain current with new trends in technology.
- A voluminous amount of technical information is available from a variety of such sources as vendor/telco/consultant presentations, conferences, technical books and magazines, and the Internet. Turn to these sources on a regular basis to help keep up with new developments in the industry.

1.4   Centralized network design

Legacy mainframe systems, dating back to the 1970s, typify centralized networks. In this environment, end users — i.e., persons or application programs — communicate with each other through communication links between devices to and from a central mainframe computer. This pattern of

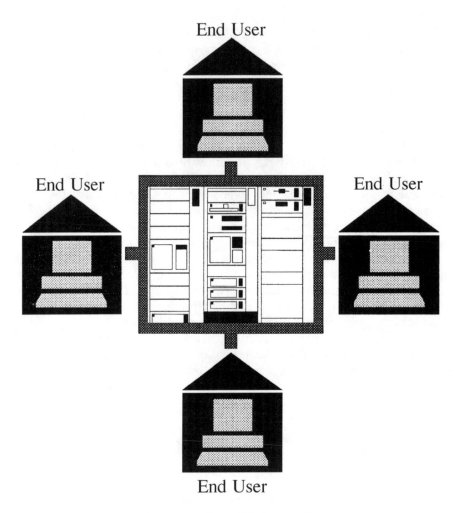

Figure 1.1   Centralized network.

connectivity forms a star-like topology illustrated in Figure 1.1. With this type of configuration, most of the application processing and associated data maintenance is performed at a central computer facility. From a network management perspective, this configuration is attractive, since it is conducive to centralized control and oversight of the network by the management information systems (MIS) department. This, in turn, makes it easier to manage daily computer operations and to control security. The star configuration may also offer economies of scale, since support for one processing site is usually cheaper than support for multiple processing sites. A major drawback of centralization is that it creates a single point of failure. A disaster at the main computer site can disable the entire network.

Historically, large IBM mainframe systems have dominated this environment. IBM's proprietary framework for supporting communication between devices (such as terminals,[1] mainframe/host processors, routers,[2] and front-end processors[3]) is called systems network architecture (SNA).

SNA provides a means to specify devices and routing paths for directing traffic between devices in the network. SNA's development predated the advent of open networking standards, personal computers, and distributed computing. As such, it was initially conceived for the centralized, mainframe environment, which it continues to support to this day. In this context, SNA defines a master-slave relationship between the host computer and terminal devices. IBM has continually refined SNA over the years to support evolving market demands and networking trends. In addition to legacy mainframe systems, SNA also supports IBM's distributed Advanced Peer-to-Peer Networking (APPN), which allows devices to communicate directly with each other without going through a central host facility. Other emerging network paradigms that SNA supports include: multiprotocol networking (such as AppleTalk,[4] TCP/IP,[5] etc.), very high bandwidth[6] networking (using frame relay,[7] ATM,[8] etc.), network management functions, and open standards (to encourage other vendors to integrate with IBM products and services). SNA remains the most prevalent network architecture in use today [Dyme94].

1.5   Distributed network design

A distributed network disperses the computing and communications capabilities throughout the network, allowing devices to communicate directly with each other. The decision to operate a distributed network rather than a centralized network is largely driven by geographic, operational, economic,

---

[1] Terminal — Any device capable of sending and/or receiving information on a communications line.

[2] Router — Protocol-specific device that finds the "best" route (i.e., the cheapest, fastest, and/or least congested route available) to send data from sender to receiver. Routers are commonly used in large networks with low bandwidth connections.

[3] Front-end processor (FEP) — A dedicated device that intercepts data before it is processed by the host computer, to perform various functions such as error control, line control, message handling, and code conversion.

[4] AppleTalk — Apple communication protocol based on seven-layer stack, similar to OSI stack.

[5] TCP/IP — Protocol suite for networking occupying layers 3 and 4 of the OSI Reference Model.

[6] Bandwidth — Range of frequencies that can be supported on a communication link. The higher the bandwidth, the higher the speed at which data can be transmitted.

[7] Frame relay — Packet-based transmission protocol operating at layer 2 of the OSI Reference Model.

[8] Asynchronous transfer mode (ATM) — Cell switching technique developed to support high-speed data transport.

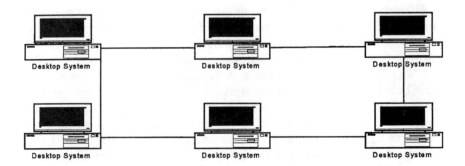

Figure 1.2   Local area network example (Source: A. Resnick).

and performance considerations that must be evaluated in the context of a particular organization's needs. Distributed networks may be used to:

- Reduce bottlenecks at high-volume usage centers by distributing workloads
- Reduce impact of a single point failure on the function of the network as a whole
- Provide improved functionality and service
- Facilitate phased, incremental system upgrades without fear that the entire network will be disrupted
- Reduce reliance on any one vendor, encouraging competitive buying opportunities

There are three major types of distributed networks:

- Local area networks (LANs)
- Metropolitan area networks (MANs)
- Wide area networks (WANs)

1.5.1   Local area networks

In contrast to WANs, LANs operate on smaller geographic distances and support higher data transmission rates. Typically, LANs connect computers that reside in the same building or campus and are usually limited to a few city blocks. LANs provide connectivity between devices operating on the network, including file-servers, printers, host processors, and personal computers. LANs are commonly used to support a wide range of applications, including: client-server[9] applications, file sharing, and centralized file backup and retrieval. An example of a LAN is provided in Figure 1.2.

---

[9] Client-server — Paradigm that defines clients (e.g., a program or computer) and how they are to be "served" by other network devices (e.g., a printer). A client can potentially be supported by many servers.

LANs are usually classified in accordance with transmission schemes (baseband and broadband), transmission media (twisted pair, coax, fiber and wireless), transmission techniques (balanced and unbalanced), topologies (star, ring, bus, and tree) and access control techniques (random control such as CSDA/CD, slotted ring, and register insertion; distributed control, such as token ring, token buys; and centralized control, such as polling, circuit switching, and time division multiplexing).

LANs employ three de facto access protocols: Ethernet,[10] token ring,[11] and FDDI.[12] The access protocols determine how the devices on the LAN share the cabling system that connects them. On a LAN, usually only one device at a time can transmit data over the network. The access protocol determines which and how each device takes turns transmitting data. Ethernet and token ring networks support transmission speeds up to 10 Mbps[13], and 16 Mbps, respectively. Fiber distributed data interface (FDDI) LANs support transmission speeds up to 100 Mbps using fiber optic cable.

1.5.2   Metropolitan area networks

The protocols used in a LAN do not support network coverage over a large, dispersed area. The need for high rates of data transport over longer distances led to the development of Metropolitan area networks (MANs). MANs are designed to support high data transmission rates (in the same range as LANs or higher) and interconnectivity of devices and/or LANs across an entire city. MANs typically support the same types of devices and applications that LANs support. In addition, MANs support transmission of voice, video, multimedia, high-quality images, and other applications demanding a very high data transfer rate. As complex applications — such as high-definition television, high-quality radiology, and videoconferencing — become increasingly common, it is expected that the demand for MANs will increase correspondingly.

MANs can be connected to other MANs and/or LANs using routers and bridges.[14] Generally, bridges are the preferred medium for interconnecting MANs, except in cases where multiple network types and protocols are to be supported. In this case, routers may be required to connect the MAN to the other network types. When planning a MAN across an entire metropolitan region, it is important that there be legal access, or right of way, to the data transmission facilities. MANs can be operated as public or private

---

[10] Ethernet — Local area networking protocol developed by Xerox and supported by a variety of vendors, including Hewlett-Packard, Intel, and others.
[11] Token ring — Local area networking protocol in which a token is passed from device to device. Devices must receive the token before they can transmit data.
[12] Fiber distributed data interface (FDDI) — Local area networking protocol for transmitting data at 100 Mbps speeds over optical fiber.
[13] Mbps = 1,000,000 bytes per second.
[14] Bridge — A device to interconnect networks using the same protocols.

networks. However, a private network, with dedicated lines,[15] is less likely to be as cost effective as a public network with switched lines.[16] A public MAN is comprised of a transport and an access network component. The transport network component is maintained by the telco or communication provider, while the access network component is maintained by the subscriber organization.

1.5.3 Wide area networks

Wide area networks (WANs) connect computers and other devices separated by very large distances. WANs support data transmission between cities, states, and/or countries. For example, a multinational corporation with offices in New York and Seattle might use a WAN to connect two or more LANs. Although WANs support transmission of many of the same types of applications as MANs, they typically do so at a much lower speed. For example, two of the more common line speeds used in WANs are 1.544 Mbps (T1)[17] and 44.746 Mbps (T3).[18]

Private[19] WANs have steadily grown in popularity because they offer flexibility, security, and control over internal data communications. During the late 1960s through the 1970s, WANs based on IBM's SNA, and TCP/IP or X.25 packet switching dominated the corporate scene. WANs built at this time were comprised of such components as terminals, processors, modems, multiplexers,[20] concentrators,[21] and communications links. Efficiency and speed were less important in these early WANs than cost-effective connectivity. The most common applications involved connecting users to remote resources to perform tasks like database retrieval and maintenance, batch file processing, on-line data entry, and so on. These applications are much less time sensitive than the demanding applications of today (e.g., videoconferencing). The most commonly used communication link was the T1, which was used to build large corporate WAN backbone networks. A "backbone" typically consists of a collection of high-speed lines — i.e., T1s — that transmit and consolidate traffic from a number of other lines operating at lower speeds. The lower-speed, dedicated lines were typically used to transmit data to and from mainframes and dumb terminals, at a rate of 9.6 Kbps or 19.2 Kbps. The consolidation of traffic onto lines with a larger capacity

---

[15] Dedicated line — Leased line usually supplied by a telephone company that permanently connects two or more locations and is for the sole use of the subscriber.

[16] Switched lines — A line through the public telephone network that is connected to a switching center.

[17] T1 — United States digital communication line operating at 1.544 Mbps.

[18] T3 — United States digital communication line operating at 44 Mbps.

[19] Private — Communications facilities that are used exclusively by the organization that owns or leases them.

[20] Multiplexer — Also known as a "mux," this is a device that allows more than one signal to be sent over a channel simultaneously.

[21] Concentrator — A device that consolidates traffic coming from a number of circuits onto a smaller number of circuits; used to reduce costs.

offered many throughput and cost advantages over many lines with smaller capacity. WAN networks such as these flourished as T1 costs steadily declined over the years.

In the 1990s and beyond, many new applications are emerging that substantially increase the traffic and demand on the network. Some of these applications include teleradiology,[22] CAD/CAM drawings, and high-definition television. In addition, networking requirements have grown larger and more complex as corporations have grown, merged, and/or increased their outreach. This has led to the need for WAN-to-WAN, LAN-to-WAN, and LAN-to-LAN interconnectivity. The most commonly used devices for interconnecting WANs and LANs include bridges, routers, extenders for remote access, brouters, and gateways.

To achieve the performance requirements mandated by WAN connectivity to LANs (which operate at much higher speeds than traditional WANs), new approaches are required. One approach is to use fast, packet-switched networking facilities. The major public switched facilities available today are frame relay, switched multimegabit data service (SMDS), and asynchronous transfer mode (ATM). These services operate at speeds ranging from 10Mbps to 2.5 Gbps. These technologies allow an organization to interconnect LANs and to consolidate all the data traffic (which might, for example, include a mix of TCP/IP, SNA, Appletalk traffic) on a single backbone network. This improves the network scalability and tolerance for growth, and also promotes the availability of new services, such as e-mail. Another appealing aspect of switched-based networking is that it essentially involves outsourcing the network maintenance to the communications carrier.

In summary, WANs may be built from a mix of public and private line facilities and from components such as routers, bridges, terminals, hosts, etc. The market for private WAN networks dominated by T1 facilities is being successfully challenged by communications carriers offering frame relay, SMDS, and ATM services. These new packet-switching services support the transport of many data types at very high speeds.

1.6   Similarities and differences between WAN and LAN design and planning

As we have discussed, there are major differences between WANs and LANs. They operate on different geographic scales and use different protocols, equipment, and facilities. LANs are usually private networks, while WANs can be either private or public. There are more choices and the choices that exist are more complex when designing a WAN as compared to designing a LAN. In the remaining chapters of this book we discuss techniques and options for designing and planning LANs and WANs, which must be guided by specific knowledge of each respective network type.

---

[22] Teleradiology — High-quality medical imaging transmitted over communications links.

One area where LAN and WAN design is converging is on the issue of how to interconnect the two. Bridges, routers, and brouters are typically used to interconnect LANs and WANs. Bridges are relatively inexpensive and are faster and easier to maintain than most routers. However, they are not suitable for large, complex networks, since they can only interconnect networks running the same protocols. They also lack the ability to reroute traffic in the event of line failures and to filter traffic that may create security or performance problems. In contrast, routers interconnect networks running multiple protocols, and they support dynamic routing schemes to redirect traffic in response to changing network conditions. Routers can also be used to implement firewalls[23] to prevent unauthorized transmissions in a network. A brouter is a hybrid of a bridge and router, and performs the functions of both. When it cannot perform the function(s) of a router, it defaults to the function(s) of a bridge.

1.7   Similarities and differences between voice and data network design and planning

Traditionally, there have been clear distinctions and choices between voice and data networks. Selection of one or the other was determined by the type of traffic carried, cost and performance characteristics, and the types of equipment and services that could be supported by each type of network. However, in the 1990s and beyond, the distinctions between voice and data networks are becoming increasingly clouded, as a variety of integrated voice and data alternatives have become available. New carrier, service, and technology options continue to proliferate. These options support not only voice and data, but also an increasingly diverse array of such applications as color facsimile, video conferencing and surveillance, high-definition television distribution, and LAN-to-LAN and MAN-to-MAN connectivity.

It is easy for an organization to install a plain vanilla voice network. To hook up to the public switched telephone network (PSTN), one has simply to call the telephone company (telco). This type of voice network provides interconnectivity between PSTN subscribers. In the early days, prior to the advent of the computer, the PSTN could only support voice traffic. The introduction of computers and the ever growing need for network connectivity led to the development of the modem (MOdulator/DEModulator). The modem is used to convert digital data signals from the computer into analog signals that can be carried on the telephone network, and vice versa. Thus, using a modem, two or more computers can send and receive data to and from each other over the PSTN. In addition to being easy to implement, this type of connectivity offers [Mull97]:

---

[23] Firewall — Router or access server used to filter transmissions to a private network for security purposes.

- High reliability
- Shared cost structure with large customer base
- Price stability and predictability
- Limited capital investment
- Few management and control responsibilities

Yet another option for implementing a voice network is to use a private branch exchange (PBX). There are two major types of PBXs: analog and digital. An analog PBX handles telephones directly and uses modems for data transmission. A digital PBX handles data transmission directly and uses codecs[24] to connect telephones. Thus, both voice and data can be carried on a PBX with the right equipment. When compared to regular telephone service, a PBX may be cost effective and may offer useful features not available otherwise. For example, some PBXs offer the capability to connect a direct high-speed line to a local area network, dynamic bandwidth allocation to support virtually any desired data rate, and connectivity to other networks, including packet-switched networks. PBXs are usually private, meaning that the network is used exclusively by the organization paying for it. This is in contrast to the public telephone network, which at any one time can be shared by many.

A key feature of the PSTN and PBX systems we have just discussed is that they employ circuit switching. Circuit switching establishes a temporary, exclusive connection between channels. Since the communications channels are dedicated for the duration of the call, if all the channels or lines are busy, any additional incoming calls will be blocked (and you will hear a busy tone). When circuit-switched networks are used to transmit data *exclusively,* they become increasingly less cost effective. One reason for this is that during a typical terminal-to-host data connection, much of the time the line is idle. A second reason is that circuit switching transmits data at a constant rate. Thus, both the sending and receiving equipment must be synchronized to transmit at the same rate. This is not always possible in a diverse computing environment with many types of computers, networks, and protocols.

Packet switching was introduced to overcome the shortcomings of circuit switching when transmitting data exclusively. In packet switching, the data is broken into small bits or "packets" before it is transmitted. In addition to containing a small portion of the data, each packet contains a source and destination address for routing purposes. The protocols developed to support packet switching are defined as X.25. Although X.25 packet-switching protocols support reliable data delivery on both analog and digital lines, the introduction of packet switching led to the development of digital networks dedicated exclusively to data.

Throughout the remainder of this book, we discuss techniques for designing data networks. As can be seen, there are many choices and options

---

[24] Codec — Device analogous to a modem that allows a digital line to process analog signals.

for designing a data network. One choice, which we discuss now, is whether the data network is to be public or private. In a private network, the organization owns or leases the network facilities (or communications links) from a communications carrier.[25] In a public network, the organization shares facilities provided by a telco or communications carrier with other users, through a switched service that becomes active at the time it is used. As was the case when deciding between a PBX and a PSTN, cost, security and performance may be factors in swaying the decision to one type of network or the other.

Telephone companies throughout the world are continually adding and upgrading their services to drive demand. An almost overwhelming array of choices for networking is available. Standard telephone service — known as narrowband communication services (i.e., subvoice grade pathways with a speed range of 100 to 200 bps) — is still available. However, an alternative to standard telephone service offered by the telcos is the integrated services digital network (ISDN).

ISDN uses digital technology to transmit data (including packet-switched data), voice, video, facsimile, and image transmissions through standard twisted pair telephone wire. Recently, ISDN has become widely available throughout the United States; however, there are still some impediments to its adoption that have not been totally resolved. One obstacle is that there are often delays of several weeks or months before ISDN lines can be installed [SKVA97, p1.]. Another potential obstacle is that some organizations have bandwidth requirements that exceed the capacity of ISDN (ISDN links come in two speeds, one capable of transmitting up to 16 Kbps and the other transmitting at speeds up to 64 Kbps[26]). Cost may be another factor to discourage use of ISDN, which is more expensive than regular phone service. In addition, the costs to upgrade to ISDN may be prohibitive. For example, ISDN does not work with standard modems operating at speeds of 28.8 Kbps. Therefore, 28.8 Kbps modem users are forced to upgrade their equipment to take advantage of ISDN.

Despite these obstacles, ISDN is increasingly being used to support telecommuting, remote access, and Internet access. As ISDN becomes more affordable, it will become especially attractive as a means to access the Internet. Banks are among the major users of ISDN, as are small businesses installing LAN-to-LAN connections. ISDN lines can be more economical than WAN leased lines — which are typically T1 lines operating at 1.544 Mbps — when data transmission is intermittent, since with ISDN you only pay for the time you use. Dedicated lines are also not needed, since ISDN allows network connections to be made on demand. ISDN connections offer higher speed then modem connections (which are too slow for LAN-to-LAN communications) at a lower cost than that of private leased lines [Skva96, p 3.].

---

[25] Communications carrier — A communications company providing circuits or lines to carry traffic, and also known as a "common carrier."

[26] Kbps — 1024 bits per second.

Broadband ISDN (BISDN) is another networking option. BISDN is an extension of ISDN that was developed by the telcos to allow very high rates of data transmission. With BISDN, channel speeds of 155 Mbps and 622 Mbps are available. BISDN is an emerging technology designed to support switched, semipermanent, and permanent broadband connectivity for point-to-point and point-to-multipoint applications.

Current packet-switching technology is simply not adequate to support such applications as broadband video telephony, corporate videoconferencing, video surveillance, high-speed digital transmission of images, TV distribution, LAN interconnection, and hi-fi audio distribution. In contrast, the deployment of BISDN relies on synchronous optical NETwork (SONET), a standard for extremely high data transmission rates using optical fiber. SONET supports transmission rates between 51.8 Mbps and 2.5 Gbps,[27] and a variety of transmission signals, including voice data, video, and metropolitan area networking (MAN). SONET also supports asynchronous transfer mode (ATM) standards. ATM is yet another standard developed to provide high-speed (150 Mbps and higher) data transport for WANs. ATM is based on a cell-switching technique that uses fixed-size 53 byte cells analogous to packets. The BISDN, SONET, and ATM standards were developed to work in conjunction with each other to support demanding data transport requirements.

The virtual private network (VPN) offers yet another option available through the telcos. A VPN is used to simulate a private network. Originally, VPNs were offered for voice traffic. However, such major players as AT&T, MCI and others are offering VPNs with switched data services that can be packaged with voice services. VPN services often provide an economical alternative to private line services, particularly for those organizations operating international networks, since charges are generally based on usage. VPN can be an attractive choice for organizations that anticipate adding new sites to the network, since it may be easier to modify the VPN than it would be to modify a private line network. Features supported by VPNs include:

- Account codes, which can be used for internal accounting and billing purposes
- Authorization codes, which can be used to restrict access and calling privileges
- Call screening based on a number of criteria, such as geographic location, trunk group, etc.
- Network overflow control
- Alternative routing capabilities when the network becomes congested.

According to [Smit97, p.20], multinational corporations are the primary users of VPN services, because VPNs offer the flexibility needed to grow and modify global networks. VPNs are particularly cost effective in providing connectivity to low-volume remote sites and to traveling employees.

---

[27] Gbps — 1 billion bits per second.

They also offer an effective alternative to a private meshed network, which for multinational companies is "an exorbitantly expensive proposition" [Smit97, p.20].

In summary, there are a variety of choices for data and voice networks. Increasingly, hybrid solutions are available, including voice and/or data networks, private and/or public networks, central office telco services, and/or private PBX services. The line between voice and data networking is becoming increasingly blurred. Traditionally, voice networks supported voice and data (with the aid of a modem). However, data-only networks specifically designed to maximize speed and efficiency have prevailed in most WAN implementations. Over time, the capability of voice networks has increased, so that they now support a wide range of services and a wide variety of transmissions, including voice, data, video, and imaging. Switched, public networks have become an attractive option to conventional private networks. However, there may be reasons — special requirements, security and control, reliability, back-up needs — that a private line network solution is best.

There are many networking options available in the marketplace. However, the pros and cons of these options are not always readily apparent without careful analysis of the network requirements. Complicating the decision is that the costs of the network options are also difficult to compute. Service providers may charge based on usage, distance, type of service, bandwidth requirements, special deals, or some combination of these. Ultimately, the decision to use a public or a private voice/data network must be based on the goals for the network and the business.

1.8   Major networking trends in industry

In the sections that follow we discuss major trends in the telecommunications industry that are shaping how networks are being designed and implemented. The common theme that unites these trends is the extraordinary pace at which the industry is changing. It is harder than ever to plan a networking strategy and to predict what the future will bring!

1.8.1   Larger, more complex networking

The world continues to grow smaller as the push to build worldwide networks continues. More and more corporations operate in international markets. As networks are developed to support these operations, there is increasing pressure to transform existing networks to global proportions.

Traditional analog lines are being phased out in favor of all digital networking facilities. The adoption of digital technology is being fueled by rapidly dropping costs for services that promise high reliability, speed, and versatility. Leading the digital revolution are technologies such as frame relay, SMDS, ATM, SONET, and cellular/cordless mobile communication. In addition to supporting a variety of data transmissions — voice, video, imaging, etc. — these technologies provide a means to interconnect a multitude of LANs, MANs, and WANs in a variety of ways to create networks that truly are global in reach. As it becomes increasingly practical and cost effective to support new applications will increase, like teleradiology for instance, it will drive the demand for these applications, which will in turn promote the offering of new digital services at competitive prices.

Frame relay networks are being used to implement high-speed virtual private networks (VPN) capable of supporting high bit-rate transmission. Frame relay provides T1/E1 access rates, typically at a substantially lower cost than comparable T1/E1 private leased-line solutions. Frame relay networks are designed to take advantage of recent improvements in transmission circuits. In the past, going back several decades, transmission circuits were far more error prone than they are today, necessitating the use of various protocols to compensate for the transmission problems. However, as optical fiber circuits have gained in popularity, the need for older, slower, more resource intensive protocols is much less important. Frame relay takes advantage of improved circuit reliability by eliminating many unnecessary error checking, correction, editing, and retransmission features that have been part of many data networks for the past two decades. Frame relay has been available for many years. However, the flexibility that Frame relay offers in allocating bandwidth on demand is somewhat new. Frame relay is one of several fast packet-switching options available today.

Switched multimegabit data service (SMDS) is yet another option for high-speed connectivity. SMDS is designed to connect LANs, MANs, and WANs. It is designed to ease the geographic limitations that currently exist with low-speed wide-area networks. SMDS is a connectionless packet-switching service that provides LAN-like performance beyond a subscriber's location. SMDS can also be used to interconnect low-speed wide area networks that are geographically dispersed. SMDS provides a convenient, high-speed interface to an existing customer network. SMDS is positioned as a *service* and not as a method for designing or building networks. SMDS is targeted for large customers and sophisticated applications that need a lot of bandwidth, but not all the time. SMDS is good at handling data applications that transfer information in a bursty manner. SMDS can also be used to support many interactive applications (with the exception of real-time, full-motion video applications, for which SMDS is too slow). For example, it takes only one to two seconds to send a high-quality color graphic image over a SMDS network. For many applications, this speed is more than adequate.

Asynchronous transfer mode (ATM) is a very new technology that provides very high-speed, low-delay, multiplexing and switching. In addition, ATM supports simultaneous transmission of any type of user traffic (such as voice, data, or video). ATM segments and multiplexes user traffic into small, fixed-length cells. Each cell contains a header, which contains virtual circuit identifiers. An ATM network uses these identifiers to relay the traffic through high-speed switches from the sending customer's premise equipment (CPE) to the receiving CPE. ATM is able to achieve very high transport speeds because it performs very little error detection and correction, and because queuing delays are minimized by the small cell size used to transmit data. ATM is most effective, in terms of speed and reliability, when used in conjunction with SONET-based fiber optic circuits.

The SONET/SDH standard is based on optical fiber technology. The term SONET is used in North America, while the term SDH is used in Europe and Japan. SONET/SDH circuits offer far superior performance and reliability than microwave and/or cable circuits. In essence, SONET/SDH is an integrated network standard that allows all types of traffic to be transported on a single fiber optic cable. Because SONET/SDH is a worldwide standard, it is now possible for different vendors to interface their equipment without conversion. SONET/SDH efficiently combines, consolidates, and segregates traffic from different locations through one facility. This ability — known as grooming — eliminates back hauling[28] and other inefficient techniques currently used in many carrier networks. SONET/SDH also eliminates back-to-back multiplexing overhead by using new techniques in the grooming process. These techniques are implemented in equipment known as add-drop multiplexers (ADM). SONET/SDH employs digital transmission schemes and supports efficient time division multiplexing (TDM) operations.

There are two major forms of mobile communications in use today: cellular and cordless. This technology is not new; however, its adoption has been somewhat slow. In part, this is due to the fact that mobile communications systems are undergoing rapid technological changes and many different protocols and standards are being used.

A cellular system usually works within a completely defined network (which operates using protocols for setting up and clearing calls and tracking mobile units) through a wide geographical area. Typically, cordless communication operates on a rather limited geographic region. With cordless personal communications, the focus is on the methods used to access a closely located transceiver — usually within a building. Cellular systems operate at a higher power than do cordless personal communications systems. Therefore, cellular systems can communicate within large cells with a radius in the kilometer range. In contrast, cordless cellular communication cells are quite small, usually on the order of 100 meters.

Cellular communication will continue to be the preferred medium for the wireless consumer and business market. Personal communications systems (PCS) are being used by telecommunications providers to help reduce

---

[28] Back hauling is a technique in which the user payload is carried past a switch that has a line to the user and to another endpoint. After the traffic destined for the endpoint is dropped off, the first user's payload is then sent back to the switch, where it is relayed to the first user. In many existing configurations, grooming eliminates the need for back hauling, but it also requires expensive configurations (such as back-to-back multiplexers that are connected with cables, panels, or electronic cross-connect equipment).

cellular churn, joining customers who are in close vendor proximity by providing end-to-end service. Wireless data connectivity is increasingly popular due to the availability of lighter and smaller equipment that can be carried by humans and in vehicles. In the near future, wireless connectivity will become an increasingly important consideration in determining corporate network infrastructure requirements.

Other emerging network technologies include digital subscriber lines, computer telephony, and integrated transmission of data and voice-over TV cables.

Digital subscriber lines (DSL) provide a means to mix data, voice, and video transmissions over phone lines. There are several different types of DSL from which to choose, each suited for different applications. All DSL technologies run on existing copper phone lines and use special, sophisticated modulation techniques to increase transmission rates.

Asymmetric digital subscriber line (ADSL) is the most publicized DSL scheme. It is commonly used to link branch offices and telecommuters in need of high-speed Intranet and Internet access. The word "asymmetric" refers to the fact that ADSL allows more bandwidth downstream (to the consumer) than upstream (from the consumer). Downstream, ADSL supports speeds of 1.5 to 8 Mbps, depending on line quality, distance, and wire gauge. Upstream rates range between 16 and 640 Kbps, depending on line quality, distance, and wire gauge. ADSL can move data up to 18,000 feet at T1 rates using standard 24-gauge wire. At distances of 12,000 feet or less, the maximum transmission speed is 8 Mbit/s.

ADSL offers other benefits. First, ADSL equipment is being installed at the carrier's central office to offload overburdened voice switches by moving data traffic from the public switched telephone network onto data networks. This is increasingly important as Internet usage continues to rise. Second, the ADSL power supply is carried over the copper wire, so ADSL works even when the local power fails. This is an advantage over ISDN, which requires a local power supply and a separate phone line for comparable service guarantees. A third benefit is that ADSL furnishes three information channels — two for data and one for voice. Thus, data performance is not impacted by voice calls. The carrier rollout plans for this service are very aggressive, and it is expected to be widely available by the end of this decade.

Rate-adaptive digital subscriber line (RADSL) provides the same transmission rates as ADSL. However, as its name suggests, RADSL adjusts its transmission speed according to the length and quality of the local line. The connection speed is established when the line synchs up or is set by a signal from the central office. RADSL applications are similar to those using ADSL and include Internet, Intranets, video-on-demand, database access, remote LAN access, and lifeline phone services.

High-bit-rate digital subscriber line (HDSL) technology is symmetric, meaning that it furnishes the same amount of bandwidth both upstream and downstream. The most mature of the xDSL approaches, HDSL has already been implemented in the telco feeder plants — which provide the lines that

extend from the central office to remote nodes — and in campus environments. Because of its speed — T1 over two twisted pairs of wiring and E1
over three twisted pairs of wiring — telcos commonly deploy HDSL as an
alternative to T1/E1 with repeaters. At 15,000 feet, the operating distance of
HDSL is shorter than that of ADSL. However, carriers can install signal
repeaters to extend HDSL's useful range (typically, by 3000 to 4000 feet).
HDSL's reliance on two or three wire-pairs makes it ideal for connecting
PBXs, interexchange carrier POPs (point of presence), Internet servers, and
campus networks. In addition, carriers are beginning to offer HDSL as a way
to carry digital traffic within the local loop, between two telco central offices
and customer premise equipment. HDSL's symmetry makes this an attractive option for high-bandwidth services like multimedia, but its availability
is still very limited.

Single-line digital subscriber line (SDSL) is essentially the same as HDSL,
with two exceptions. SDSL uses a single wire pair and has a maximum
operating range of 10,000 feet. Since it is symmetric and needs only one
twisted pair, SDSL is suitable for applications like video teleconferencing or
collaborative computing, which require identical downstream and upstream
speeds. The standards for SDSL are still under development.

Very high-bit rate digital subscriber line (VDSL) is the fastest DSL technology. It delivers downstream rates between 13 to 52 Mbps and upstream
rates between 1.5 to 2.3 Mbps over a single wire pair. The maximum operating distance, however, is only 1000 to 4000 feet. In addition to supporting
the same applications as ADSL, VDSL, with its additional bandwidth, can
potentially enable carriers to deliver high-definition television (HDTV).
VDSL is still in the definition stage.

A number of critical issues must be resolved before DSL technologies
achieve widespread deployment. First, the standards are still under development and must be firmly established. Issues that must be resolved relate
to interoperability, security, eliminating interference with ham radio signals,
and lowering the power system requirements from the present 8 to 12 watts
down to 2 to 3 watts. A nontechnical, but nonetheless important factor in
the success of DSL technology is how well the carriers can translate the
successes they have realized in their technology trials to success in market
trials and commercial deployment.

Cable systems provide an attractive, high-bandwidth channel to carry
data as well as video signals. However, the implementation of cable systems
has been slow. Cable can be used in both business and residential areas.
Currently, entertainment transmissions are unidirectional. In order to support communication in both directions, cable modems are needed. In North
America, there is an agreement between suppliers establishing how CATV
systems will carry digital video and data within standard 6 MHz cable
channels. The reliability of cables, their power supply, and their management
are critical issues that must be addressed before this emerging technology
is widely used in commercial applications.

Computer-telephone integration (CTI) technology takes many forms. Essentially, this technology can be used to establish a link between a computer and phone, enabling such functions as unified messaging or having a single PC-based interface to all of voice, e-mail, and fax messages. On the high end, a call center can receive incoming calls, identify callers either through caller identification or by asking them to use their touch-tone dial to enter an account number. The interest in this technology is great, but it is not widely deployed.

1.8.2   High-speed LANs

In this section, we summarize recent technological developments that have made high-speed local area networks possible. Some of these technological innovations are being used in metropolitan and wide-area networks as well as in high-speed LANs.

FDDI/CDDI (fiber or copper distributed data interface)

FDDI is an accepted standard that has been implemented in a wide range of products, including bridges, routers, concentrators, wiring hub cards, and virtually all network interface cards for any bus. FDDI supports transmission over two-pair category-5 unshielded twisted-pair, multimode fiber, and single-mode fiber. In the next decade, FDDI will supplant Ethernet as a de facto LAN standard.

FDDI is based on a token passing scheme that employs two rings of fiber optic cable. Packets rotate around the rings in opposite directions. An FDDI network can achieve burst speeds of 100 Mbps and sustained data transmission speeds of around 80 Mbps. An FDDI network can stretch 100 km and connect over 500 stations spaced as far as 2 km apart.

Despite its high cost, FDDI is attractive due to the many benefits it offers that cannot be found with any other communication medium. These benefits include:

- *Speed* — First and foremost among FDDI's advantages over first-generation, copper-based LANs is the speed at which data can be transferred over the network. Ethernet's 10 Mbps and IBM's token ring speed of 16 Mbps offer considerably less throughput in comparison to FDDI's speed of 100 Mbps.
- *Security* — Security is always a concern with any network. It is especially critical when the network carries sensitive data. Copper-based medium emits electromagnetic interference patterns that can be monitored. This problem is completely eliminated with fiber optics. Metal cables are also much easier to tap into than fiber optic cables. Tapping into a coaxial cable is almost as easy as attaching a clamp to the cable. To tap into a fiber optic cable, the cable must be cut and very precisely

spliced back together or the transmission will be impaired and the line break will be detected.

- *Noise immunity* — Unlike lightwave communications, metal-based (coaxial cables or twisted pair wiring) networks are subject to EMI interference. Using proper installation techniques, FDDI networks can stretch significantly beyond the distances achieved by Ethernet. For example, Ethernet networks are limited to a distance of no greater than 2.8 km, while FDDI can cover a distance of 100 km.

- *Fault tolerance* — The FDDI specification is designed to address the need for fault tolerance. Fault tolerance refers to the network's ability to recover from problems in the transmission equipment. This is a critical need in high-performance applications. In a fully configured FDDI network with two fiber optic rings, a break or failure in one of the rings will not disable the network. Using built-in station management and network management functions, in the event of a ring failure, an FDDI network can reconfigure itself so that all nodes maintain normal communications.

- *Determinism* — In certain real-time applications, network designers must know the time needed to transfer data from one node on the network to another or from a supercomputer to a peripheral device. The FDDI token passing scheme is based on a timed token protocol designed for just these types of applications. The protocol guarantees a maximum token rotation time, as determined by a bidding process between the nodes when the network is initialized. The node requiring the fastest transfer time can dictate the token rotation time for the ring. Network developers can determine the maximum time needed for each node on the network to acquire a token and, consequently, the maximum time needed to communicate with any other node on the network.

FDDI node processors (i.e., which provide connectivity to the FDDI and handle much of the protocol processing) are responsible for connecting bridges, routers, computers, and other devices to the FDDI backbone. Some of the characteristics to carefully consider in a node processor are its design flexibility, internal bandwidth, host bus interface, and its ability to handle FDDI's demanding protocol processing requirements. A special consideration is the use of encapsulation or a transparent protocol. Protocol encapsulation limits the use of FDDI backbones to the same protocol. A transparent protocol implementation does not have the same restriction, enabling users to construct hybrid FDDI backbone networks. Many FDDI networks use a combination of single and dual attachment FDDI stations, depending on the application(s) being run and the objectives of the network designer.

Given their role in connecting the FDDI network, the capabilities of node processors today and over the next several years should be of extreme importance to network implementers and OEM manufacturers, who will be incorporating FDDI into their product lines.

DQDB (dual queue dual bus)

DQDB uses two optical fiber buses to connect basically the same types of network components as does FDDI, but at a higher transmission rate (140 Mbps). The buses are separate and unidirectional, and use the same signaling frequency. Network nodes are tied to each other as a logical bus, and thus, each node is located both upstream and downstream from other bus nodes. DQDB technology revolves around the fact that all stations have knowledge of the frames queued at all other stations. Under DQDB, the end nodes continuously transmit empty data frames around each ring. Whenever a station on one of the network nodes has something to send, it generates a frame request on the bus that is carrying traffic away from its node.

If redundancy is required, a DQDB network can be configured as a looped bus in which the two ends of the buses are co-located. If a fault occurs in one of the buses, the nodes at either end of the fault can act as both beginning and end points. Most typically, DQDB has been used to implement MAN backbones.

Full duplex LANs

Network throughput can be increased when transmission is supported in both directions simultaneously. In most cases, the LAN cabling does not need to be changed at all to support full duplex transmission.[29] For example, if two cable pairs are available between the stations and hubs, full-duplex operation can be supported, thus doubling the potential throughput. The higher speed does not cause transmission problems, because the individual cables are operated at the "old" speed. This is a tactical solution for network managers that is very helpful in offloading heavily used network segments. In particular, server-hub-connections can benefit from this type of solution. To migrate to this type of solution requires the use of adapter cards, hub boards, and bridge parts supporting full duplex operations. Figure 1.3 illustrates this type of solution for an Ethernet. Other possible implementation options include full duplex fast Ethernet, full duplex FDDI, and full duplex token ring.

Asynchronous transfer mode

ATM provides very high speed transport of data, voice, and LAN communications over local, metropolitan, and wide-area networks. ATM also offers bandwidth-on-demand up to a committed rate for specific connections. ATM is a promising technology for interconnecting LANs.

The ability of ATM to integrate existing Ethernet and token ring networks is a key to its acceptance in industry. ATM is available at speeds of

---

[29] Full duplex — Full duplex lines allow data to be transmitted in both directions simultaneously. This is in contrast to half duplex lines, which allow data to be transmitted in only one direction at a time.

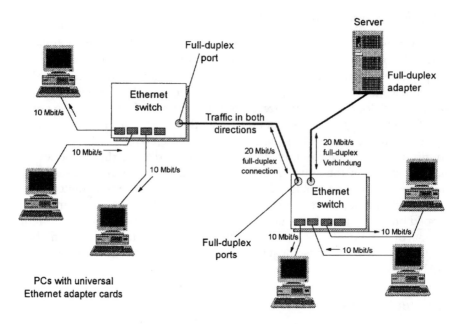

Figure 1.3   Example of half duplex Ethernet.

44.74, 51.84, 100, 155, and 622 Mbps, and many users expect that 25 Mbps transmission rates will soon be available. ATM supports many cabling types, including two-pair and four-pair category-3 unshielded twisted pair, multimode-fiber, DS-3, and T1/E1 copper circuits.

100Base-T

100Base-T provides 100 Mbps of bandwidth using existing Ethernet media access control (MAC) sublayers. The standard bodies are busy working to further improve this level of efficiency and transmission speed. This technology does not represent a major innovation. Rather, it represents a way of combining shared media using a relatively inefficient access method. The majority of the market leaders support this technology.

The cabling supported in the first release of this standard includes four-pair Category-3 unshielded twisted pair, two-pair Category-5 unshielded twisted pair, two-pair shielded twisted pair, and multimode fiber. 100Base-T cabling is configured in a star, which is supported by centrally located repeater hubs. Between any end user devices, there is a limit of at most two hubs. The distance of 100Base-T is limited to 210 meters, and to 100 meters between hubs and end-user devices.

100VG AnyLAN

The name AnyLAN refers to the fact that it supports both Ethernet and token ring frame formats at 100 Mbps. AnyLAN is limited to a distance of

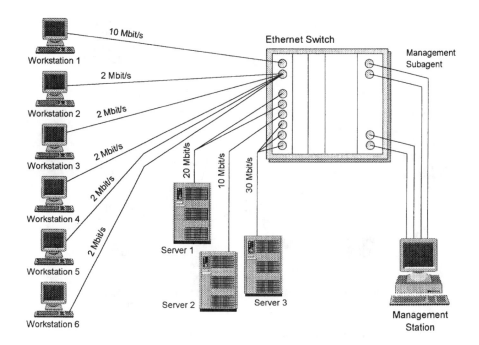

Figure 1.4   Example of full duplex Ethernet.

approximately 100 meters between hubs and end user devices. Manufacturer support of AnyLAN is still very limited.

It operates over four-pair Category-3 unshielded twisted pair, two-pair shielded twisted pair, and multimode fiber. The physical topology is a star, whereby stations are connected to a central hub of network segments. The hubs function as repeaters. In addition, they also provide a level of network security by filtering incoming and outgoing packets based on the source and destination addresses of the transmission traffic.

The hub polls each station to determine whether they want to send data or not. Usually, the polling is sequential; however, various stations and applications can be assigned higher transmission priorities (by changing the hub polling table). By virtue of its ability to alter demand priorities, the utilization levels that can be achieved using AnyLAN are expected to be higher than those that can be achieved using 100Base-T.

Switched and virtual LANs

Over the last few years, a new class of switched and virtual LAN products has emerged that can increase the bandwidth of overloaded Ethernet and token ring networks, while using conventional cabling and adapters. More of these products are available for Ethernet than for token ring, but the technology used is the same in both cases. Figure 1.4 shows the example with Ethernet and Figure 1.5 for token ring.

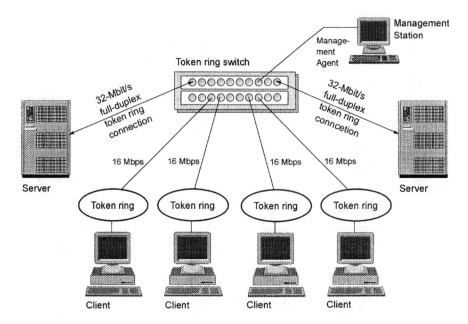

Figure 1.5   Example of virtual token ring LAN.

In the case of Ethernet, switching may be implemented with both 10Base-T and 100Base-T. Once switched, the connection between the hub with the switch and workstations or servers may receive the full bandwidth. There are four basic technologies used within the switch:

- Software switch with shared-memory-interface
- Switch with high-speed backbone
- Switch with high-speed backplane
- Switch with high-speed crossbar

There are two core technologies available for Ethernet and switching:

- Static Ethernet switching (also called dedicated switching)
- Dynamic Ethernet switching (also called bandwidth-on-demand)

Static switching was designed to simplify the job of change management. Implementing the switching policies into software supports automation of configuration management. The network manager is authorized to make assignments between servers, segments, and workstations. Static port switching allows network managers to move users, via software, from one of a hub's shared Ethernet buses to another, thus facilitating configuration management. Ports can be individually selected and added to different shared Ethernet buses. Moving individual ports helps to maximize port usage, reducing the number of modules that need to be purchased for the hub. The static module switch also operates as a software-controlled system for implementing additions, moves, and changes to the network. However, the static module switch moves the entire hub module, including all its parts, from one shared bus to another.

Dynamic switching was designed to increase LAN bandwidth on an as-needed basis. The switch examines packets originating in a workstation connected to a port on a dynamic switch. The switch examines the packet's source and destination addresses. The switch then opens a dedicated circuit (10 or 100 Mbps) that carries the packet through the switching fabric from source to destination. The collisions normally associated with Ethernet LANs are substantially reduced or even eliminated, giving each user access to the LAN's full bandwidth capability. Within the switch, many source-to-destination connections can be maintained simultaneously. Once transmission requests have been fulfilled, the switch drops the circuit and redirects its attention to other requests. In this way, the dynamic switch allocates bandwidth on-demand.

Dynamic port switching allows each port to be connected to a single end station or server. Since the switch gives each port the maximal available bandwidth-on-demand, its function is to allocate much greater network bandwidth to individual end users and servers than is possible using shared Ethernets. Changes are requested only in hubs that accommodate the switches.

The dynamic segment switch functions internally exactly the same way as the dynamic port switch. It allocates private, point-to-point 10 Mbps or 100 Mbps circuits to its ports on demand across the switching fabric. However, each dynamic segment switch port is able to connect to an entire network segment and is not limited to an individual end station or server. The dynamic segment switch accomplishes this by being able to identify a large number of MAC addresses on each port. The ability to string entire networks from each port often allows the dynamic segment switch to substitute for bridges and routers to segment LANs.

In summary, LAN switching offers the following unique features:

- Full bandwidth to each user, while preserving existing infrastructures
- Scalable, cost-effective solutions based on cost/performance requirements
- Standards-based connectivity and network management
- Congestion management
- High-port density
- Flexibility of combining various switching techniques
- Fault tolerance
- Port prioritization
- Multicasting
- Support of port-level, multi-hub virtual networking

Software control of switched LANs offers a diverse range of solutions based on the creation of virtual LANs. A virtual LAN is a secure, software defined group of workstations and servers that function as a networked workgroup of users that are logistically close together although physically the nodes may actually be located in diverse places, such as rooms, floors, buildings, or sites. The virtual LAN is called virtual because it is defined entirely in software of the hub and may be rapidly deleted or altered at any time by the network manager to meet business needs.

Virtual LANs constitute a new kind of network design tool created to help LAN managers quickly adapt the network to fit a competitive organization's frequently changing workflow requirements. The interconnected virtual LAN also includes shared resources, such as print servers and E-mail servers. These resources may be accessed by all subnetworks.

The management of virtual LANs is not trivial. Not only the physical layout, but also the virtual topology must be maintained and visualized. The network management instruments must be aware of the virtual components and any changes made in real-time. Segment monitoring using RMON-probes must be redefined as well.

1.8.3   Emergence of Internet and Intranet technologies

"The use of Internet and Intranet technologies is rapidly exploding in the corporate scene. As further evidence of this trend, the winners of an annual *Datapro* survey of 'Hot Products' were products that manage, secure, or boost the performance of intranets" [Saun97].

The Internet was originally designed for the Department of Defense to provide robust and inexpensive networking capabilities. These advantages are in large part why the Internet is so popular today and is available in even the most remote corners of the world. Since the Internet is a packet-switched network, it can exchange data with any other TCP/IP network. Many companies are using, or are thinking of using, the Internet to augment their WAN capabilities. In comparison to dial-up lines, or dedicated T1 or T3 lines, or frame relay connections, or other commonly used networking options, the Internet is very economical. The Internet does not require a lot of start-up investment or maintenance, and the ongoing usage costs are usually very low. Both dedicated and switched lines can be used to connect to the Internet. For example, using an analog line, a traveling employee with access to a local phone line can connect into the Internet. Thus, some companies see the Internet as a cost-effective way of providing WAN access to remote locations with little traffic. The most significant concerns with the Internet are usually associated with security and transmission delays. Firewalls and encryption are the two most common ways of addressing security concerns. Transmission delays on the Internet can be caused by many things and are harder to control. For applications with high performance, it is possible that a private WAN solution will be preferable to an Internet solution.

Like the Internet, Intranets are being implemented everywhere. An Intranet is a company-specific network that uses software programs based on the Internet's TCP/IP protocol. TCP/IP is platform independent and works with Macintosh, PC, or UNIX-based computers. As the popularity of Intranets increases, so has the demand for new tools and Web-based solutions. This has increased competition among software manufacturers which, in turn, has resulted in better and cheaper products.

1.8.4   Security

Security is vital for all networks, no matter how small or large. However, as networks become larger, and the applications they run become more critical to the organization, it becomes essential to have a comprehensive approach to managing the network security. There are numerous threats to the security of a network, and it takes skill, imagination, and constant vigilance to ward off potentially serious network intrusions. One of the first steps in developing a plan to secure the network is to recognize the types of security breaches that may occur. For example, this includes unauthorized transmission, interception, or alteration of data on the network.

It is not hard to see that there are many vulnerable points in any network. To list a few potential vulnerable points: (a) devices — such as sniffers — can be attached to a network to intercept line transmissions, (b) some LAN network interface cards (NICs) and workstations can be modified to intercept transmissions that are not intended for them, (c) unattended workstations that are connected to the network can be accessed by unauthorized people, (d) network management and monitoring equipment can be used to access sensitive information on the network, (e) leased lines can be tapped ("Find a circuit number beginning with FDDC or FDDN, and all you need is a modem and a recording device to steal information.") [Data8_96].

One strategy to deal with network security is to physically secure sensitive access points to the network and to limit access to authorized users. This limits the access outsiders have to the network; however, it does not eliminate the threat from insiders within the organization. The ways of compromising a network's security are too numerous to fully enumerate, and new techniques to thwart network security are continually being invented. A variety of strategies need to be used in even a basic network security plan to ensure that an adequate level of security is in place to ward off the most likely forms of attack. This is not easy, and sometimes it is made harder by the network itself. For example, LANs have certain features that compound the difficulty of securing them. First, by its very design, a LAN can transmit to any device on the network, and thus any device on the network can potentially be used to intercept information. Second, the addressing method used by LANs is not secure, and it is fairly easy for a knowledgeable culprit to modify a LAN transmission so it appears to be coming from someplace else. Thus, the opportunities for unauthorized interception, modification of data, and masquerading are compounded in a LAN environment.

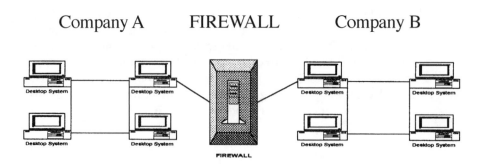

Figure 1.6 Implementing a firewall between inside and outside network users (Source: A. Resnick).

In LAN and WAN environments, one way security can be improved is to determine from where the potential threats are most likely to come. If internal traffic is perceived as trustworthy, while outside traffic is not, then secure routers or bridges can be used as firewalls between the two. An example of this type of solution is illustrated in Figure 1.6. It should be noted that a firewall does not lock out intruding computer viruses, since it does not actually inspect the contents of the data. If internal traffic is perceived to be potentially untrustworthy, then other security measures must be implemented at each end terminal. This can be accomplished by using data encryption schemes, which encode the data in such a way that it can only be decoded by an authorized person (or a very clever, unauthorized person using lots of computing power to decrypt the message). The use of routers and encryption are also useful in securing Internet or Intranet transmissions. We discuss this at greater length in Chapter 2 and in Chapter 5.

A caution is in order with regard to encryption. A story illustrates our point. In 1996, when Qualcomm needed to secure its transmissions from its San Diego headquarters to a branch office in Singapore over its virtual private network WAN, it wanted to use a 56-bit DES data encryption scheme. However, under United States law this is illegal. Any encryption scheme employing a key greater than 40 bits is considered a potential threat to national security, and cannot be used outside the United States. When Qualcomm asked the State Department for an exception, it was denied [Star96]. There are a number of approaches around this problem, but none of them are wholly adequate for secure transmission overseas. One obvious strategy is to use a 40-bit encryption scheme for international transmissions and a stronger (i.e., longer and harder to decrypt) key within the United States.

It should be emphasized that most security violations are not the result of high-tech break-ins. More likely, they are the result of "low-tech" opportunistic events, like finding a system password taped to a monitor, or they are the result of unintentional error(s). In any event, since there are many potential places where a network can be compromised, the security plan must be *comprehensive* if it is to be effective. Ideally, proactive measures should be used to prevent problems before they occur. These should be supplemented with reactive measures that involve system monitoring and automatic alert generation when security breaches are detected. Finally, passive measures can be used after a security breach to try to track and identify the problem source. An example of a passive measure is to collect data on network usage that can be reviewed as needed.

1.8.5 Regulation trends

In February 1996, Congress passed the Telecommunications Act of 1996. This is one of the most important pieces of legislation since the Communications Act of 1994. The primary intent of the legislation is to promote competition among various providers of communications services. We outline a number of the key provisions in the new legislation, which are expected to have significant impacts on the telecommunications industry [Levi96]:

1. New obligations are mandated for local exchange carriers (LECs) requiring them to interconnect to competitors, unbundle their services, and establish wholesale rates so that competitors can resell LEC services to fill gaps in their own service offerings.
2. State and municipalities are barred from restricting any provider offering interstate or intrastate communications services. It encourages the removal of barriers that would discourage small businesses and entrepreneurs from offering services in the telecommunications market.
3. All remaining restrictions in the 1982 consent decree that broke up the Bell System were lifted. This will allow the Bell Operating Companies (BOCs) to provide telephone service (e.g., private lines, local calling, etc.) outside and within their own regions, if certain conditions are met. It would also allow the BOCs to manufacture terminal and other networking equipment and to enter new markets that were previously forbidden (e.g., electronic publishing).
4. Barriers separating the cable and the phone industries were removed, and provisions were made to encourage cable companies to enter the local exchange and local access markets.
5. Section 502 of the Act makes it illegal to send indecent communications to minors, to send obscene material through communication devices, or to use a communications device to harass.

State governments are recognizing the fact that the telecommunications industry provides a powerful stimulus for economic growth. There are many state initiatives to attract and promote the telecommunications industry. To ward off potential federal interference with state rights, many state legislatures are taking a proactive role in passing measures to deregulate local telephone services. States are directing public utility commissions to establish rules and conditions for local competition. For information on specific regulatory initiatives by state, the interested reader is referred to [Mull96].

In summary, many of the traditional barriers to competition in the local phone markets have been removed. In this environment, the role of the public utility commission is changing dramatically. It is now being encouraged to actively support fair competition in the local phone market, and it will no longer be required to "establish rates of return or to ensure protection against losses" [Mull96]. States are seizing the initiative to create a regulatory environment that protects consumers and promotes competition. These initiatives are expected to provide powerful incentives to the phone companies to "cut costs, invest in new technologies, and introduce new services" [Mull96].

1.9 Summary

In this chapter, we introduced a systems methodology for developing the network planning and implementation approach. This methodology involves:

- Defining the business needs and objectives
- Assessing the major risks and dependencies
- Defining the project requirements
- Designing an implementation strategy

We surveyed the major network types, and identified similarities and differences in the planning and implementation of these networks.

We examined emerging trends that will have substantial impacts on the design and planning of networks. These trends include:

- Increasing demand for fast, high-capacity networks to support larger and more complex networks handling larger and more complex applications
- Increasing adoption of open-based standards, such as the Internet and TCP/IP based protocols
- Emerging concerns with security and control over network activity
- Evolving regulatory environment to promote growth in the telecommunications industry

Bibliography

[Cole93] Cole, G., IEEE 802.6 metropolitan area network standard, *Datapro*, McGraw-Hill, New York, 7, May 1993.

[Coop97] Ramo, J. C., Welcome to the wired world as the globe's political and business elites meet in Davos, Switzerland, to ponder the implications of the information revolution, the future is glowing — and growing — all around them, *TIME*, Vol. 149, No. 5, February 3, 1997.

[Data95] Datapro Information Services Group, Financial planning for network managers, *Datapro*, McGraw-Hill, February 1997.

[Data8_96] Datapro Information Services Group, Sources of leaks in lans and wans, *Datapro*, McGraw-Hill, Inc., August 1996.

[Data96] Datapro Information Services Group, High speed local area networks, *Datapro*, McGraw-Hill, November 1996.

[Data97] Datapro, Switched Multimegabit Data Service (SMDS), *Datapro*, McGraw-Hill, March 1997.

[Dyme94] Dymek, W., IBM systems network architecture (SNA), *Datapro*, McGraw-Hill, April 1994.

[Keis89] Keiser, G., *Local area networks*, McGraw-Hill Book Company, New York, 1989.

[Levi96] Levine, H., Blaszak, J., Block, E., and Boothby, C., The telecommunications act of 1996, *Datapro*, McGraw-Hill, June 1996.

[Mart94] Martin, J., with Chapman, K., and Leben, J., *Local area networks*, Second Edition, Prentice-Hall, Englewood Cliffs, NJ, 1994.

[Mino93] Minoli, D., *Broadband network analysis and design*, Artech House, Boston, 1993.

[Mull96] Muller, N., Federal and state telecom reform initiatives, *DataPro*, McGraw-Hill, April 1996.

[Mull97] Muller, N., and Costello, R., Planning voice networks, *DataPro*, McGraw-Hill, March, 1997.

[Owen91] Owen, J., Data networking concepts, *Datapro*, McGraw-Hill, New York, 14, 1997.

[Saun97] Saunders, S., The brightest ideas in networking, *Data Communications*, McGraw-Hill, New York, 1997.

[Scha95] Schatt, S., Local routers and brouters, *Linking LANs*, McGraw-Hill, New York, 1995, pp. 91-101.

[Skva97] Skvarla, C., ISDN services in the U.S: overview, *Datapro*, Mc-Graw-Hill, April 1996.

[Smit97] Smith, M., Virtual private network (VPN) services: overview, *Datapro*, Mc-Graw-Hill, New York, 1997.

[Spoh93] Spohn, D., *Data network design*, McGraw-Hill, New York, 1993.

[Stall94] Stallings, W., *Data and computer communications*, Fourth Edition, Macmillan Publishing Company, New York, 1994.

[Star96] Stark, M., Encryption for a small planet, *BYTE*, McGraw-Hill, March 1996.

[Stei96] Steinke, S., The internet as your wan, Lan Magazine, Miller Freeman, Vol. 11, No. 11, pp. 47-52, October 1996.

[Wall95] Waller, D., Spies in cyberspace. *TIME*, 145, 11, March 20, 1995.

chapter two

Technical considerations in network design and planning

Contents

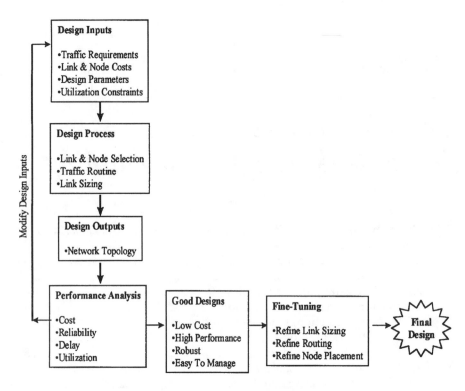

Figure 2.1 Overview of network design process.

2.1   Overview of the network design process

Network design is a painstaking, iterative process. The first step of this process is to define the requirements the network must satisfy. This involves collecting information on anticipated traffic loads, traffic types (e.g., data, video, etc.), and sources and destinations of traffic. This information is used, in turn, to estimate the network capacity needed. These requirements are used as input to the second step, the design process. In this step, various design techniques and algorithms are used to produce a network topology. The design process involves specifying link and node placements, traffic routing paths, and equipment sizing. After a candidate network solution is developed, it must be analyzed to determine its cost, reliability, and delay characteristics. This third step is called performance analysis. When these three steps are completed, the first design iteration is finished. Then the entire process is repeated, either with revised input data (e.g., using revised traffic estimates, etc.) and/or by using a new design approach. This process is summarized in Figure 2.1.

The basic idea of this iterative process is to produce a variety of networks from which to choose. Unfortunately, for most realistic design problems, it is not possible from a mathematical perspective to know what the optimal network should look like. To compensate for this inability to derive an analytically perfect solution, the network designer must use a judicious form of trial and error to determine his or her best options. After surveying a variety of designs, the designer can select the one that appears to provide the best performance at the lowest cost.

Because network design involves exploring as many alternatives as possible, automated heuristic design tools are often used to produce quick, approximate solutions. Once the overall topology and major design aspects have been decided, it may be appropriate to use additional, more exact solution techniques to refine the details of the network design. This fine-tuning represents the final stage of the network design process.

In this chapter, we take the reader through each of the major steps involved in network design: requirements analysis, topological design, and performance analysis.

2.2   Data collection

The network requirements must be known before the network can be designed. However, it is not easy to collect all the information needed to design a network. Often, this is one of the most time consuming aspects of the design process. Data must be collected to determine the sources and destinations of traffic, the volume and character of the traffic flows, and the types and associated costs of the line facilities to transport the traffic. It is rare that these statistics are readily available in the succinct summary form needed by network design algorithms and procedures.

Most typically, a considerable volume of data is collected from a variety of sources. For existing networks, it may be possible to collect information on:

- Session type(s)
- Length of session(s) — average, minimum, and maximum time
- Source and destination of data transmissions
- Number of packets, characters sent, etc.
- Application type
- Time of session
- Traffic direction and routing

To relate these findings to the user requirements, interviews and data collection activities may be needed to determine the following:

- Required response time
- Current and planned services and applications to be supported (e.g., E-mail, database transactions and updates, file transfer, etc.)
- Anticipated future services and application needs
- Communications protocols, including network management protocols to be supported
- Network management functions to be supported

- Network reliability requirements
- Implementation and maintenance budget

These data must then be related, culled, and summarized if they are to be useful. Ultimately, the traffic data must be consolidated to a single estimate of the source-destination traffic for each node. Between each source and destination of traffic, a line speed must be selected that will have sufficient capacity[1] to transport the traffic requirement.

The line speed and the line end points are used in turn to determine the line cost. Usually only one line speed is used when designing a network (or a significant subnetwork portion), since multiple line speeds complicate network manageability and may introduce incompatibilities in transporting the data. Table 2.1 below illustrates the data in the summarized form needed by most network design procedures.

Table 2.1 Sample Traffic and Cost Data Needed for Network Design

Traffic source (1)	Traffic destination (2)	Estimated traffic (bytes) (3)	Usable line capacity (bytes) (4)	Estimated line cost ($ monthly) (5)
City A	City B	80,000	1,000,000 (T1)	1,000.00
City A	City C	770,000	1,000,000 (T1)	3,500.00
City B	City N	500,000	1,000,000 (T1)	6,006.00
City B	City C	30,500	1,000,000 (T1)	5,135.00

Thus far, we have not addressed the issue of node[2] costs. As shown in Chapter 3, the reason for this is that these costs seldom play a major role in the design of the network topology. *After* a network has been designed, the node costs are factored into the total network cost (see Section 2.6.8).

When collecting traffic and cost data, it is helpful to maintain a perspective on the level of design needed. The level of design — i.e., be it high-level or finely detailed — helps to determine the amount of data that should be collected, and when more detailed data are required and when not. It may be necessary to develop multiple views of the traffic requirements and candidate node sets in order to perform sensitivity analysis. It is easier to develop strategies for dealing with missing and/or inconsistent data when the design objective is clear.

To the extent practical, the traffic and cost data collected should be current and representative. In general, it is easier to collect static data than it is to collect dynamic data. Static data remains fairly constant over time,

---

[1] Capacity — The capacity of a line refers to the amount of traffic it can carry. Traffic is usually expressed in "bits per sec," which is abbreviated as *bps*. The actual carrying capacity of a line is dependent upon technology, since the technology determines the amount and nature of "overhead" traffic, which must be carried on the line.

[2] Node — In the context of the network topology, a node is a connection point between lines. It is a very general term for a terminal, a processor, a multiplexer, etc.

while dynamic data may be highly variable over time. If the magnitude of the dynamic traffic is large, it makes sense to concentrate more effort in trying to estimate it more accurately, which may have a substantial impact on the network performance. Wherever possible, automated tools and methods should be used to collect the data. However, this may not be feasible, and sometimes the data must be collected by hand. However, a manual process increases the likelihood of errors and limits that amount of data that can be analyzed.

2.2.1   Automated data generation

There may be situations where no data are available on existing traffic and cost patterns. This may be the case when an entirely new network is being implemented. When actual data are unavailable, an option is to use traffic and cost generators. Traffic and cost generators may also be used to augment actual data (particularly when they are missing and/or inconsistent) or to produce data for benchmark studies.

For a traffic and cost matrix similar to Table 2.1, for a network with n sources and destinations, the number of potential table entries is[3]:

$$\binom{n}{2}$$

As stated in [Cahn98]:

There is only one thing certain about a table with 5,000 or 10,000 entries. **If you create such a table by hand it will contain thousands and thousands of errors and take weeks of work.**

Thus, there is strong motivation for using automated tools to generate and/or augment the traffic and cost data needed to design a network.

2.2.2   Traffic generators

A traffic generator, as its name implies, is used to automatically produce a traffic matrix (i.e., the first three columns of Table 2.1) based on a predetermined model of the traffic flow. Many design tools have traffic generators built into them. Alternatively, stand-alone software routines are available that can be used to produce traffic matrixes. A traffic generator can easily produce matrixes representative of increasing or decreasing traffic volumes. Each of these matrixes can be used to design a network, and the results can be studied to analyze how well a design will handle increasing traffic loads

---

[3] The notation refers to the number of combinations of n sources and destination nodes taken two at a time, with each set containing two different nodes and no set containing exactly the same two nodes.

over time. The traffic generator might also be used to produce traffic matrixes based on a uniform or random traffic model. While these are not realistic traffic distributions for most networks, they may, nonetheless, provide useful results for benchmarking studies. Other more realistic traffic models can be used to produce traffic matrixes that more accurately conform to observed or expected traffic flows. For example, it may be appropriate to use a model that adjusts traffic flows as a function of node population, geographic distance between other sites, and anticipated link utilization levels [Kersh89]. Other models might be based on the type of traffic — i.e., e-mail, World Wide Web traffic, and client/server traffic — that is to be carried by the network. The interested reader is referred to [Cahn98], which lists a number of exemplar software routines for producing traffic matrixes to conform to a variety of model conditions and assumptions. The main decision when using a traffic generator is the selection of the traffic flow model that will be used to produce the traffic matrix.

2.2.3 Cost generators and tariff data

Cost generators are similar in concept to traffic generators. Cost generators are used to produce a cost matrix (i.e., columns (1), (2), and (5) of Table 2.1). In an actual design situation, it may be necessary to produce several cost matrixes representing the various line and connectivity options. In this case, each circuit or line option should be represented in a separate traffic/cost matrix.

A major challenge in creating cost matrixes is that the tariff[4] structures (which determine the costs for lines between two points) are complex and contain many anomalies and inconsistencies. For instance, tariffs are not based solely on a simple distance calculation. Thus, it is possible that a longer circuit may actually cost less than a shorter circuit. Two circuits of the same length may have different costs, depending on where they begin and terminate. In the United States, circuits within a LATA[5] may cost less than circuits that begin and end across LATA boundaries. Depending on the geographic scope of the network, domestic (United States) and/or international tariffs may apply. Given the complexities of the tariff structures, the "ideal" way of keeping track of them is to have a single look-up table containing all the published tariff rates between all points.

Although comprehensive, accurate tariff information is available, it is very expensive, costing thousands of dollars. Thus, the "ideal" solution for gathering tariff data may be too costly to be practical. NAC, Inc. — which produces the WinMIND™ design tool discussed in Chapter 3 — is an example of a commercial supplier of tariff information.

---

[4] Tariff — A tariff is a published rate for a specific communications service, equipment, or facility that is legally binding on the communications carrier or supplier.

[5] LATA — A local access and transport area defines geographic regions within the United States within which the Bell operating companies (BOCs) can offer services. Different LATAs have different tariff rates.

Access to a comprehensive tariff database does not guarantee that the desired cost data are obtainable for all points. For example, tariff information is only available for direct services that are actually provided. If no direct service is available between two points, then obviously a tariff will not be published for these points. Furthermore, even if accurate tariff information is available, it may still be too complex to use easily.

When the tariff data are either too costly or complex to use directly, alternative methods must be used to estimate the line costs. It should be noted that while these alternative methods may be easier and cheaper than the purchase of a commercial tariff tool, they are not as accurate. The need for accuracy must be weighed against the trade-off of using a simpler scheme to estimate line costs. In making this determination it is helpful to bear in mind that the single largest expense in most data networks is the cost of the communication lines, since line costs may exceed node costs by a factor of 2 to 6 [Pili97].

One simple cost model is based on a linear distance function. In this model, line costs between two nodes i and j are estimated by a function containing a fixed cost component, f, and a variable distance based component, v [Kersh89]. It should be noted that the f and v cost components will vary by link type. The linear cost function can be summarized as:

$$\text{Cost}_{ij} = f + v \ (\text{dist}_{ij}) \tag{2.1}$$

where Cost_{ij} = cost for line between two nodes i and j
   f   = fixed cost component
   v   = variable cost based on distance
   dist_{ij} = distance between nodes i and j

When the locations of nodes i and j are expressed as V & H[6] coordinates (i.e., (V_i, H_i), and (V_j, H_j), respectively) the distance, dist_{ij}, between the nodes is easily calculated using a standard distance formula [Kersh89]:

$$\text{dist}_{ij} = \sqrt{\left(V_i - V_j\right)^2 / 10 + \left(\left(H_i - H_j\right)^2 / 10\right)} \tag{2.2}$$

This simple model can also be used to simplify a complex tariff structure. Linear regression can be used to transform selected points from the tariff table into a linear cost relationship. The fixed cost component, f, and the variable cost component, v, can be derived by taking the partial derivatives of the cost function with respect to f, and with respect to v, respectively

---

[6]   V&H — The vertical and horizontal coordinate system was developed by AT&T to provide a convenient method of computing the distance between two points using a standard distance formula.

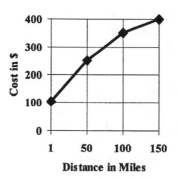

Figure 2.2   Example of piecewise linear cost function.

[Cahn, p.151]. This simplified model may perform well in special cases where the tariff structure is highly linear.

A somewhat more realistic estimate of cost may be possible using a piecewise linear function. The piecewise linear cost function is very similar to the linear cost function, except that the f (fixed) and v (variable) cost components vary according to distance. An example of a piecewise linear function is presented below and in Figure 2.2:

$$\text{Cost}_{ij} = \$100 + \$3/(\text{dist}_{ij}) \text{ (for dist}_{ij} \text{ between 0–50 miles)}$$

$$\text{Cost}_{ij} = \$100 + \$2/(\text{dist}_{ij}) \text{ (for dist}_{ij} \text{ between 50–100 miles)}$$

$$\text{Cost}_{ij} = \$100 + \$1/(\text{dist}_{ij}) \text{ (for dist}_{ij} \text{ between 100–150 miles)} \qquad (2.3)$$

where Cost_{ij} = cost for line between two nodes i and j
        dist_{ij} = distance between nodes i and j

Many service providers use this model to price private lines. Note that there are no additional usage fees in the model we have presented. With this type of cost model, there is an economic incentive to fill the line with as much traffic as possible for as much time as possible.

A step-wise linear function is illustrated in Figure 2.3. A hypothetical step-wise linear function is given below. Note that in this function the fixed costs are only constant within a given range, and there is no longer a variable cost component.

$$\text{Cost}_{ijk} = \$100 \text{ (for dist}_{ijk} \text{ between 0–50 miles)}$$

$$\text{Cost}_{ijk} = \$200 \text{ (for dist}_{ijk} \text{ between 50–100 miles)}$$

$$\text{Cost}_{ijk} = \$300 \text{ (for dist}_{ijk} \text{ between 100–150 miles)} \qquad (2.4)$$

where Cost_{ijk} = cost for line between two nodes i and j up to point k
        dist_{ijk} = distance between nodes i and j

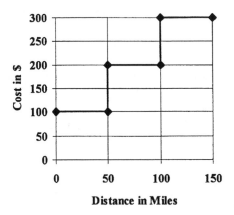

Figure 2.3 Example of stepwise linear cost function.

When international circuits must be priced, the cost models may need to be extended to provide more realistic estimates. For example, if a line is installed across international boundaries, adjustments may be needed in the cost model to account for differences in the tariff structures of each country involved. In addition, lines installed across international boundaries are usually priced as the sum of two half-circuits. A communication supplier in one country supplies half of the line, while a communication supplier in the other country supplies the other half of the line. In addition, the cost of a connection between any two countries is usually the same for *all* end points. Thus, a connection from "Tokyo to Sao Paolo … will cost the same as a circuit from Kyoto to Rio de Janeiro" [Cahn98, p.159]. The interested reader is referred to [Cahn98] for examples of cost generators for both domestic and international tariffs.

2.3   *Technical requirements specification*

2.3.1   *Node selection, placement, and sizing*

In this section we discuss matters relating to the selection, placement, and sizing of network nodes. We begin this discussion with a review of the devices commonly used as network nodes:

- Bridge — This device is used to interconnect networks or subnetwork components that share the same protocols. Bridges are fairly simple and do not examine the data passing across them.
- Concentrators — These devices allow multiple devices to dynamically share a smaller number of lines than would otherwise be possible. Concentrators are sometimes called MAUs for multiple access units. Concentrators only allow one device at a time to use the communications channel, although many devices can connect to the concentrator. This is a significant difference from multiplexers.

- Digital service units (DSUs) and channel service units (CSUs) — These devices are used to connect digital devices to digital circuits.
- Gateways — Gateways are used to interconnect otherwise incompatible devices using protocol conversion techniques.
- Hub — A device used to interconnect workstations or devices.
- Modems — A MODulator/DEModulator device that is used to connect digital devices to analog circuits.
- Multiplexers — These devices are also known as MUXs. They allow multiple devices to transmit signals over a shared channel, thereby reducing line and hardware costs. In general, multiplexers do not require the devices on the line to operate using the same protocols. There are two types of multiplexers. One type uses frequency division multiplexing (FDM). This type of multiplexing divides the channel frequency band into two or more narrower bands which each act as separate channels. The second type of multiplexer uses time division multiplexing (TDM). This type of multiplexing divides the line into separate channels by assigning each channel to a repeating time slot.
- Intelligent multiplexers or statistical time division multiplexers (STDM) — These devices use statistical sampling techniques to allocate time slots for data transmission based on need, thereby improving the line efficiency and capacity. The design of STDMs is based on the fact that most devices transmit data for only a relatively small percentage of the time they are actually in use. STDMs aggregate (both synchronous and asynchronous) traffic from multiple lower speed devices onto a single higher speed line.
- Router — A protocol-specific device that transmits data from sender to receiver over the best route, which may mean the cheapest, fastest, or least congested route.
- Switches — These devices are used to route transmissions to specific destinations. Two of the most common types are circuit switches and packet switches.

The selection of a specific device depends on many factors. These factors may include the node cost and the requirements that have been established for protocol compatibility and network functionality. This is a context-specific decision that must be made on a case by case basis.

As discussed earlier, the selection of a particular type of node device has little impact on the design of the network topology. However, the placement of nodes within the network does impact the network topology. Typically, nodes are placed near major sources and destinations of traffic. However, this is not always true, as sometimes node placements are based on organizational or functional requirements that do not strictly relate to traffic flow. For instance, a node may be placed at the site of a corporate headquarters, which may or may not be a major source of traffic. If the node locations must be taken as given and cannot be changed, then the decisions on node placement are fairly easy.

However, in other cases, the network designer is asked to suggest optimal node placements. In Chapter 3, we introduce several node placement algorithms. One algorithm — the center of mass algorithm (COM) — suggests candidate locations based on traffic and cost considerations. A potential shortcoming of the center of mass algorithm is that is may suggest node placements in areas that are not feasible or practical.

The add and drop algorithms may be used to select an optimal subset of node locations based on a predefined set of candidate nodes. Thus, potential sites for node placements must be known in advance when using these latter two algorithms, which are described in detail in Chapter 3.

Once the node placements and the network topology have been established, the traffic flows through the node can be estimated. This is needed to size the node. Measuring the capacity of a node is generally harder than measuring the capacity of a link. Depending on the device, the node capacity may depend on the processor speed, the amount of memory available, the protocols used, and software implementation. It may also reflect constraints on the type, amount, and mix of traffic that can be processed by the device. These factors should be considered when estimating the rated vs. the actual usable node capacity.

Queuing models can be used to test whether or not the actual node capacity is sufficient to handle the predicted traffic flow. Queuing analysis allows the network designer to estimate the processing times and delays expected for traffic flowing through the node, for various node capacities and utilization rates. In Section 2.6.2, we provide an introduction to the queuing models needed to perform this analysis. The node should be sized so that it is adequate to support current and future traffic flows. If the node capacity is too low, or the traffic flows are too high, the node utilization and traffic processing times will increase correspondingly. Queuing analysis can be used to determine an acceptable level of node utilization that will avoid excessive performance degradation.

In sizing the node's throughput requirement, it may also be necessary to estimate the number of entry points or ports needed on the node. A straightforward way of producing a preliminary estimate is given below:

$$\text{Number of ports} = \frac{\left(\text{Total traffic through ports in bps}\right)}{\left(\text{Usable port capacity in bps}\right)}$$

This estimate is likely to be too low, because it does not allow for excess capacity to handle unanticipated traffic peaks. Queuing models similar to those used for node sizing can be used to adjust the number of ports upward to a more appropriate figure. A queuing model allows one to examine the cost of additional ports vs. the improvements in throughput. Queuing analysis is a very useful tool for port sizing.

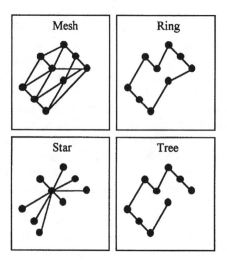

Figure 2.4 Sample network topologies.

2.3.2 Network topology

The network topology defines the logical and/or physical configuration of the network components. The basic types of network topologies are listed below. An illustrative example of each network type is provided in Figure 2.4.

- Star — In this topology, all links are connected through a central node.
- Ring — In this topology, all the links are connected in a logical or physical circle.
- Tree — In this topology, there is a single path between all nodes.
- Mesh — In this topology, nodes can connect directly to each other, although they do not necessarily all have to interconnect.

The selection of an appropriate network topology depends on a number of factors, including protocol and technology requirements. In Chapter 3, we discuss in detail the issues involved in selecting a network topology.

2.3.3 Routing strategies

A number of schemes are used to determine the routing of traffic in a network. These routing schemes can be described in the following terms [Cahn98, p. 216]:

Table 2.2 Taxonomy of Network Routing

Routing Characteristics
One route between nodes vs. multiple routes between nodes
Fixed routing vs. dynamic routing
Minimum hop routing vs. minimum distance routing vs. arbitrary routing
Bifurcated routing vs. non-bifurcated routing

The meaning of "one route between nodes," and "multiple routes between nodes," and "arbitrary routing" is self-explanatory. Fixed routing means the traffic routing is predetermined and invariant, irrespective of conditions in the network. In contrast, dynamic routing may change depending on network conditions (i.e., traffic congestion, node or link failures, etc.). Minimum hop routing attempts to send the traffic through the least possible number of intermediate nodes. Minimum distance routing is used to send traffic over the shortest possible path. Bifurcated routing splits traffic into two or more streams that may be carried over different paths through the network. Nonbifurcated routing requires all the traffic associated with a given transmission to be sent over the same path.

The actual routing scheme used depends on the characteristics of the devices and the technology employed in the network. This is demonstrated in the examples below [Cahn98, p. 217]:

- Router networks frequently use a fixed minimum routing or minimum hop routing scheme.
- SNA uses a static, arbitrary, multiple, bifurcated routing scheme.
- Multiplexer-based networks generally use minimum distance or minimum hop routing schemes.

2.3.4   Architectures

Network architectures define the rules and structures for the network operations and functions. Before the introduction of network architectures, application programs interacted directly with the communications devices. This meant that programs had to be written explicitly to work with specific devices and network technologies. If the network was changed, the application programs needed to be modified accordingly. The purpose of a network architecture is to shield application programs from specific network details and device operations. As long as the application programs adhere to the standards defined by the architecture, the architecture can handle specific device and network implementation details.

Most networks are organized as a series of layers, each built upon its predecessor. The number of layers used and the function of each network layer varies by protocol and by vendor. The purpose of each layer is to off-load work from successively higher layers. This divide and conquer strategy is designed to reduce the complexity of managing the various network functions. Layer n on one machine operates with layer n on another machine. The rules and conventions used in this interaction are collectively known as the layer n protocol. Peer processes represent entities operating at the same layer on different machines. Interacting peer processes pass data and control information to the layer immediately below, until the lowest level is reached. Between each adjacent pair of layers is an interface that defines the operations and services the lower layer offers to the upper one. The network architecture defines these layers and protocols. Among other things, the network architecture provides a means for establishing and terminating device connections, specifying traffic direction (i.e., simplex, full-duplex, half-duplex), error control handling, and methods of controlling congestion and traffic flow.

There are both open and proprietary network architectures. IBM's proprietary network architecture — Systems Network Architecture (SNA) — is an example of one of the earliest network architectures. It is based on a layered architecture. Although SNA was originally designed for centralized mainframe communications, it has been continually updated over the years and supports distributed and peer-to-peer communications. It is estimated that approximately 50,000 data centers worldwide use the SNA network architecture [Data94]. Digital Equipment Corporation's digital network architecture (DNA) is another example of a proprietary networking architecture. It too uses a layered approach. Novell's NetWare and Banyan's VINES are two popular examples of proprietary LAN network architectures.

One of the primary issues in using a proprietary-based architecture is that it tends to lock the user into a single vendor vision or solution. This may also make it more difficult to incorporate other third-party products and services into the network. In response to strong market pressures and technological evolution, all the above-named proprietary architectures have gradually evolved toward more open standards.

The International Systems Interconnection (ISO) protocol was developed by the International Organization for Standardization to provide an open network architecture. The OSI model is based on seven layers. The lowest level is the physical layer, which specifies how bits are physically transmitted over a communications channel. The next layer is the data link layer. This layer creates and converts data into frames so that transmissions between adjacent network nodes can be synchronized. The third layer is the network layer. This layer determines how packets received from the transport layer are routed from source to destination within the network. The fourth layer is the transport layer. This layer is responsible for providing end-to-end control and information exchange. It accepts data from the session layer and passes it to the network layer. The next highest layer is the session layer. The session layer allows users on different machines to establish sessions between each other. The presentation layer performs syntax and semantics checks on the data transmissions, and structures the data in required display and control formats. It may also perform cryptographic functions. The highest layer, the application layer, provides an interface to the end user. It also employs a variety of protocols (e.g., a file transfer protocol).

Network architecture considerations come into play when the network is being implemented because the network architecture has profound impacts on the types of devices, systems, and services the network can support and on how the network can be interconnected with other systems and networks. The network architecture also has a significant impact on what and how new products and services can be integrated into the network, since any additions must be compatible with the architecture in use. Some of the key decisions involved in selecting a network architecture include:

1. Open or proprietary architecture — In making this decision, it is helpful to keep in mind that the full promise of open architectures has yet to be achieved. Although there is steady progress toward open architectures, they are not fully implemented in the marketplace. An expedient compromise might be to select a network or network components that encompass a subset of OSI functionality.
2. Selection of network management protocol — Network management protocols come in both proprietary and open varieties. The requirements for network management may dictate the selection of one over the other. For example, it may be necessary to manage a diverse array of third-party devices. This might influence the decision to adopt a network management protocol that can successfully integrate device management across multiple platforms.
3. Specification of application requirements — Depending on the requirements at hand, a particular network architecture may be selected because it facilitates important applications that must be supported by the network.
4. Special device requirements — A requirement for specific devices supported by the network architecture may influence its selection.
5. Selection of communication services — All network architectures support traditional digital (e.g., T1, fractional T1, and T3 lines) and analog lines. However, the use of various other network services may dictate specific network architecture requirements. For example, the network architecture must explicitly support satellite data links, if satellite services are to be used. As another example, networks offering frame relay, SMDS, or ATM services also require specialized network architectures.
6. Future plans and expected growth — Plans for future network migrations may influence the selection of a network architecture, particularly if the network architecture under consideration is moving in a direction consistent with the evolution of the organization's needs.

2.3.5 Network management

The importance of being able to manage the network after it is implemented is gaining increasing recognition in the marketplace. This is reflected in the emergence of both proprietary and open (e.g., SNMP[7] and CMIP[8]) network

---

[7] SNMP — This stands for simple network management protocol. This protocol is designed to work with TCP/IP and establishes standards for collecting information and for performing security, performance, fault, accounting, and configuration functions associated with network management.
[8] CMIP –The common management information protocol was designed, like SNMP, to support network management functions. However, it is more comprehensive in scope and is designed to work with all systems conforming to OSI standards. It also requires considerably more overhead to implement than does SNMP.

management protocols. These protocols provide a means to collect information and to perform network management functions relating to:

- Configuration management — This involves collecting information on the current network configuration, and managing changes to the network configuration.
- Performance management — This involves gathering data on the use of network resources, analyzing this data, and acting on the insights provided by the data to maintain optimal system performance.
- Fault management — This involves identifying system faults as they occur, isolating the cause of the fault(s), and correcting them, if possible.
- Security management — This involves identifying locations of sensitive data, and securing the network access points as appropriate to limit the potential for unauthorized intrusions.
- Accounting management — This involves gathering data on resource utilization. It may also involve setting usage quotas and generating billing and usage reports.

The selection of a network management protocol can have significant impacts on the network costs and on the selection of the network devices and systems. For example, IBM's proprietary network management system — NetView — is expensive and requires an IBM operating system; however, it provides comprehensive network management functionality for both SNA and non-SNA networks [Dyme94]. In selecting a network management approach, the benefits must be weighted against costs, compatibility issues, and network requirements.

In general, network management is easier when the network is simple, homogeneous, and reliable. Designing a network with this in mind means that complexity is to be avoided unless it serves a good purpose. Network complexity should reflect a requirement for services and functions that cannot be provided by simpler solutions. All other things being equal, it is better to have a network comprised of similar, compatible components and services. A network that is robust, reliable, and engineered to support growth will be easier to maintain than a network with limited capacity. Thus, a network with good manageability characteristics should be given preference over designs that are harder to manage, particularly when these benefits can be achieved without incurring significantly higher costs. Network manageability may also be enhanced by careful vendor selection. In this context, vendors who guarantee the quality and continuity of network products and services are preferable to those who do not.

Network management encompasses all the processes needed to keep the network up and running at agreed upon service levels. Network management involves the use of various management instruments to optimize the network operation at a reasonable cost. Network management is most effective when a single department or organization controls it. The major players and functions in the network management process are:

- Clients
- Client contact point(s)
- Operations support
- Fault tracking
- Performance monitoring
- Change control
- Planning and design
- Billing and finance

Clients represent internal or external customers or any other users of network management services. Clients may report problems, request changes, order equipment or facilities, or ask for information through an assigned contact point. Ideally, this should be a single point of contact. The principal activities of the contact point include:

- Receiving problem reports
- Handling calls
- Handling and processing inquiries
- Receiving change requests
- Handling orders
- Making service requests
- Opening and referring trouble tickets
- Closing trouble tickets

The contact point forwards trouble tickets (i.e., problem reports) to operations support. In turn, operations support may respond with the following types of activities:

- Problem determination by handling trouble tickets
- Problem diagnosis
- Corrective actions
- Repair and replacement of software and/or equipment
- Referrals to third parties
- Backup and reconfiguration activities
- Recovery processes
- Logging and documenting events and actions

It is possible that various troubleshooting activities by clients and/or operations support may result in change control requests. Problem reports and change requests should be managed by a design group (usually in operations support) assigned to fault monitoring. The principal functions of fault monitoring include:

- Manual tracking of reported or monitored faults
- Tracking progress on status of problem resolution and escalating the level of intervention, if necessary

- Information distribution to appropriate parties
- Referral to other groups for resolution and action

Fault monitoring is a key aspect of correcting service and quality-related problems. Fault monitoring often results in requests for various system changes. These requests are typically handled by a change control group. Change control deals with:

- Managing, processing, and tracking service orders
- Routing service orders
- Supervising the handling of changes

After the change requests have been processed and validated, they should be reviewed and acted on by the group designated to perform planning and design. Planning and design performs the following tasks:

- Needs analysis
- Projecting application load
- Sizing resources
- Authorizing and tracking changes
- Preparing purchase orders
- Producing implementation plans
- Establishing company standards
- Quality assurance

The recommendations made by planning and design are generally then passed on to finance and billing and to implementation and maintenance. Implementation and maintenance makes changes and processes work orders approved by planning and design and by change control. In addition, this area is in charge of:

- Implementing change requests and work orders
- Maintaining network resources
- Performing periodic inspections
- Maintaining database that tracks various network components and their configuration
- Performing network provisioning

Network status and performance information should be continuously monitored. Ideally, fault monitoring should be proactive in detecting problems and in opening and referring trouble tickets to the appropriate departments. Performance monitoring deals with:

- Monitoring the system and network performance
- Monitoring service level agreements and how well they have been satisfied

- Monitoring third party and vendor performance
- Performing optimization, modeling, and network tuning activities
- Reporting usage statistics and trends to management and to users
- Reporting service quality status to finance and billing

Security management is also a vital part of network management. It is responsible for ensuring secure communication and protecting the network operations. It supports the following functions:

- Threat analysis
- Administration (access control, partitioning, authentication)
- Detection (evaluating services and solutions)
- Recovery (evaluating services and solutions)
- Protecting the network and network management systems

Systems administration is responsible for administering such functions as:

- Software version control
- Software distribution
- Systems management (upgrades, disk space management, job control)
- Administering the user-definable tables (user profiles, router tables, security servers)
- Local and remote configuring resources
- Name and address management
- Applications management

Finance and Billing is the focal point for receiving status reports regarding service-level violations, network plans, designs, and changes, and invoices from third parties. Finance and Billing is responsible for:

- Asset management
- Analysis and assignment of service costs
- Billing clients
- Collection of usage and system outage data
- Calculating rebates to clients
- Bill verification
- Software license control

The instruments available to support each of the network management functions are highly varied in sophistication, scope, and ease of use. The selection of the appropriate tools and organizational processes to support the network management functions is heavily influenced by the business context in which the network is being operated.

2.3.6   *Security*

Network security requirements are not explicitly considered during the execution of topological network design algorithms. Nonetheless, security considerations may have considerable impact on the choice of network devices and services. For example, an often cited reason for a private network, as opposed to a public network, is the need for control and security.

There are many ways to compromise a network's security, either inadvertently or deliberately. Therefore, to be effective, network security must be comprehensive and should operate on several levels. Threats can occur from both internal and external sources, and can be broadly grouped into the following categories:

- *Unauthorized access to information* — This type of threat includes wiretapping and people correctly guessing a password to gain access to a system they are not authorized to use.
- *Masquerading* — This type of threat occurs when someone gains access to the network by pretending to be someone else. An example of this type of threat is a Trojan horse. An example of a Trojan horse is a software routine that appears to be benign and legitimate but is not. A Trojan horse masquerading as a log-on procedure can prompt a network user to supply the password required to gain entry to the system. The network user may never even know that they have given away their password to the Trojan horse!
- *Unauthorized access to physical network components* — This type of threat might occur if someone were to cut through a communications link while making building repairs or if a bomb were to be exploded where it could disrupt the network.
- *Repudiation* — This threat occurs when someone denies having used the network facilities in an inappropriate or improper manner. An example of this is someone sending harassing e-mail to another person, while denying it.
- *Unauthorized denial of service* — This threat occurs when a user prevents other users from gaining access to the network to which they are entitled. This may occur if the network is inundated with traffic flooding the network, thus blocking entry to the system. This type of threat can be caused by intentional or unintentional acts.

One level of security is offered by protocol security, and thus it is important to assess the level of vulnerability posed by the presence or lack of good protocol security in the network. For example, SNMP and other network management protocols that have been designed with security in mind can be used to identify and protect the network against unauthorized use. IP[9]

---

[9] IP — This stands for Internet protocol. It controls the network layer protocol of the TCP/IP protocol suite.

networks, on the other hand, are potentially vulnerable to source address spoofing. Spoofing is a form of masquerading where packets appear to come from a source that they did not. IP networks are also susceptible to packet flooding caused by an open connection. This creates system overloads that may lead to a denial of service on the network. A good defense against this type of attack is to configure routers and firewalls in the network to filter out incoming packets that are not from approved sources. In future versions of IP, new security provisions will undoubtedly become available. For example, in IP Version 6, recommendations have been made to provide for IP authentication headers and IP encapsulating security payload. Other emerging security protocols being developed to protect Internet traffic include secure sockets layer (SSL) and secure hypertext transport protocol (SHTTP).

Operational security provides a second level of network security. Operational security involves disabling network services that are not necessary or appropriate for various types of users. In this context, remote login and file transfer protocols may be disabled or controlled so that viruses and unauthorized personnel are prevented from gaining entry to the network. Operational security also involves such good practices as changing passwords regularly, constant use of updated antivirus programs, ongoing monitoring of anonymous and guest system access, and enabling and reviewing security logs and alerts.

Network security can also be implemented at the physical level. This approach attempts to safeguard access to the network by securing network components and limiting access to authorized personnel only.

Network security can be implemented at the data level. This involves the use of encryption technology to protect the confidentiality of data transmissions. Use of encryption technology implies that both the sender and the receiver must employ compatible procedures to encrypt and decrypt data. This, in turn, has implications on the management and implementation of the network services.

There are two major forms of encryption: single-key and public/private key. An overview of single-key cryptography is provided in Figure 2.5. DES is a widely used single-key encryption scheme. With DES, the data to be encrypted is subjected to an initial permutation. The data is then broken down into 64-bit data fields that are in turn split. The resulting two 32-bit fields are then further permuted in a number of iterations. Like all secret key encryption schemes, DES uses the same key to perform both encryption and decryption. The DES key, by convention, is 56 bits long.

Seminal work by Diffie and Hellman and by Rivest, Shamir, and Adleman (RSV) led to the development of public/private key cryptography. In this scheme, data is encrypted with a public key that can be known by many and is decrypted by a private key known only to one. The beauty of public key encryption is that it is computationally infeasible to derive the decipherment algorithm from the encipherment algorithm. Therefore, dissemination

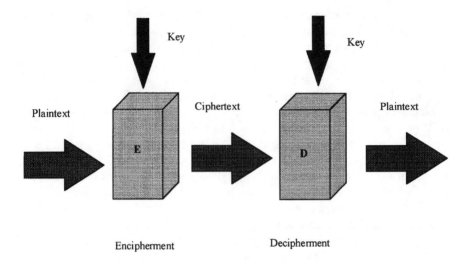

Figure 2.5   Secret key example.

Legend:

Plaintext   = Data before encryption
Ciphertext   = Data after encryption
E   = Encryption function
D   = Decryption function
Key   = Parameter used in cipher to ensure secrecy
Random Seed = Randomly selected number used to generate public and secret keys

of the encryption key does not compromise the confidentiality of the decryption process. Since the encryption key can be made public, anyone wishing to send a secure message can do so. This is in contrast to secret key schemes that require both the sender and receiver to know and safeguard the key. Public key encryption is illustrated in Figure 2.6.

One application of public key cryptography is the generation of digital signatures. A digital signature assures the receiver that the message is authentic, i.e., the receiver knows the true identity of the sender and that the contents of the message cannot be modified without leaving a trace. A digital signature is very useful for safeguarding contractual and business related transmissions, since it provides a means for third party arbitration and validation of the digital signature. Public and private keys belong to the sender, who creates keys based on an initial random number selection (or random seed). The message recipient applies the encipherment function using the sender's public key. If the result is plaintext, then the message is considered valid. Digital signatures are illustrated in Figure 2.7.

In summary, comprehensive network security involves active use of protocol, operational, and encryption measures. Good management oversight and employee training complement these security measures.

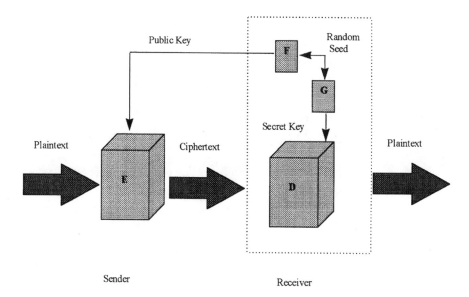

Legend:

Plaintext	= Data before encryption
Ciphertext	= Data after encryption
E	= Encryption function
D	= Decryption function
Key	= Parameter used in cipher to ensure secrecy
Random Seed	= Randomly selected number used to generate public and secret keys

Figure 2.6 Public key cryptography.

2.4 Representation of networks using graph theory

There are numerous rigorous mathematical techniques for solving network design problems. Many of these techniques are based on graph theory. Graph theory provides a convenient and useful notation for representing networks. This, in turn, facilitates the mathematical and computerized implementation of network design algorithms.

When we introduce network design algorithms in later sections, we will need the following definitions and nomenclature relating to graph theory.

Definition 2.4.1: Graph

A graph G is defined by its vertex set (nodes) V and its edges (links) E.

Definition 2.4.2: Link

A link is a *bidirectional* edge in which the ordering of the nodes attached to the link does not matter. A link can be used to represent network traffic

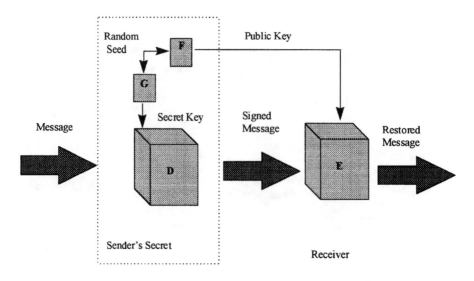

Legend:

Plaintext	= Data before encryption
Ciphertext	= Data after encryption
E	= Encryption function
D	= Decryption function
Key	= Parameter used in cipher to ensure secrecy
Random Seed	= Randomly selected number used to generate public and secret keys (i.e., F and G)

Figure 2.7   Digital signature.

flowing in either direction. Full duplex lines in a communications network support traffic in both directions simultaneously and are often represented as links in a graph.

Definition 2.4.3: Undirected graph
An undirected graph contains only bidirectional links. See Figure 2.8 for an illustration.

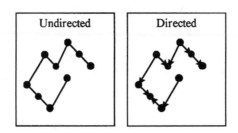

Figure 2.8   Examples of undirected and directed graphs.

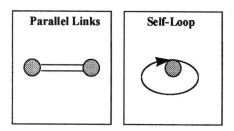

Figure 2.9 Parallel links vs. self-loop.

Definition 2.4.4: Arc
An arc is a link with a specified direction between two nodes. Half-duplex lines in a communications network handle traffic in only one direction at a time and can be represented as arcs in a graph.

Definition 2.4.5: Directed graph
A directed graph is a graph containing arcs. See Figure 2.8 for an illustration and comparison with an undirected graph.

Definition 2.4.6: Self-loop
A self-loop is a link that begins and ends with the same node. See Figure 2.9 for an illustration of a self-loop.

Definition 2.4.7: Parallel link
Two links are considered parallel if they start and terminate on the same nodes. See Figure 2.9 for an illustration.

Definition 2.4.8: Simple graph
A simple graph is a graph without parallel links or self-loops. Most network design algorithms assume that the network is represented as a simple graph.

Definition 2.4.9: Adjacency
Two nodes i and j are adjacent if there exists a link (i, j) between them. Adjacent nodes are also called *neighbors*.

Definition 2.4.10: Degree
The degree of a node is the number of links incident on the node or the number of neighbors the node has.

Definition 2.4.11: Incident link
A link is said to be incident on a node if the node is one of the link's end points.

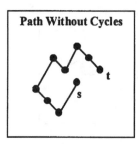
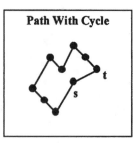

Figure 2.10 Examples of path without cycles and path with cycles.

Definition 2.4.12: Path
A path is a sequence of links which begins at an initial node, s, and ends at a specified node, t. A path is sometimes designated as (s, t).

Definition 2.4.13: Cycle
A cycle exists if the starting node, s, in a path (s, t) is the same as the terminating node, t.

Definition 2.4.14: Simple cycle
A simple cycle exists if the starting node s is the same as the terminating node t and all intermediate nodes between s and t appear only once. See Figure 2.10 for an example of a graph with a cycle and a graph with no cycles.

Definition 2.4.15: Connected graph
A graph is considered connected if at least one path exists between every pair of nodes.

Definition 2.4.16: Strongly connected graph
A directed graph with a directed path from every node to every other node is considered a strongly connected graph.

Definition 2.4.17: Tree
A tree is a graph that does not contain cycles. Any tree with n nodes will contain (n-1) edges. See Figure 2.11 for an example of a tree graph.

Definition 2.4.18: Minimal spanning tree
A minimal spanning tree is a connected graph that links all nodes with the least possible total cost or length and does not contain cycles. (Assume that a weight is associated with each link in the graph. This weight might represent the length or cost of the link.).

Definition 2.4.19: Star
A graph is considered a star if only one node has a degree greater than 1. See Figure 2.11 for an example of a star graph and a comparison with a tree graph.

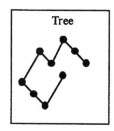 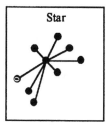

Figure 2.11 Tree graph vs. star graph.

2.5 Introduction to algorithms

Many network design problems are solved using special techniques called algorithms. Given the importance of algorithms in solving network design problems, we take time at this juncture to formally define and review important properties of algorithms.

2.5.1 Definition of algorithms

An algorithm is a well-defined procedure for solving a problem in a finite number of steps. An algorithm is based on a model that characterizes the essential features of the problem. The algorithm specifies a methodology to solve the problem using the model representation.

Algorithms are characterized by a number of properties. These properties are necessary to ensure that the algorithm correctly solves the problem for which it is intended, or correctly identifies when a solution is impossible to find, in a finite number of steps. These properties include:

- Specified inputs: The inputs to an algorithm must be from a prespecified set.
- Specified outputs: For each set of input values, the algorithm must produce outputs from a prespecified set. The output values produced by the algorithm comprise the solution to the problem.
- Finiteness: An algorithm must produce a desired output after a finite number of steps.
- Effectiveness: It must be possible to perform each step of the algorithm exactly as specified.
- Generality: The algorithm should be applicable to all problems of the desired form. It should not be limited to a particular set of input values or special cases.

To illustrate, we now present an example of a greedy algorithm. The algorithm is considered greedy because it selects the best choice immediately available at each step, without regard to the long-term consequences of each selection in totality and in relation to each other.

Table 2.3 List of Potential Link Costs

From node	To node	Link cost
A	B	1
A	C	6
A	D	+∞
B	C	5
B	D	+∞
C	D	2

We will use the greedy algorithm to find the cheapest set of links to connect all of a given set of terminals. A graphical representation is used to model our network design problem. Using this representation, all terminal devices are modeled as nodes (a, b, c, and d), and all communications lines are modeled as links. Associated with each possible link (i, j), — where i is the starting node and j is the terminating node — is a weight, representing the cost of the link if it is used in the network. A cost of +∞ is used to indicate when a link is prohibitive in cost or is not available. The link costs for our sample problem are summarized in Table 2.3.

Greedy algorithm (also known as Kruskal's algorithm)

1. Sort all possible links in ascending order and put in a link list.
2. Check to see if all the nodes are connected.
   - If all the nodes are connected, terminate the algorithm, with the message "Solution complete."
   - If all the nodes are not connected, continue to next step.
3. Select the link at the top of the list.
   - If no links are on the list, then terminate the algorithm. Check to see if all nodes are connected, and if not, terminate the algorithm with the message "Solution cannot be found."
4. Check to see if the link selected creates a cycle in the network.
   - If the link creates a cycle, remove it from the list. Return to step 2.
   - If the link does not create a cycle, add it to the network, and remove link from link list. Return to step 2.

We now solve the sample problem using the algorithm specified above:

1. Sort all possible links in ascending order and put in a link list. The results are summarized in Table 2.4.

Table 2.4 Sorted Link List

From node	To node	Link cost
A	B	1
C	D	2
B	C	5
A	C	6
A	D	+∞
B	D	+∞

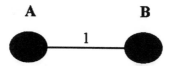

Figure 2.12 First link selected by greedy algorithm.

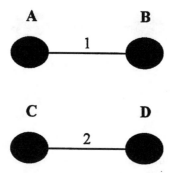

Figure 2.13 Second link selected by greedy algorithm.

2. Check to see if all the nodes are connected. *Since none of the nodes are connected, we proceed to the next step.*
3. Select the link at the top of the list. *This is link AB.*
4. Check to see if the link selected creates a cycle in the network. *It does not, so we add link AB to the solution and remove it from the link list. We obtain the partial solution shown in Figure 2.12 and proceed with the algorithm.*
5. Check to see if all the nodes are connected. *They are not, and we proceed to the next step of the algorithm.*
6. Select the link at the top of the list. *This is link CD.*
7. Check to see if the link selected creates a cycle in the network. *It does not, so we add link CD to the solution, and remove it from the link list. We obtain the partial solution shown in Figure 2.13 and proceed with the algorithm.*
8. Check to see if all the nodes are connected. *They are not, and we proceed to the next step.*
9. Select the link at the top of the list. *This is link BC.*
10. Check to see if the link selected creates a cycle in the network. *It does not, so we add link BC to the solution, and remove it from the link list. We obtain the partial solution shown in Figure 2.14 and proceed with the algorithm.*
11. Check to see if all the nodes are connected. *All the nodes are now connected, and we terminate the algorithm with the message "Solution complete." Thus, Figure 2.14 represents the final network solution.*

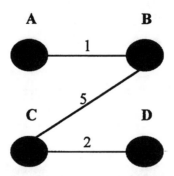

Figure 2.14 Third and final link selected by greedy algorithm.

Using the checklist below, we verify that the greedy algorithm exhibits all the necessary properties we have defined:

☑ Specified inputs: The inputs to the algorithm are the prespecified nodes, potential links, and potential link costs.

☑ Specified outputs: The algorithm produces outputs (link selections) from the prespecified link set. The outputs produced by the algorithm comprise the solution to the problem.

☑ Finiteness: The algorithm produces a desired output after a finite number of steps. The algorithm stops when all the nodes are connected, or after all the candidate links have been examined, whichever comes first.

☑ Effectiveness: It is possible to perform each step of the algorithm exactly as specified.

☑ Generality: The algorithm is applicable to all problems of the desired form. It is not limited to a particular set of input values or special cases.

The greedy algorithm just described provides an optimal network solution when there are no restrictions on the amount of traffic that can be placed on the links. In this form, the algorithm is called an "unconstrained optimization technique." However, it is not realistic to assume that links can carry an indefinite amount of traffic. When steps to check the line capacity restrictions are added to the algorithm, the algorithm becomes a constrained optimization technique. When the algorithm is constrained, and the restrictions on the algorithm are active (e.g., this would occur if the traffic limit is reached at some point, and therefore a link initially selected for inclusion in the design must be rejected) there is no longer a guarantee that the algorithm will produce an optimal result.

Many studies have been conducted on the constrained form of Kruskal's greedy algorithm. These studies show that although the greedy algorithm does not necessarily produce an optimal, best case result, in general it produces very good results that are close to optimal [Kersh93].

In summary, an algorithm is considered good, if it always provides a correct answer to the problem or indicates when a correct answer cannot be found. A good algorithm is also efficient. In the next section, we discuss what it means to be an efficient algorithm.

2.5.2   Introduction to computational complexity analysis

The efficiency of an algorithm can be measured in several ways. One estimate of efficiency is based on the amount of computer time needed to solve the problem using the algorithm. This is also known as the *time complexity* of the algorithm. A second estimate of efficiency is the amount of computer memory needed to implement the algorithm. This is also referred to as the *space complexity* of the algorithm. Space complexity is very closely tied to the particular data structures used to implement the algorithm.

In general, the actual running time of an algorithm implemented in software will largely depend upon how well the algorithm was coded, the computer used to run the algorithm, and the type of data used by the program. However, in complexity analysis we seek to evaluate an algorithm's performance *independent of its actual implementation*. To do this, we must consider factors that remain constant irrespective of the algorithm's implementation.

Since we want a measure of complexity that is not dependent upon processing speed, we ignore space complexity, because it is so closely tied with implementation details. Instead, we will use time complexity as a measure of an algorithm's efficiency. We will measure time complexity in terms of the number of *operations* required by the algorithm instead of the actual CPU time the algorithm requires. Expressing time complexity in these units allows us to compare the efficiency of algorithms that are very different.

For example, let N be the number of inputs to an algorithm. If we are told that algorithm A requires a number of operations proportional to N^2, and algorithm B requires a number of operations proportional to N, we can see that algorithm B is more efficient. If N is 4, then algorithm B will require approximately four operations, while algorithm A will require 16. As N becomes larger, the difference in efficiency between algorithms A and B becomes more apparent.

When we are told that algorithm B requires time proportional to f(N), this also implies that given any reasonable computer implementation of the algorithm, there is some constant of proportionality C such that algorithm B requires no more than (C * f(N)) operations to solve a problem of size N. Algorithm B is said to be of order f (N) — which is denoted O (f (N)) — and f (N) is called the algorithm's growth rate function. Because this notation uses the capital letter O to denote *order*, it is called the Big O notation.

Some examples and explanations for the Big O notation are given below:

- If an algorithm is O(1), this means it requires a constant time that is independent of the problem's input size N.
- If an algorithm is O(N), this means it requires time that is directly proportional to the problem's input size N.
- If an algorithm is O(N²), this means it requires time that is directly proportional to the problem's input size N².

In Table 2.5, we summarize commonly used terminology describing computational complexity. The terms listed below are sorted from low to high complexity. In general, network design algorithms are considered efficient if they are of O(N²) complexity or less. The greedy algorithm we examined in the last section can be shown to be O(N log N), where N is the number of edges examined, and is considered very computationally efficient [Kersh93]. Most of the effort expended in executing the greedy algorithm presented in the last section goes toward creating the sorted link list created in the first step.

Table 2.5 Computational Complexity Terminology

Complexity	Terminology
O(1)	Constant complexity
O(log n)	Logarithmic complexity
O(n)	Linear complexity
O(n log n)	n log n complexity
$O(n^b)$	Polynomial complexity
$O(b^n)$, where b > 1	Exponential complexity
O(n!)	Factorial complexity

Brute force and exhaustive search algorithms are considered strategies of last resort. Using these methods, all the potential solution candidates are examined one by one, even after the best one has been found. This is because these methods do not recognize the optimal solution until the end, after all the candidates have been compared. Brute force and exhaustive search methods are usually $O(b^N)$ or worse. The worst computational complexity is factorial complexity, which is generally associated with n-p (i.e., nonpolynomial time) complete problems. Public key cryptography and traveling salesman problems are two examples of n-p complete problems. In general, when the input size is large, n-p complete problems are exceedingly difficult to solve and require very large amounts of computing time. Figure 2.15 shows the effect of increasing computational complexity on the number of operations required by the algorithm to solve the problem.

A big O estimate of an algorithm's time complexity expresses how the time required to solve the program changes *as the input grows in size.* Big O

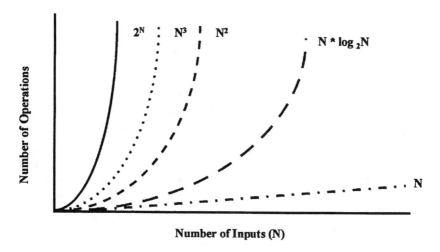

Number of Inputs (N)

Figure 2.15   Growth rate function comparison.

estimates do not directly translate into actual computer time, since the big O method uses a simplified reference function to estimate complexity. The simplified reference function *omits* constants and other terms that may affect actual computer time. Thus, the big O method provides a *lower* bound on computer time. We illustrate this in the examples that follow.

Example 1
If algorithm is: $O(N^3 + N^2 + 3N)$, we can ignore the low order terms in the growth rate function. Therefore, the algorithm's complexity can be expressed by a simplified growth function: $O(N^3)$.

Example 2
If algorithm is: $O(5N^3 + 2N^2 + 3N)$, we can ignore multiplicative constants in the high order terms of the growth rate function. Therefore, the algorithm's complexity can be expressed by a simplified growth function: $O(N^3)$.

Example 3
If algorithm is: $O(N^3) + O(N^3)$, we can combine the growth rate functions. Therefore, the algorithm's complexity can be expressed by a simplified growth function: $O(2 * N^3) = O(N^3)$.

A formal generalization of these examples is given below.

If: we are given $f(x)$ & $g(x)$, functions from the set of real numbers
Then: we can say $f(x) = O(g(x))$
If and only if: there are constants C and k such that $| f(x) | \le C | g(x) |$
                              whenever $x > k$

We provide an example to illustrate:

Show: $f(x) = O(x^2 + 2x + 1)$ is $O(x^2)$

Solution: Since $0 \le (x^2 + 2x + 1) \le (x^2 + 2x^2 + x^2) = 4x^2$,

                when ever $x > 1 = k$, and $C = 4$

                Therefore, it follows: $f(x) = O(x^2)$

Complexity analysis can be used to examine worst case, best case, and average scenarios. Worst case analysis tries to determine the largest number of operations needed to guarantee a solution will be found. Best case analysis seeks to determine the smallest number of operations needed to find a solution. Average case analysis is used to determine the average number of operations used to solve the problem, assuming all inputs are of a given size.

It should be noted that time complexity is not the only valid criterion to evaluate algorithms. For example, other important criteria might include the style and ease of the algorithm's implementation. The appeal of time complexity, as we have described here, is the fact that it is independent of specific implementation details and provides a very robust way of characterizing algorithms so they can be compared.

It should also be noted that different orders of complexity do not always matter much in an actual situation. For example, a more complex algorithm may run faster than a less complex algorithm with a high constant of proportionality, especially if the number of inputs to the problem is small. Time complexity, as presented here, is particularly important when big problems are to be solved. When there are very few inputs, an algorithm of any complexity will usually run quickly.

In summary, computational complexity is important because it provides a measure of the computer time required to solve a problem using a given algorithm. If it takes too long to solve a problem, the algorithm may not be useful or practical. In the context of network design, we need algorithms that can provide reasonable solutions in a reasonable amount of time. Consider the fact that if there are n potential node locations in a network, then there are $2^{(n * (n-1)/2)}$ potential topologies. Thus, brute force and computationally complex design techniques are simply not suitable for network problems of any substantial size.

2.5.3   Network design techniques

Three major types of techniques are used in network design: heuristic, exact, and simulation techniques. Heuristic algorithms are the preferred method for solving many network design problems, because they provide a close to optimal solution in a reasonable amount of time. For this reason, we concentrate the bulk of our attention in this book on heuristic solution techniques or algorithms.

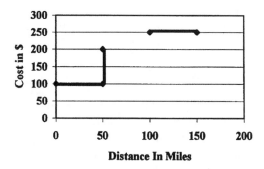

Figure 2.16 Graph with discontinuities.

Linear programming is a powerful, exact solution technique that is also used in network design. Linear programming methods are based on the simplex method. The simplex method guarantees that an optimal solution will be found in a finite amount of time. Otherwise, the simplex algorithm will show that an optimal solution does not exist or is not feasible given the constraints that have been specified. Linear programming models require that both the objective function and the constraint functions be linear. Linear programming also requires that the decision variables be continuous, as opposed to discrete.

In the context of network design, we might want to find a low-cost network (i.e., we want to minimize a cost function based on link costs or tariffs), subject to constraints on where links can be placed and constraints on the amount of traffic that the links can carry. Although linear functions may provide useful approximations for the costs and constraints to be modeled in a network design problem, in many cases they do not. The tariffs that determine the line costs are usually nonlinear and may exhibit numerous discontinuities. A discontinuity exists when there is a sharp price jump from one point to the next, or when certain price/line combinations are not available, as illustrated in Figure 2.16. Linear programming may be used successfully when the cost and constraint functions can be approximated by a linear or piecewise linear function that is accurate within the range of interest. This implies that the linear programming approach is best applied, in the context of network design, when the neighborhood of the solution can be approximated a priori.

When the decision variables for designing the network are constrained to discrete integer values, the linear programming problem becomes an integer programming problem. Integer programming problems, in general, are much harder to solve than a linear programming problem, except in selected special cases. In the case of network design, it might be desirable to limit the constraints to zero (0) or one (1) values to reflect whether or not a link is being used. However, these restrictions complicate the problem a great deal. By complicate, we mean that the computational complexity of the integer programming technique increases to the point where it may be impractical to use. For more information on linear programming and integer programming techniques, the reader is referred to [Mino93] and [Wagn75].

Simulation is a third commonly used technique in network design. Simulation is often used when the design problem cannot be expressed in an analytical form that can be solved easily and/or exactly. A simulation tool is used to build a simplified representation of the network. Experiments are then conducted to see how the proposed network will perform as the traffic and other inputs are modified. In general, simulation approaches are very time consuming and computationally intensive. However, they are helpful in examining the system performance under a variety of conditions to test the robustness of the design. There are a number of software packages available that are designed exclusively for simulation and modeling studies. Two of the better known commercially available simulation packages are SIMSCIPT and GPSS.

In summary, as network size increases, so does the computational complexity of the network design process. This occurs to such an extent that for most problems of practical interest, exact optimal solution techniques are impractical, except in special cases. Heuristic design techniques providing approximate solutions are usually used instead. Although in many cases, good solutions can be obtained with heuristic techniques, they do not necessarily guarantee that the best solution will be found. After an approximate solution is found using a heuristic method, it may be helpful to use an exact solution technique or simulation to fine-tune the design.

2.6   Performance analysis

Once a network has been designed, its performance characteristics can be analyzed. The most common measures of network performance are cost, delay, and reliability. In this section, we present an overview of the methods used to calculate these performance measures. However, other measures of network performance — for example, link, memory, and node utilization — are possible and may also be helpful in evaluating the quality of the network design. For more information on other network performance measures and their calculation, the reader is referred to [Cahn98] and [Kersh93].

2.6.1   Queuing essentials

In this section, we summarize some of the major results from queuing theory that are applicable to the analysis of network delay and reliability. We concentrate on results that are easily grasped using basic techniques of algebra and probability theory. We refer readers who want more information on this subject to [Kersh93] and [Gross74]. In addition to providing detailed theoretical derivations and proofs, these references also discuss other more complex queuing models that are beyond the scope of this book.

Queuing theory was originally developed to help design and analyze telephone networks. Since the rationale for a network is to provide a means to share resources, there is an implicit assumption that the number of resources available in the network will be less than the total number of potential system users. Therefore, it is possible that some users may have to wait until others relinquish their control over a telecommunications facility or device. Queuing analysis is used to estimate the waiting times and delays that system users will experience, given a particular network configuration.

Queuing theory is very broadly applicable to the analysis of systems characterized by a stochastic[10] input process, a stochastic service mechanism, and a queue discipline. The key descriptive features of the input process are the size of the input population, the source of the input population, interarrival times between inputs, and how the inputs are grouped, if at all. The features of interest for the service mechanism are the number of servers, the number of inputs that can be served simultaneously, and the average service time. The queue discipline describes the behavior of the inputs once they are within a queue. Given information on these characteristics, queuing theory can be used to estimate service times, queue lengths, delays while in the queue and while being serviced, and the required number of service mechanisms.

Some queuing notation is needed for the discussion that follows. We now present essential definitions and nomenclature.

Standard queuing notation

The following notation specifies the assumptions made in the queuing model:

Arrival process/Service process/Number of parallel servers/Optional additions

Optional additions to this notation are:

/Limit on number in the system/Number in the source population/Type of queue discipline

---

[10] Stochastic process — A process with events that can be described by a probability distribution function. This is in contrast to a deterministic process whose behavior is certain and completely known.

The arrival and service processes are defined by probability distributions that describe the expected time between arrivals and the expected service time. The number of servers or channels operating in parallel must be a positive integer. Table 2.6 summarizes abbreviations and assumptions commonly used in queuing models.

Table 2.6 Summary of Standard Queueing Nomenclature

Standard abbreviation	Meaning
M	Markovian or exponential distribution
E_k	Erlangian or gamma distribution with K identical phases
D	Deterministic distribution
G	General distribution
c	Number of servers or channels in parallel
FCFS	First come, first served
LCFS	Last come, first served
RSS	Random selection for service
PR	Priority service

To illustrate these abbreviations, we provide the following examples:

Table 2.7 Queuing Abbreviations

Queuing model abbreviation	Meaning
M/M/c	Markovian input process and Markovian service distribution with c parallel servers
M/M/1/n/m	Markovian input process and Markovian service distribution with 1 server and a maximum system capacity of n and a total potential universe of m customers
M/C/3/m/m	Markovian input process and constant service distribution with 3 servers and a maximum system capacity of n and a total potential universe of m customers

A Markovian distribution is synonymous with an exponential distribution. This distribution has the interesting property that the probability a system input that has already waited T time units must wait another X time units before being served is the same as the probability that an input just arriving to the system will wait X time units. Thus, the system is "memoryless" in that the time an arrival has already spent in the system does not in any way influence the time the arrival will remain in the system.

Other widely used queuing notation is summarized below in Table 2.8.

Table 2.8 Queuing Notation

Notation	Meaning
$1/\lambda$	Mean inter-arrival time between system inputs
$1/\mu$	Mean service time for server(s)
λ	Mean arrival rate for system inputs
μ	Mean service rate for server(s)
P	Traffic intensity = $(\lambda/(c * \mu))$, where c = number of service channels = utilization factor measuring maximum rate at which work entering system can be processed.
L	Expected number in the system, at steady state, including those in service
L_q	Expected number in the queue, at steady state, excluding those in service
W	Expected time spent in the system, including service time, at steady state
W_q	Expected time spent in the queue, excluding service time, at steady state
N(t)	Number of units in the system at time t
$N_q(t)$	Number of units in the queue at time t

Using the notation above, we introduce Little's law. This is a very powerful relationship that holds for many queueing systems. Little's law says that the average number waiting in the queuing system is equal to the average arrival rate of the inputs to the system multiplied by the average time spent in the queue. Mathematically, this is expressed [Gross74, p. 60]:

$$L_q = \lambda * W_q \qquad (2.5)$$

Another important queuing relationship derived from Little's law says that the average number of inputs to the system is equal to the average arrival rate of inputs to the system multiplied by the average time spent in the system. Mathematically, this is expressed [Gross74, p. 60]:

$$L = \lambda * W \qquad (2.6)$$

The intuitive explanation for these relationships goes along the following lines. An arrival, A, entering the system will wait on average W_q time units before entering service. Upon being served, the arrival can count the number of new arrivals behind it. On average this number will be L_q. The average time between each of the new arrivals is $1/\lambda$ units of time, since by definition this is the interarrival rate. Therefore, the total time it took for the L_q arrivals to line up behind A must equal A's waiting time W_q. A similar logical analysis holds for the calculation of L in Equation 2.6.

A number of steady-state models have been developed to describe queuing systems that are applicable to network analysis. A steady-state model is used to describe the long-run behavior of a queuing system after it has stabilized. It also represents the average frequency with which the system will occupy a given state[11] over a long period of time.

2.6.2   Analysis of loss and delay in networks

In the following discussion, we provide examples of queues relating to network analysis, and we illustrate how the queuing models can be used to calculate delays and service times. Before using any queuing model, it is important to first identify the type of queue being modeled (e.g., is it an M/M/2 queue or some other type of queue?). Next, the assumptions of the queue need to be examined to see if they are reasonable for the situation begin modeled. A fundamental assumption inherent in the models we present is that the system will eventually reach a steady state. This is not a reasonable assumption for a system that is always in flux and changing. One way that this assumption can be checked is to examine the utilization factor ρ (see definition above in Table 2.8). As ρ approaches 1, a queuing system will become congested with infinitely long queues and will not reach a steady state. The steady-state queuing models we present below also assume that the interarrival times and the service times can be modeled by exponential probability distributions.

Sometimes not enough is known about the arrival and service mechanism to specify a probability distribution, or perhaps what is known about these processes is too complex to be analyzed. In this case, you may have to make do with the first and second moments of the probability distributions. This corresponds to the mean and variance, respectively. These measures allow one to calculate the squared coefficient of variation of the distribution. The coefficient of variation, C^2, is defined as:

$$C^2 = V(X)/(\overline{X})^2 \quad \textit{where } V(X) \textit{ is variance of probability distribution}$$
$$\textit{and } \overline{X} \textit{ is mean of probability distribution}$$

When the arrival rate is deterministic and constant, C^2 is equal to zero. When the arrival rate is exponentially distributed, C^2 is equal to 1. When C^2 is small, the arrival rate tends to be evenly spaced. As C^2 approaches 1, the distribution approaches a Poisson distribution, with exponential interarrival time, and the process is said to be random. When C^2 exceeds 1, the probability distribution becomes bursty, with large intermittent peaks.

Except where explicitly stated otherwise, all the steady-state queuing models we present assume an infinite population awaiting service. In reality, this is seldom the case. However, these formulas still provide a good approximation when the population exceeds 250 and the number of servers is small.

---

[11] State — The state of a queuing system refers to the total number in the system, both in queue and in service. The notation for describing the number of units in the system (including those in queue and those in service) is $N(t)$. Similarly, the notation for describing the number of units in the queue at time t is: $N_q(t)$.

When the population is under 50, then a finite population model should be used [Mino96, Data94, p.18].

The steady-state models we introduce also assume that traffic patterns are consistent and do not vary according to the time of day and the source of traffic. This, too, is an unrealistic assumption for most telecommunication systems. Despite the fact that this assumption is rarely satisfied in practice, the steady-state models still tend to give very good results.

The steady-state models also assume that all the inputs are independent of each other. In a network, it is entirely likely that problems in one area of the network will contribute to other failures in the network. This might occur, for example, when too much traffic from one source creates a system overload that causes major portions of the network to overload and fail as well. Despite the fact that this assumption is also rarely satisfied in practice, the models still provide useful results in the context of network design.

One of the compelling reasons for using steady-state queuing models is that despite their inherent inaccuracies and simplifications of reality they often yield robust, good results. The models are also useful because of their simplicity and closed-form solution. If we try to interject more realism in the model, the result is often an intractable formula that cannot be solved (at least as easily). The requirements for realism in a model must always be weighed against the resulting effort that will be required to solve a more complex model.

2.6.2.1 *M/M/1 model*

This type of queue might be used to represent the flow of jobs to a single print server, or the flow of traffic on a single T1 link. Using standard queuing notation, the major steady-state relationships for M/M/1 queues are given below. For these relationships to hold, λ/μ must be less than 1. When the (λ/μ) ratio equals or exceeds 1, the queue will grow without bound and there will be no system steady state.

Probability that the system will be empty	$= 1 - (\lambda/\mu)$	(2.7)
Probability that there will n inputs in the system	$= \rho^n (1-\rho)$	(2.8)
Expected number of inputs in the system	$= L = \rho/(1-\rho)$	(2.9)
Expected number of inputs in the queue	$= L_q = \rho^2/(1-\rho)$	(2.10)
Expected total time in system	$= W = 1/(\mu - \lambda)$	(2.11)
Expected delay time in queue	$= W_q = \rho/((1-\rho) * \mu)$	(2.12)

M/M/1 example of database access delay

A very large number of users request information from a database management system. The average interarrival time of each request is 500 milliseconds. The database look-up program requires an exponential service time averaging 200 milliseconds. How long will each request have to wait on average before being processed?

Answer

As given in the problem, $\lambda = 1/.50$ s $= 2$ seconds; $\mu = 1/.20$ s $= 5$ seconds; $\lambda/\mu = \rho = 2/5$.

Therefore, the expected delay waiting in queue is $W_q = \rho/((1-\rho) * \infty) = .4/((1-.4) * 5) = .1333$ seconds.

M/M/1 example of delay on communications link

A very large number of users share a communications link. The users generate on average one transmission per minute. The message lengths are exponentially distributed with an average of 10,000 characters. The communications link has a capacity of 9600 bps. We want to know: (a) what is the average service time to transmit a message?, (b) what is the average line utilization?, (c) what is the probability that there are no messages in the system?

Answer

(a) The average service time is the message length divided by the channel speed:

$1/\mu = (10{,}000$ characters $* 8$ bits per character$)/9{,}600$ bits per second $= 8.3$ seconds

(b) The average line utilization is $\lambda/\mu = \rho = (.0167)/(.12)$ or the line is utilized at an average rate of 13.9%.

$\lambda = 1$ message per minute $* 60$ seconds per minute $= .0167$ messages per second

$\mu = 1/8.3 = .12$ messages per second

(c) The probability that the system will be empty is $1 - (\lambda/\mu) = 1 - (.139) = .861$ or 86.1% of the time the line is empty.

2.6.2.2   M/M/1/k model

This type of queue is used to represent a system that has a maximum total capacity of k positions, including those in queue and those in service. This type of queue might be used to model a network node with a limited number of ports or buffer size. Using standard queuing notation, the major steady-state relationships for a M/M/1/k queue are summarized below.

Probability that the system will be empty	$= (1-\rho)/(1-\rho^{k+1})$	(2.13)
Probability that there will n inputs in the system	$= p_k$	
	$= \rho^n (1-\rho))/(1-\rho^{k+1})$	(2.14)
Expected number of inputs in the system	$= L = \rho/(1-\rho)$	(2.15)
Expected number of inputs in the queue	$= L_q = L - \rho (1- p_k)$	(2.16)
Expected total time in system	$= W = L/(\lambda *(1- p_k)$	(2.17)
Expected delay time in queue	$= W_q = L_q/(\lambda *(1- p_k)$	(2.18)

M/M/1/5 example of jobs lost in front-end processor due to limited buffering size

Assume the buffers in a front-end Processor (FEP) can handle at most 5 input streams (i.e., $k = 5$). When the FEP is busy, up to a maximum of four jobs are buffered in queue. The average number of jobs arriving per minute is 5, whereas the average number of jobs the FEP processes per minute is 6. How many jobs on average are lost or turned away, due to inadequate buffering capacity:

Answer

A job will be turned away when it arrives at the system and there are already four jobs (one in service and three in the queue) ahead of it. Thus, to find the number of jobs that are turned away, calculate the probability that a job will arrive when there are already four jobs in the queue and one in service, and multiply this by the arrival rate.

$$\lambda = 5 \text{ per minute; } \mu = 6 \text{ per minute; } \lambda/\mu = \rho = 5/6 = .833$$

Probability that there will 5 inputs in the system $= p_5 = \rho^5(1-\rho))/(1-\rho^6)$
$$= (.833)^5/1-(.833)^6) = .10$$

Therefore: $\lambda * p_5 = (5 \text{ per minute}) * (.1) = .5$ jobs per minute are turned away.

2.6.2.3 M/M/c model

This type of queue might be used to represent the flow of traffic to c dial-up ports, or the flow of calls to a PBX with c lines, or traffic through a communications link with c multiple trunks. Using standard queuing notation, the major steady-state relationships for M/M/c queues are listed below. For these relationships to hold, $\lambda/(c * \mu)$ must be less than 1. When this ratio equals or exceeds 1, the queue will grow without bound and there will be no system-steady state. Although the equations are more complicated than the ones we have introduced thus far, they are nonetheless easily solved using a calculator or computer.

Probability that the system will be empty $= P_0 =$

$$\left[\sum_{n=0}^{c-1} (\lambda/\mu)^n /n! + \frac{(\lambda/\mu)^c}{c!\left(1 - \frac{\lambda}{c\mu}\right)} \right]^{-1} \tag{2.19}$$

The probability that the system will have n in the system $= P_n =$

$$\begin{cases} \left(\dfrac{(\lambda/\mu)^n}{n!} \right) * P_0 & \text{for } 0 \leq n \leq c \\[4mm] \left(\dfrac{(\lambda/\mu)^n}{c! c^{n-c}} \right) * P_0 & \text{for } n > c \end{cases} \tag{2.20}$$

Expected number of inputs in the system $= L \quad = \rho + L_q$ (2.21)
Expected number of inputs in the queue $= L_q \quad = \rho_2/(1-\rho)$ (2.22)
Expected total time in system $= W \quad = 1/\mu + W_q$ (2.23)
Expected delay time in the queue $= W_q \quad = L_q/\mu$ (2.24)

M/M/c example of print server processing times

Two print servers are available to handle incoming print jobs. Arriving print jobs form a single queue from which they are served on a first come first served (FCFS) basis. The accounting records show that the print servers handle an average of 24 print jobs per hour. The mean service time per print job has been measured at 2 minutes. It seems reasonable, based on the accounting data, to assume that the interarrival times and service times are exponentially distributed. The network manager is thinking of removing and relocating one of the print servers to another location. However, the system users want a third print server added to speed up the processing of their jobs. What service levels should be expected with one, two, or three print servers?

Answer

The appropriate model to use for the current configuration is (M/M/2) with $\lambda = 24$ per hour, and $\mu = 30$ per hour. The probability that both servers are expected to be idle is equal to the probability that there are no print jobs in the system. From Equation 2.19, this is calculated:

$$P_0 = \left[\sum_{n=0}^{1} (24/30)^n / n! + \frac{(24/30)2}{2!\left(1 - \frac{24}{60}\right)} \right]^{-1} = [1 + .8 + .5333]^{-1} = .43$$

One server will be idle when there is only one print job in the system. The fraction of time this will occur is:

$$P_1 = \left(\frac{24}{30}\right) * P_0 = .344$$

Both servers will be busy whenever two or more print jobs are in the system. This is computed:

$$P(Busy) = 1 - P_0 - P_1 = .226$$

The expected number of jobs in the queue is given in Equation 2.22:

$$L_q = \frac{\left(\frac{24}{30}\right)^2 \left(\frac{24}{60}\right)(.433)}{2\left(1 - \frac{24}{60}\right)} = .153$$

The expected number in the system is calculated according to Equation 2.21:

$$L = \rho + L_q = .153 = .8 = .953$$

The corresponding W and W_q waiting times, as computed from Equation 2.23 and Equation 2.24 are:

W $= 1/\mu + W_q = .953/24 = .397$ hours * 60 minutes per hour
$= 2.3825$ minutes
W_q $= L_q/\mu = .153/24 = .006375$ hours * 60 minutes per hour
$= .3825$ minutes

Similar calculations can be performed for a single print server and for three print servers. The results of these calculations are summarized in Table 2.9 below.

Table 2.9 Comparison of M/M/1, M/M/2, and M/M/3 Queues

	1 Print server	2 Print servers	3 Print servers
P_0	.2	.43	.44
W_q	8 minutes	.3825 minutes	.1043 minutes
W	10 minutes	2.3825 minutes	2.1043 minutes

M/M/1 vs. M/M/c example and comparison of expected service times

Given the same situation presented in the previous example, what would happen to the expected waiting times if a single, upgraded print server were to replace the two print servers currently used? The network manager is thinking of installing a print server that would have the capacity to process 60 jobs per hour.

Answer

The new option being considered equates to an M/M/1 queue. The new print server has an improved service rate of $\mu = 60$ jobs per hour. The calculations for the expected waiting times are shown below.

P_0 $= 1 - (\lambda/\mu) = 1 - (24/60) = .6$
W $= 1/(\lambda - \mu) = 1/(60 - 24) = .0277$ hours * 60 min./hour $= 1.662$ min.
W_q $= \rho/((1-\rho) * \mu) = .4/((1-. 4) * 60) = .1111$ hours * 60 min./hour $= .666$ min.

These calculations demonstrate a classic result; that it is always better to have a single more powerful server whose service rate equals the sum of c servers than it is to have c servers with c queues. Another important implication of this is that when a network is properly configured from a queuing perspective it may be possible to provide better service at no additional cost.

2.6.2.4 *M/G/1 model*

In the models presented thus far, we have assumed that all the message lengths are randomly and exponentially distributed. In the case of packet-switched networks, this is not a valid assumption. We will now consider the effects that packet-switched data have on the delay in the network. Let us consider an M/G/1 queue, in which arrivals are independent, a single server is present, and the arrival rate is general. Using the Pallaczek-Khintchine formula, it can be shown that the average waiting time in an M/G/1 system is [Giff78, p. 185]:

$$W_q = \frac{\lambda E\left[1/\mu^2\right]}{2(1-\rho)} \tag{2.25}$$

where $E[1/\mu^2]$ is the second moment of the service distribution.

The second moment is defined as:

$$E\left[1/\mu^2\right] = \sum_{j+1}^{M} P_j \mu_j^2 \tag{2.26}$$

where P_j = probability of a message being type j
         $1/\mu$ = service time for message of type j.

The variance V of the service distribution is given by:

$$V = E\left[(1/\mu)^2\right] - \left(E[1/\mu]\right)^2 \tag{2.27}$$

M/G/1 example of packet switch transmission delay

We are given two networks. Both networks use 56Kbps lines that are on average 50% utilized. Both networks transmit 1000 bit messages on average. The first network transmits exponentially distributed message lengths. The second network transmits packets of constant message length. Compare the waiting times for message processing in the two network configurations.

Answer

First, consider the case when the message length is constant. From the data, the average arrival rate can be computed:

$$\lambda = \frac{(56000)\,(.5)}{1000} = 28 \; messages \, / \, \sec$$

The mean service time is computed from the data as given:

$$\left[(1/\mu)^2\right] = .00648 \frac{\sec^2}{msg}$$

The waiting time can now be computed from Equation 2.25 as:

$$W_q = \frac{\lambda E\left[1/\mu^2\right]}{2(1-\rho)} = \frac{(28m/s).00648}{2(1-.5)} = .18144 \; \sec/msg$$

Now, consider the case when the message length is exponentially distributed. This corresponds to an M/M/1 queue. The waiting time calculation for this queue is given by Equation 2.12 and is calculated below:

$$\mu = \frac{1000 \; bits}{56000 \; bits/\sec} = .018 \; \sec$$

$$W_q = \frac{\rho}{\mu(1-\rho)} = \frac{.5}{.018(1-.5)} = 55.5 \; \sec/msg$$

Thus, the delay when the message lengths vary according to an exponential distribution is considerably longer than when the message lengths are constant. Note that in both of these cases the average message length is the same. This is an important result that demonstrates why, all other things being equal, packet-switched networks are more efficient than networks that transmit messages of varying length.

2.6.2.5   M/M/c/k model

We now present results for a M/M/c/k model. We assume that c is greater than k. This model corresponds to the situation where the system only has room for those in service. It has no waiting room. When all the servers are busy, the arrivals will be turned away and denied service. Certain types of telephone systems that cannot buffer calls exhibit this behavior. The steady-state equations for this queue are given below.

The probability that the system will be empty = $P_0 =$

$$\left[\sum_{n=0}^{c-1}(\lambda/\mu)^n/n!+\frac{(\lambda/\mu)^c}{c!}\left(\frac{1-\rho^{k-c+1}}{1-\rho}\right)\right]^{-1} \tag{2.28}$$

where $\mu_n = n\mu, 1 \leq n \leq c$

   $\rho = \lambda/c\mu$

The probability that the system will contain n inputs = $P_n =$

$$\begin{cases} \left(\dfrac{(\lambda/\mu)^n}{c!n^{n-c}}\right)P_0 & for\ c < n \leq k \\ \\ \left(\dfrac{(\lambda/\mu)^n}{n!}\right)P_0 & for\ 0 \leq n \leq c \end{cases} \tag{2.29}$$

The effective arrival rate is less than the service rate under steady-state conditions, and is calculated as:

$$\lambda_e = \lambda/(1 - P_k) \tag{2.30}$$

The expected queue length, L_q, is calculated from the use of sum calculus and its definition as:

$$=\frac{(\lambda/\mu)^c(\rho \times P_0)}{c!(1-\rho)^2}\left\{1-[(k-c)](1-\rho)+1]\rho^{k-c}\right\} \tag{2.31}$$

where $\rho = \lambda/c\mu$

Since the carried load is the same as the mean number of busy servers, we can calculate the expected number in the system as:

$$L = L_q + \frac{\lambda}{\mu}(1-p_k) \tag{2.32}$$

From Little's law, it follows that the waiting times are:

$$W_q = L_q/\lambda(1-P_k) \tag{2.33}$$

and

$$W = L_q + 1/\mu \tag{2.34}$$

We now consider a special case of the M/M/C/K model: the M/M/c/c queue. For this queue, the effective arrival and service rates are:

$$\lambda_n = \begin{cases} \lambda, & 0 \le n < c \\ 0 & n \ge c \end{cases} \tag{2.35}$$

$$\mu_n = \begin{cases} n\mu, & 1 \le n < c \\ 0 & \textit{elsewhere} \end{cases} \tag{2.36}$$

Probability that the system will be empty = P_0 =

$$\left[\sum_{n=0}^{c} (\lambda/\mu)^n / n! \right]^{-1} \tag{2.37}$$

The probability that the system will have n in the system = P_n =

$$\frac{(c\rho)^n / n!}{\sum_{j+0}^{c} (c\rho)^j / j!} \tag{2.38}$$

where: $\rho = \lambda / c\mu$

2.6.3   *Erlang's loss formula*

In the M/M/c/c model, the system is saturated when all the channels are busy, i.e., for P_c. By multiplying the numerator and the denominator of Equation 2.38 by $e^{-c\rho}$, we can obtain a truncated Poisson distribution with parameter values ($c * \rho = \lambda/\mu$). These values can be obtained from tables of the Poisson distribution, or from an Erlang B table, simplifying the calculations for P_c below. This formula is perhaps better known as Erlang's loss formula. In the context of a telephone system, it describes the probability that an incoming call will receive a busy signal and will be turned away. This formula was used to design telephone systems that satisfy predefined levels of acceptable service.

$$P_c = \frac{e^{-c\rho} (c\rho)^c / c!}{\sum_{j+0}^{c} e^{-c\rho} (c\rho)^j / j!} \tag{2.39}$$

M/M/c/c/example of blocked phone lines

An office has four shared phone lines. Currently, half the time a call is attempted, all the phone lines are busy. The average call lasts two minutes. What is the probability that a call will be attempted when all four lines are busy?

Answer

In the problem we are given the fact that $P_4 = .5$. Therefore, using Equation 2.39: From an Erlang loss or Poisson table, the value for P_4 is obtained when $4 \rho = \lambda/\mu = 6.5$. Therefore, the implied arrival rate is:

$$\lambda = 6.5 * \mu = 3.25 \text{ calls per minute.}$$

2.6.4   Total Network Delay

Thus far, we have used queuing models to estimate delay on a single network component. In this section, we illustrate how these calculations can be extended to estimate the overall network delay or response time. Note that, in general, multiple calculations are needed to compute the overall network delay, as we demonstrate below. These calculations can be involved and tedious — especially for a large network — and in actual practice it is best to use software routines to automate these calculations. Many network design tools — such as WinMind™ — offer built-in routines to estimate delay using techniques similar to those described here.

By way of example, we will construct a delay model for a packet-switched network. In packet-switched networks, propagation delay, link delay, and node delay are the major components of network delay. This can be summarized in the following equation:

Delay total = Total average link delay + Total average node delay

+ Total average propagation delay   (2.40)

Assuming an M/M/1 service model to approximate traffic routing on the network, total average link delay on the network can be estimated by:

$$D_{link} = \frac{1}{\sum\limits_{i=1}^{I} \sum\limits_{j=1}^{J} R_{ij}} \sum\limits_{l=1}^{L} (D_l * F_l) \qquad (2.41)$$

where R_{ij} = traffic requirement from node i to node j
         F_l = flow on link l

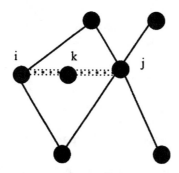

:::::: The shortest path from (i,j) is through the shortest path
from (i,k) and the shortest path from (k,j).

Figure 2.17   Nested shortest paths.

$$D_l = \text{delay on link } l = \frac{\left(P/S_{ij}\right)}{1-\left(F_l/C_{ij}\right)}$$

P = packet length
S_{ij} = link speed from node i to node j
C_{ij} = link utilization from node i to node j

Thus, total average network delay is the sum of the expected delay on all the links. The unknown variable in the above equation for D_{link} is link flow. A shortest-path algorithm can be used at this stage to assign link flows to solve for this variable.

A shortest-path algorithm computes the shortest path between two given points and is based on the insight that the shortest paths are contained within each other. This is illustrated in Figure 2.17. Thus, as described in [Kers93, p. 157]:

> If a node, k, is part of the shortest path from i to j, then the shortest i, j-path must be the shortest i, k-path followed by the shortest j, k-path. Thus we can find shortest paths using the following recursion:

$$d_{ij} = \min_{k}\left(d_{ik} - d_{kj}\right) \text{ where } d_{xy} \text{ is the length of the shortest path from x to y.}$$

Dijkstra's algorithm and Bellman's algorithm, two well-known examples of shortest-path algorithms, specify ways to initiate the recursion equation given above so that a solution can be found.

Node delay is a function of node technology. Assume, for the purposes of illustration, that this delay is a constant 120 milliseconds. Thus, total average node delay is estimated by:

$$D_{node} = C_{node} * A \qquad\qquad (2.42)$$

where D_{node} = total average node delay
C_{node} = 120 milliseconds/node
A = average number of nodes per shortest routing path

Propagation delay is the time it takes electromagnetic waves to traverse the transmission link. Propagation delay is proportional to the physical distance between the node originating a transmission and the node receiving the transmission. If the tariff database is used to compute link costs, it is also likely that V (vertical) and H (horizontal) distance coordinates for each node are available so that the mileage between nodes can be estimated. This mileage multiplied by the physical limit of electronic speed — 8 microseconds per mile — can be used to obtain an estimate of the total average propagation delay, as indicated below. In general, propagation delays are small compared to the queuing and transmission delays calculated in the previous section.

$$D_{prop} = \frac{C_{prop} * \sum_{l=1}^{L} F_l * M_l}{\sum_{i=1}^{I} \sum_{j=1}^{J} R_{ij}} \qquad\qquad (2.43)$$

where D_{prop} = total average propagation delay
R_{ij} = traffic requirement from node i to node j
C_{prop} = 8 microseconds/mile
F_l = flow on link l
M_l = length in miles of link l

2.6.5 Other types of queues

There are other types of queues that we have not covered which are useful in network analysis. These queues include:

- Queues with state dependent services — In this model, the service rate is not constant and may change according to the number in the system, either by slowing up or by increasing in speed.
- Queues with multiple traffic types — These types of queues are common in network applications. To handle the arriving traffic, you may wish to employ a priority service regime, to give faster service to some kinds of arrivals relative to the others.

- Queues with impatience — This type of queue is designed to model the situation where an arrival joins the queue but then becomes impatient and leaves before it reaches service. This is called reneging. Another type of impatience, called balking, occurs when the arrival leaves on finding that a queue exists. A third type of impatience is associated with jockeying, where an arrival will switch from one server to the next to try to gain an advantage in being served.

These models interject more realism in the network analysis at the expense of simplicity. The interested reader is referred to [Kersh93], and [Klie75] for more information in this area.

2.6.6   Summary

Queuing is broadly applicable to many network problems. In this section we studied a few examples, including:

- Delay in a T1-based WAN
- Transmission speed in a packet-switched network
- Printer-server on a lan with limited buffering
- Expected call blockage in a PBX
- Total network delay in an APPN packet switched network

In our presentation, we used queuing models based on steady-state assumptions. These assumptions were made to keep the models simple and easy to solve. It is important to maintain a perspective on what degree of accuracy is needed in estimating the network delays. If further refinements to the delay estimates will require substantially more time and effort, then this may well temper the decision to attempt to introduce more realism into the models. The models introduced here provide a high-level approximation of reality, and thus there may well be discrepancies between the anticipated (i.e., calculated) system performance and actual observed performance.

2.6.7   Analysis of network reliability

There are numerous ways that network reliability can be estimated. Here we survey three commonly used methods for analyzing reliability: (1) component failure analysis, (2) graphical tree analysis, and (3) k-connectivity analysis. The interested reader is referred to [Kersh93] for a comprehensive treatment of this subject.

One measure of network reliability is the fraction of time the network and all of its respective components are operational. This can be expressed as:

$$\text{Reliability} = 1 - (\text{MTTR})/(\text{MTBF}) \tag{2.44}$$

where   MTTR = mean time to repair failure
        MTBF = mean time before failure

For example, if the above equation is used to compute a reliability of .98, this means that the network and its components are operational 98% of the time.

Since the network is comprised of multiple components, each component contributes to the possibility that something will fail. One commonly used model of component failure assumes that failures will occur according to the exponential probability distribution. An exponential random variable with parameter λ is a continuous random variable whose probability density function is given for some $\lambda > 0$ by:

$$f(x) = \left\{\lambda e^{-\lambda x} \text{ if } x \geq 0 \quad \text{and} \quad f(x) = \left\{0 \text{ if } x < 0 \right. \right. \tag{2.45}$$

The cumulative distribution function F of an exponential variable is given by:

$$F(a) = 1 - e^{-\lambda a} \quad \text{for} \quad a \geq 0 \tag{2.46}$$

We will now use these definitions in an illustrative sample problem. Assume that the failure of a network component can be modeled by an exponential distribution with parameter $\lambda = .001$, and X = time units in days. This means that for this component:

$$f(x) = .001 e^{-.001X} \quad \text{and} \quad F(x) = 1 - e^{-.001X}$$

Using the cumulative probability density function F (x) above, one can compute the probabilities of failure over various time periods:

Probability that the network component will fail within 100 days $= 1 - e^{-1} = .1$
Probability that the network component will fail within 1000 days $= 1 - e^{-1} = .63$
Probability that the network component will fail within 10,000 days $= 1 - e^{-10} = .99$

This model describes the probability of failure for a single network component. A more generalized model of network failure is given in the simple model below. This model says the network is connected only when all the nodes and links in the network are working. It also assumes that all the nodes have the same probability of failure p, and all the links have the same probability of failure p′. A final assumption made by this model is that the network is a tree containing n nodes. This model of the probability that the network will fail can be written as [Cahn98]:

$$\text{Probability (failure)} = (1 - p)^n \times (1 - p')^{n-1} \tag{2.47}$$

When the networks involved are more complex than a simple tree, this formula no longer holds. When the network is a tree, there is only one path between the nodes. However, in a more complex network there may be more than one path, even many paths, between nodes that would allow the network to continue to function if one path were disconnected. With the previous approach, all combinations of paths between nodes have to be examined to determine the probabilities of link failures associated with a network failure. For a network of any substantial size, this gives rise to many combinations, i.e., there is a combinatorial explosion in the size of the solution space. This type of problem is, in general, very computationally intensive to solve.

In the text that follows, we discuss alternative strategies for estimating the reliability of complex networks containing cycles and multiple point-to-point connections. Except for the smallest of networks, the calculations are sufficiently involved as to require a computer.

Graph reduction is one technique used to simplify the reliability analysis of complex networks. The idea of graph reduction is to replace all or part of the network with an equivalent, yet simpler graphical tree representation. For instance, one type of reduction is parallel reduction. If there are parallel edges between two nodes and the two edges are operational with probabilities p and p', respectively, then it is possible to replace the two edges with a single edge whose probability is equal to the probability that either or both of the edges are operational. Other transformations are possible — e.g., to reduce a series of edges, etc. — and are likely necessary to sufficiently transform a complex network so that its reliability can be calculated. For more information on graph reduction techniques the reader is referred to [Shoo1991] and [Kersh93].

K-connectivity analysis is also useful in assessing network reliability. K-connectivity strives to characterize the survivability of a network in the event of a single component failure, a two-component failure, or a k-component failure. If a network contains k separate, disjoint paths between every pair of nodes in the network, then the network is said to be k-connected. Disjoint paths have no elements in common. Paths are edge-disjoint if they have no edges in common. Likewise, paths are node-disjoint if they have no nodes in common. An example of a one- and a two-connected graph is provided in Figure 2.18.

It is possible to test for k-connectivity, either node or edge, by solving a sequence of maximum flow problems. This is a direct result of work by Kleitman who showed that [Klei69]:

Given a graph G with a set of vertices and edges (V, E), G is said to be k-connected if for any node $v \in V$ there are k node-disjoint paths from v to each other node and the graph G' formed by removing v and all its incident edges from G is (k-1) connected.

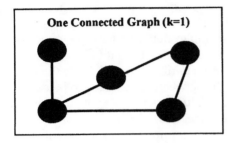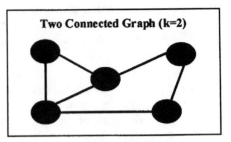

Figure 2.18 Example of 1-connected and 2-connected graphs.

Thus, it is possible to determine the level of k-connectivity in a network by performing the following iterative procedure described in [Kersh93]. The computational complexity of this process is O (k^2N^2).

> it is only necessary to find *k* paths from any node, say v_1, to all others, k-1 paths from another node, say v_2 to all others in the graph with v_1 removed, and k-2 paths from v_3 to all others in the graph with v_1 and v_2 removed, etc. [Kersh93]

Thus far, we have focused our discussion on link failures. Clearly, if there is a node failure, the network will not be fully operational. However, a network failure caused by a node cannot be corrected by the network topology, since the topology deals strictly with the interconnections between nodes. To compensate for possible link failures, we can design a topology that provides alternative routing paths. In the case of a node failure, if the node is out of service the only way to restore the network is to put the node (or some replacement) back in service. In practice, back-up or redundant node capacity is designed into the network to compensate for this potential vulnerability.

We now demonstrate how k-connectivity can be used to assess the impact of node failures. We begin by transforming the network representation to an easier one to analyze. Suppose we are given the undirected graph in Figure 2.19. To transform the network, we begin by selecting a target node. We then transform all the incoming and outgoing links from that node into directed links, as shown in Step 2 in Figure 2.19. Finally, we split the target node into two nodes, i and i'. We connect nodes i and i' with a new link. All the incoming links stay with node i and all the outgoing links stay with node i'. This is shown in Step 3 of Figure 2.19. Once the nodes are represented as links, the k-connectivity algorithm presented above can be used to determine the level of k-node connectivity in the network

We conclude this section with some guiding principles. Single points of failure are to be avoided in the network. To prevent a single line failure from disabling the network, the network should be designed to provide 2-k edge-disjoint connectivity or better. This will provide an alternative route to transmit

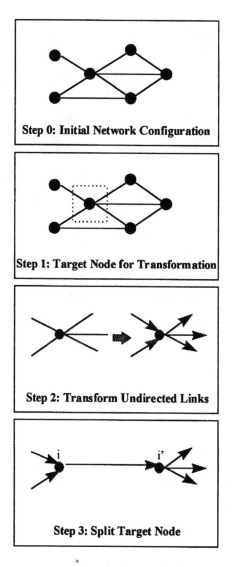

Step 0: Initial Network Configuration

Step 1: Target Node for Transformation

Step 2: Transform Undirected Links

Step 3: Split Target Node

Figure 2.19 K-connectivity reliability analysis.

traffic if one link should fail. However, multiple k-connectivity does not come cheap. In general, a multiconnected network is substantially more expensive than a similar network of lower connectivity. A common target to strive for is 2-connectivity, and to compensate for weakness in the topology by using more reliable network components. However, k-connectivity alone does not guarantee that the network will be reliable. Consider the network illustrated in Figure 2.20. In this network, there are two paths for routing traffic between any two pairs of nodes. However, should the center node fail, the entire network is disconnected. A single node whose removal will

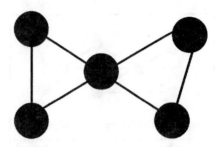

Figure 2.20   Example of 2-connected graph with single-node point of failure.

disconnect the network is called an articulation point. One solution to this problem is to avoid a design where any one link or node would have this impact. It is apparent that the failure of some nodes and links may have more impact on the network than the failure of others. To the extent that is possible, we want to design networks with excess capacity strategically located in the network. While it is desirable to have excess capacity in the network — for performance reasons so that traffic of varying intensity can be easily carried without excessive delays — we also want to add network capacity where it will make the most difference on the overall network reliability [Kersh93].

2.6.8   Network costs

Many of the costs associated with the network are obtainable from manufacturers and organizations leasing equipment and providing services. Important costs associated with the network include:

- Tariff costs for all network links (these are location specific charges)
- Monthly charges for other network expenses (including device costs, software costs, etc.)
- Installation charges
- Usage sensitivity charges

All network expenses should be reduced to the same units and time scale. For example, one-time costs — such as installation costs and one-time purchases — should be converted to monthly costs by amortization. For example, if a network device costs $9,000, this lump sum should be converted to a monthly charge. Thus, a $9,000 device with an expected useful life of three years has a straight-line amortized monthly cost of $250.

Likewise, usage sensitive charges should be converted to consistent time and dollar units. Usage charges, as the name implies, vary according to the actual system usage. When calculating the network costs, a decision must be made as to whether or not an average cost or a worst case cost calculation is needed. In the former case, the average monthly usage fee, and in the latter case the largest possible monthly usage fee should be used in the final cost calculation as shown in Equation 2.49.

A tariff database gives the cost of individual links in the network, based on their type and distance. It is the most accurate source of information on monthly line charges. However, the fees for accurate, up-to-date tariff information may be substantial. Alternatively, monthly link costs can be estimated using the techniques described in Section 2.2.3. Once the individual link costs have been tabulated, by whatever means, they are summed to obtain the total link operating cost:

$$\text{Total monthly line costs} = \sum_{n=1}^{N} \sum_{i=1}^{I} \sum_{j=1}^{J} O_{nij} * N_{nij} \qquad (2.48)$$

where O_{nij} = cost of link type n, from node i to node j

M_{nij} = number of type n links between nodes i and j

A similar calculation can be performed for the node costs. The total monthly network cost is computed as the sum of all the monthly charges, as indicated below:

Total monthly costs = Monthly line costs + Monthly amortized costs +

Monthly usage costs + Other monthly service costs (2.49)

Bibliography

[Cahn98] Cahn, R., *Area network design: concepts and tools for optimization,* Morgan Kaufmann, San Francisco, 1998.

[Data94] Datapro, network architectures: overview, *Datapro,* McGraw-Hill, November 1995.

[Dyme94] Dymek, W., IBM Netview, *Datapro,* McGraw-Hill, September 1994.

[Giff78] Giffen, W., *Queuing basic theory and applications,* Grid Series in Industrial Engineering, Columbus, OH, 1978.

[Gross74] Gross, D., and Harris, C., *Fundamentals of queuing theory,* John Wiley & Sons, New York, 1974.

[Kersh89] Kershenbaum, A., Interview with T. Rubinson on April 27, 1989.

[Kersh93] Kershenbaum, A., *Telecommunications network design algorithms,* McGraw-Hill, New York, 1993.

[Klei69] Kleitman, D., Methods of investigating connectivity of large graphs, *IEEE Transactions on Circuit Theory (Corresp.),* CT-16: 232-233, 1969.

[Klie75] Kleinrock, L., *Queuing systems,* Volumes 1 and 2, Wiley-Interscience, New York, 1975.

[Mino93] Minoli, D., *Broadband network analysis and design,* Artech House, Boston, 1993.

[Mino96] Minoli, D., Queuing fundamentals for telecommunications, *Datapro,* McGraw-Hill, June 1996.

[Pili97] Piliouras, B., Interview with T. Rubinson on August 6, 1997.

[Ross80] Ross, S., *Introduction to probability models*, Second Edition, Academic Press, New York, 1980.

[Rubi92] Rubinson, T., *A fuzzy multiple attribute design and decision procedure for long term network planning*, Ph.D. dissertation, June 1992.

[Shoo91] Shooman, A., and Kershenbaum, A., Exact graph-reduction algorithms for network reliability analysis, *IEEE proceedings from Globecomm '91*, August 19, 1991.

[Wagn75] Wagner, H., *Principles of operations research*, Second Edition, Prentice-Hall, Englewood Cliffs, NJ, 1975.

chapter three

WAN network design

Contents

3.1   Management overview of WAN network design

In this chapter, we focus on the topologic design of WAN data networks. The ultimate result of this process is a set of recommendations for placing specific nodes and links in a network.

In Chapter 2, we discussed considerations involved in selecting node equipment. This selection is largely predicated on the functionality that must be supported at a particular network location. For instance, a firewall may be needed at a strategic network access point, and therefore a router may be selected as a node type. Once the functionality needed at a particular site has been determined, specific equipment options can be evaluated based on their performance and cost characteristics. In this chapter, we present techniques for determining where nodes should be placed in the network to minimize line costs. These techniques do not explicitly consider the node functionality or capacity.

Until fairly recently, only a few line choices were available. One could choose between private lines (operating at DDS,[1] DS1,[2] or DS3[3] speeds) or public lines (operating X.25 at 56 Kbps or 64 Kbps). Now, there are many high bandwidth options available, including frame relay, switched multimegabit data service (SMDS), asynchronous transfer mode (ATM), and ISDN. In the sections that follow, we discuss the pros and cons of these choices for various network scenarios and ways that they can be used in combination to provide a total network solution.

All these choices compound the difficulty of designing the network. The network designer must generate several alternative designs for *each* link type and technology considered to achieve a sense of the price/performance trade-offs associated with each option. The decisions and comparisons involved in planning all but the smallest networks are sufficiently complex as to warrant the aid of automated network design tools. We review Win-MIND™, and other network design tools, which offer a variety of design techniques for producing networks using different link types and technologies.

3.1.1   Outsourcing

When implementing the network, the organization must decide whether or not to use in-house and/or outside staff and resources, or an appropriate mix. After identifying several promising designs, an organization may wish to solicit bids from various service providers. The major carriers will develop bids according to an organization's needs, which may include

---

[1]   DDS — This stands for digital data service, a transmission service that supports line speeds up to 56 Kbps.

[2]   DS1 — This stands for digital signal 1, a transmission service that supports 24 voice channels multiplexed onto one 1.544 Mbps T1 channel.

[3]   DS3 — This stands for digital signal 3, a transmission service that supports 44 Mbps on a digital T3 channel.

design, implementation, and even operation of the network. These bids provide very useful information on the expected network performance and cost. The organizational budget, in-house skills, security requirements, and the role of the telecommunication/MIS functions in the survival of the company are major decision factors when deciding to use outside services to implement all or part of the network.

The first-year benefits of outsourcing the network implementation and management are usually positive because the difficult task of coordinating these functions is shifted to another party. Upper management sees a drop in payroll and related employee benefits, and the telecom manager does not need to tie up all of his or her capital on one-time purchases. Since technology is always changing, contracts with outsourcers are likely to be subject to ongoing negotiation and mutual misunderstandings. However, outsourcing may help position the company to adopt new technologies as they become available and to grow to meet expanding network demands.

When a company elects to buy network equipment, the one-time purchase costs are typically amortized over a 3- or 5-year period. However, network technology is often outdated within 12 to 18 months. Thus, many telecom managers find they need updated customer premise equipment (CPE) before they have fully depreciated it. One way to avoid this situation is to lease the network equipment with the stipulation that upgrades will be made available on a continual basis. This way network managers no longer need to maintain an obsolete depreciable asset on the network accounting budget. At the end of the lease, the equipment can usually be bought outright if the network manager so desires. However, leasing can be expensive. In addition, if the equipment is leased, there may be a cyclical need to replace the equipment as the lease period expires. If the equipment is owned, the communications/IS department can continue to function while plans are made to update the equipment.

From the outsourcer's point of view, the first year of network support requires a high outlay of capital to cover start-up costs. Typically, the outsourcer does not realize a profit until several years into the contract. Coincidentally, this is about the same time that the customer is growing tired of paying monthly costs for outdated equipment and services. At this time, the customer may want to renegotiate the contract or determine their liability for breaking or breaching the contract. Conversely, the outsourcer wants to keep the contract active since this is when the majority of the return on investment (ROI) is realized.

Outsourcing offers many potential benefits:

- Reduced communication overhead costs
- Reduced labor and training costs
- Improved availability of skilled network management personnel
- Economies of scale in buying services and equipment
- Competitive edge due to the outsourcer's reliance on customer satisfaction and bottom-line performance

- Reduced demands on general management for day-to-day operational decision making
- Better control of large, fixed-cost expenditures through variable cost contracts tailored to the organization's yearly activities

However, outsourcing may also be associated with these problems:

- It may be difficult to accurately determine the outsourcer's qualifications and track record
- There may be limited options for rectifying a nonworking alliance
- There may be judgment differences between the outsourcer and the organization due to rapid technology changes and corporate culture issues
- The outsourcer may lack the critical mass and capital market assets needed to properly service large contracts

In summary, many factors must be weighed when choosing a network design. The final network selection will reflect the following decisions.

3.1.2   Organizational goals

The role of the network in the organization must be clearly defined. If the applications supported by the network dominate the organizational requirements, it may be necessary to develop specialized networks, or subnetworks, to provide optimal support for each type of application. This approach may involve trading simplicity and cost for increased network performance. In contrast, when simplicity and operational control are paramount, price and performance may matter less. This change in emphasis may have significant impacts on the type of design that is favored. Therefore, the organization must choose the design strategy carefully to ensure that it is consistent with the business goals. Enterprise networking may be strongly influenced by the need and/or desire for cost reductions, improved utilization of existing network facilities, implementation of new applications and services, simplified network infrastructure, support of diverse network/internetworking requirements, and network reliability.

3.1.2.1   Use of public vs. private leased lines
The organization's need for control, specialized services, and cost economies may dictate the choice between public and private lines. For example, private lines are notable for their high degree of reliability and predictable performance. However, X.25 public lines (which use statistical multiplexing to increase the efficiency of the line utilization, in contrast to private lines, which use the less efficient, static, time-division form of multiplexing) may offer an attractive, cost-effective solution for certain transmission requirements.

3.1.2.2 Type of service

The mix of traffic to be carried by the network helps to determine the type of service needed (e.g., DDS lines, frame relay, SMDS, etc.). For example, frame relay only supports data transmissions, and thus, it may not be suitable for an organization transmitting voice, data, and possibly video traffic. The respective strengths and limitations of each type of service must be weighed relative to the organization's needs.

3.1.2.3 Carrier Selection

If most of the network locations are in the same city, it may be advisable to use a local carrier. However, if most of the network locations are geographically dispersed over state or country boundaries, then an IEC[4] will be needed. Some carriers only offer a limited number of services. Depending on the requirements, this may impact the selection of the network carrier.

3.1.2.4 Use of in-house vs. outside staff and resources

The organization's skill sets, budget, future growth, and specialized needs must be considered when determining the appropriate mix of internal and outside staffing and resources. Suggestions for evaluating which choices are more appropriate in a given context are provided in Chapter 6.

3.2 Technical overview of WAN network design

The secret to managing the complexity of network design is: Divide and conquer! Most networks can conceptually be divided into hierarchical levels. Recognizing the number of operational levels needed in the network is a critical first step in the design process. For example, a backbone network is composed of one or more links that interconnect clusters of lower-speed lines. The clusters of lower-speed lines are called access networks. The backbone can be thought of as one level of the network topology. The access networks can be thought of as a second level of the network topology.

After decomposing the network design problem into the appropriate number of levels, the next challenge is to decide what type of topology is appropriate to support the traffic that must be carried on that portion of the network. Sometimes, more than one topology is possible and should be explored. Selecting a topology for each network segment is important because the design techniques are classified according to the type of network they produce.

For example, a backbone network is designed to carry substantially greater traffic than the access network(s), and therefore it is not surprising that the respective topologies of each may be very different. Backbone networks are usually designed with a mesh[5] topology, whereas local access

[4] IEC — This stands for interexchange carrier.
[5] Mesh — A mesh network provides point-to-point connectivity between various nodes in the network.

networks are often designed as trees. A mesh topology provides greater routing diversity and reliability than a tree. However, for low-volume traffic, the cost savings offered by a tree topology may more than compensate for the loss of reliability and performance. When designing a network comprised of a backbone and a number of access networks, the standard approach is to design the backbone separately from each of the access networks. After the design of each respective component has been optimized, the total network solution can be put together.

It is possible that several networks will need to be interconnected. The strategy for designing this type of internetwork is very similar to what we have just described. Each subnetwork should be designed and optimized with respect to its particular requirements. Once the subnetworks have been designed, the interconnection between the networks can be treated as a separate design problem. When all these problems have been solved, the design solution is complete.

In the sections that follow, we describe methods for decomposing and solving centralized and distributed network design problems.

3.2.1   Centralized network design

Centralized networks are perhaps best characterized by legacy mainframe configurations. In a centralized network, all communications are controlled centrally by a single site or processing unit. All communications from one device to another go through the central site using a static, predetermined routing path. An example of a centralized network is given in Figure 1.1. Typically, devices, or nodes, in this type of network include the central processing unit, concentrators, and computer terminal devices.

Centralized network design can be decomposed into three general subproblems:

- Concentrator location
- Terminal assignment
- Multi-point line layout

Each of these subproblems is solved in turn using any of a number of heuristic design algorithms. In this chapter, we present some of the better known and most popular heuristic design algorithms. The results of each subproblem are then combined to form an overall design solution. The heuristic design algorithms provide an initial low-cost solution, which can be fine-tuned as necessary. In practice, it is advisable to use several different algorithms to solve the same problem(s) to get as much insight as possible on what the "best" solution looks like.

The concentrator location problem, as its name implies, involves placing concentrators at key points in the network. Likewise, the terminal assignment problem involves connecting terminals to the closest or most central concentrator. The multipoint line layout problem involves clustering terminal devices on a shared line to minimize line costs. After the multipoint lines are designed, they are treated as a single terminal, and terminal assignment algorithms can be used to connect them to the closest concentrator. Thus, the interconnection of the devices to a concentrator is solved separately from the interconnection of the concentrators to each other or to the central site. The order in which these problems are solved is likely to have a significant bearing on the final outcome. It may be useful to design the network assuming in a first pass that the concentrator locations are taken as given for the purposes of terminal assignment and multipoint line layout. Conversely, the problem can be resolved, assuming that the terminal locations and multipoint lines are taken as given, in order to determine optimal concentrator placements.

After all the pieces have been put together, the resulting network can be analyzed. Node and line utilization can be checked. There may be nodes or links with too little traffic going through them to be cost effective, and thus, it may make sense to remove them from the final design. It may also be advisable to place links in strategic places to provide additional capacity or redundancy in the network.

We now review solution techniques for each of the three major subproblems associated with centralized network design. As stated earlier, a centralized network uses a static path for routing traffic between devices. Recalling an earlier definition in Chapter 2, a minimal spanning tree is the cheapest way to connect all the nodes in a network. Therefore, the lowest-cost routing path in a centralized network corresponds to a minimal spanning tree. In Chapter 2, we introduced Kruskal's algorithm as a means to construct a minimal spanning tree. The only problem with Kruskal's algorithm as we presented it was that it did not take constraints on line capacity into account.

Obviously, the multipoint line layout should reflect the capacities of the lines to be used (which typically correspond to DDS lines operating at speeds up to 56 Kbps). When capacity constraints must be observed, the multipoint layout problem becomes a capacitated minimal spanning tree (CMST) problem. In the following sections, we introduce four CMST algorithms: Prim's algorithm, a revised CMST version of Kruskal's algorithm, Esau-William's (E-W) CMST algorithm, and the Unified algorithm. In Section 3.2.4.2, we demonstrate how these techniques can be used to compute a CMST for multipoint line layout. Briefly summarizing, these algorithms take the terminal and concentrator locations as given, as well as the line speed being considered for the multipoint line. From this information, the algorithms construct a multipoint line layout that minimizes costs and avoids violating capacity and cycling constraints (a minimal spanning tree, by definition, does not allow cycling, since cycling would introduce redundant links).

CMST algorithms assume that there is no constraint on the number of lines coming into the terminal or concentrator. In actual practice, there may be restrictions on the number of incoming lines that can be handled. In addition, CMST algorithms rely on differences in link costs when designing the multipoint line. However, there may be situations where large numbers of terminals in the same building or location have similar link costs. In these situations, CMST algorithms are not suitable, and bin-packing algorithms should be used instead to design the multipoint line. Bin-packing algorithms are used to "pack" as many terminals as possible on a single line to minimize the number of multipoint lines created. In Section 3.2.4.2, we demonstrate three bin packing algorithms for designing multipoint lines:

- First fit decreasing
- Best fit decreasing
- Worst fit decreasing

The terminal assignment problem, which involves associating a multi-point line or a terminal to a concentrator, may be easy to solve if there is only one natural center to which to connect. If more than one concentrator or center location is possible, a simplistic approach is to assign each terminal to the nearest *feasible* center. This is a type of greedy algorithm that may, unfortunately, strand the last terminals connected far from the concentrator or center. One way to compensate for this problem is to use a modified form of the Esau-William's algorithm. In this approach, the value of a trade-off function is computed for each terminal. This trade-off calculates the difference in cost between connecting a terminal to the first-choice concentrator and the second nearest choice concentrator. When this difference is large, preference should be given to the first-choice connection. This form of the terminal assignment algorithm is described in more detail in Section 3.2.4.2. When a design contains stranded terminals, far from a natural center, it may also indicate that there is insufficient concentrator coverage in the network. It may be possible to better solve this problem by adding new or more strategically placed concentrators in the network.

The concentrator location problem, as its name implies, is concerned with determining where concentrators should be placed and which terminals should be associated with each concentrator. There are several types of concentrator location algorithms. The center of mass (COM) algorithm is used when there are no candidate sites for concentrators. The COM algorithm proposes concentrator sites based on natural traffic centers. The results of the COM algorithm must be checked for feasibility, because the algorithm may recommend a location that is not practical. We review the COM algorithm in detail in Section 3.2.5.2. In addition to the COM algorithm, in Section 3.2.5.2, we introduce two other concentrator location procedures — the ADD and the DROP algorithms. Both the ADD and the DROP algorithms assume that the following information is available:

- Set of terminal locations (i)
- Set of potential concentrator locations (j)
- Cost matrix specifying the cost, c_{ij}, to connect terminal i to concentrator j for all terminals i and concentrators j
- Cost matrix of d_j, the cost of placing a concentrator at location j

From this information, these algorithms construct a set of concentrator locations and a set of terminal assignments for each concentrator location.

In addition to selecting concentrator sites in a centralized network, the ADD and the DROP algorithms can also be used to identify where backbone nodes should be placed in a distributed network. Although the ADD and the DROP algorithms may yield similar results, there is no guarantee or expectation that they will.

3.2.2 Distributed network design

In a distributed network, traffic is no longer routed in a static pattern through a central host processor (as it is in centralized networks). Instead, traffic may flow between nodes, over various possible routes. This communication pattern gives rise to a mesh topology, which is illustrated in Figure 3.17. In general, distributed network design is harder analytically and more computationally complex than centralized design.

Distributed network design can be decomposed into three major subproblems:

1. Develop network topology (which specifies node and link connections).
2. Assign traffic flows over the network links.
3. Size the line capacities, based on the topology and estimated traffic flows.

Solving these problems, especially the latter two, is difficult, because the link flows are inextricably tied to the line capacities. In addition, in a mesh topology, there may be many possible routing paths for the traffic flows. Without explicit knowledge of the traffic flows, it is difficult to estimate the required line capacities, and vice versa. Mathematically, the simultaneous solution of both (2) and (3) above is very complex. According to [Kersh93], the expected computational complexity of any mesh design algorithm is at best "$O(N^5)$ since there are $O(N^2)$ candidate links to consider, and consideration of a link would involve doing a routing, which requires finding paths between all pairs of nodes, itself an $O(N^3)$ procedure." As we discussed in Chapter 2, a procedure with a computational complexity of $O(N^5)$ is considered very complex indeed and is impractical for solving problems with a large number N of nodes.

Integer programming techniques can be used to find exact, optimally distributed design solutions in cases where the network does not exceed more than a few dozen nodes [Kersh93, p.305]. In this context, an optimal solution is one that can be mathematically proven to be the lowest cost design that does not violate cycling and capacity constraints. Important references on mesh topology procedures include [Fran72], [MGE74], [MAR78], and [MS86]. For a more comprehensive survey of distributed network design methods, the reader is referred to [Kersh93] and [Cahn98].

Within the literature, one mesh topology algorithm stands out for its low computational complexity and high-quality results — the Mentor algorithm. Remarkably, the computational complexity of Mentor is $O(N^2)$. Mentor's low computational complexity is achieved by *implicitly* routing traffic as the algorithm proceeds. We explain in Section 3.2.6.2 how this implicit routing is achieved, when we review Mentor in detail. We focus considerable attention on the Mentor algorithm, because it is so powerful and because it provides consistently good results (we note some exceptions in Section 3.2.6.2 and Section 7.2.3.7). Consistent with standard methodology, the Mentor design algorithm decomposes the design problem into several smaller problems. The first problem it tackles is selecting locations for the network backbone. The algorithm then proceeds to assign links, based on certain traffic routing assumptions. The simplifying assumptions that Mentor uses are based on the following insights about the characteristics of "good" distributed network designs [Kersh91]:

- Traffic flows follow short, direct paths as opposed to long, circuitous ones.
- Links should be highly utilized, as long as performance is not impacted adversely.
- Long, high-capacity links are used in preference to short- and/or low-capacity links to achieve cost economies of scale.

Although these characteristics are not wholly compatible with each other, Mentor attempts to achieve a reasonable balance between them. Briefly summarizing, the Mentor algorithm takes as given the terminal and/or concentrator locations, a maximum limit on the amount of usable line capacity that can be used, and a cost matrix specifying the cost of links between each of the terminal and concentrator locations. The output of the algorithm is a design layout that minimizes cost without violating the prespecified line capacity restrictions. We discuss several variations of the Mentor algorithm that are used to reflect different uses of technology (e.g., multiplexer-based networks, router-based networks, etc.).

3.2.3   Star topology networks

3.2.3.1   Overview

The star topology is the simplest and easiest type of network to design. It was one of the first network topologies to emerge in early centralized networks.

A star topology features a single internetworking central hub/switch that provides access from leaf internetworks into the backbone with access to each other only through the core networking device. Each station in a wide area network (WAN) star topology is directly connected to a common central hub/switch that uses circuit-switched technology to communicate with each end station or device. Star networks commonly employ digital private branch exchange (PBX) and digital data switches as the central networking device. Often, a time division multiplexer (TDM) will be used to aggregate the PBX and data switch traffic onto larger bandwidth facilities in order to reduce network costs.

Other examples of a star topology include local area networks (LANs) and metropolitan area networks (MANs) that employ packet broadcasting through a central node that connects to each end station by means of two unidirectional point-to-point links. In this case, all transmissions to the central switch are transmitted to all of the other endstations. Although this LAN/MAN topology physically resembles a star, logically it performs like a bus.

Star topologies are recommended for:

- Small, centralized networks
- Voice networks
- Message switching
- Switched networks accessing backbone networks

The advantages of a star approach include:

- Straightforward topology for centrally controlled networks
- Simplified network management
- Isolates line failures to single site
- Predictable network performance
- Easy to expand

Some of the disadvantages associated with star topologies are:

- High circuit cost
- Requires central switching point
- Central switch presents single point of failure
- Topology is not scalable

3.2.3.2 Design considerations and techniques

Star topologies help ensure that network delay is kept to a minimum, and are good for voice and fax applications that are sensitive to time delays. Delay is also a critical design factor when using satellite facilities that can themselves introduce signal delays of up to 250 milliseconds, resulting in annoying voice echoes. To combat this phenomenon, many popular multiplexer manufacturers provide built-in echo cancellation circuitry. Digital signal processors (DSPs) are often used to perform voice compression, to reduce the bandwidth needed between sites. However, DSP algorithms can add up to 100–150 milliseconds of delay (50–75 milliseconds for each end) and can produce a pseudo "echo." Over the years, this delay has been reduced by more efficient DSP design and sampling techniques and digitized bit reassembly. This has reduced the need, in some cases, for a star topology in voice-sensitive applications.

In a star topology, the main design decision is where to place the center node. In many cases, this decision is determined by pragmatic considerations, such as the location of existing data processing and telecommunication facilities. In the event that there are no guidelines for selecting the center location, clustering and center of mass algorithms (see Section 3.2.5.2 for examples) can be used to find the most central location for the node placement.

One such algorithm is Prim's algorithm, which we introduce at this time. The first step of Prim's algorithm involves computing a network center, based on estimated traffic flows. The second step involves assigning links to the center node. Depending on a prespecified design parameter, Prim's algorithm produces a starlike or treelike topology (thus, Prim's algorithm is also useful for designing shared, multipoint lines). The difference between a starlike and a treelike topology is that in the former, all the links converge toward a central location, whereas in the latter, they do not. Prim's algorithm takes as given the locations of the sources and destinations of traffic, estimated traffic flows between these locations, and line costs to connect the sources and destinations.

Prim's algorithm

Step 1: Find the network median.

(1a) Calculate a weight for each node i, as the sum of all traffic leaving and entering the node:

$$\text{Node\_Weight}_i = \sum_{j=1}^{n} \left(T_{j,i} + T_{i,j}\right) \tag{3.1}$$

where   i is a selected node from a total set of nodes n
        $T_{i,j}$ is the traffic flowing from node i to node j
        $T_{j,i}$ is the traffic flowing from node j into node i.

(1b) Calculate a figure of merit for each node i as:

$$\text{Figure of Merit}_i = \sum_{\text{for } j \neq i, j=1}^{n} \left(\text{Node\_Weight}_j \times \text{Cost}_{i,j}\right) \tag{3.2}$$

where   i is a selected node from a total set of nodes n
        Node_Weight_j is as defined above
        Cost_{ij} is the cost of a link between node i and j.

Note that in Equation 3.2, i *cannot* equal j.

(1c) Sort the figures of merit from low to high.
(1d) Select the node with the *smallest* number as the network center, designated c.

Step 2: Assign links.

(2a) Given all links (i,j) with i in the tree and j outside the tree, compute L'(i,j) as:

$$L'(i,j) = \text{Cost}_{i,j} + \alpha\left(\text{Cost}_{i,c}\right) \qquad (3.3)$$

where $\text{Cost}_{i,j}$ is the cost of a link between nodes i and j
$\text{Cost}_{i,c}$ is the cost of a link between nodes i and c
and α, a design parameter, is set to a number between 0 and 1.

(2b) Sort L'(i,j) values from low to high.
(2c) Check to see if all nodes have been connected into the network. If not, continue to the next step. If yes, then stop the algorithm. The solution has been found.
(2d) Select the lowest L'(i,j) value and remove it from sorted list. Add node j and link (i,j) to the network. Check to see if cycling or capacity constraints have been violated. If the constraints have been violated, remove node j and link (i, j) from the solution. Return to step (2a) above.
(2e) Update L(i,j) to reflect the addition of a new link into the network. Return to step (2a) above.

We now present a sample problem to illustrate Prim's algorithm. As stated above, the node locations, estimated traffic flow, and line costs must be collected before the algorithm can be initiated. We summarize the data for our sample problem in Table 3.1. Assume for this problem that the maximum traffic flow on a line is 10,000 units. Because the line capacity is very high relative to the traffic flows in this sample problem, the traffic constraint is not active. The design parameter, α, is set to 1 to encourage the selection of a star topology (recall that when α is set to 0, Prim's algorithm produces a tree topology).

Solution to sample star topology problem using Prim's algorithm

Step 1: Find the network median.

(1a) Calculate a weight for each node i as the sum of all traffic leaving and entering the node using Equation 3.1.

Table 3.1   Traffic and Cost Data for Star
Topology Design Problem

Node pair	Traffic requirements	Line cost
1,2	22	80
1,3	23	1307
1,4	6	1074
2,3	7	1387
2,4	10	1154
3,4	13	1528

Table 3.2   Node Weight Calculations
for Sample Star Topology Problem

Node	Node weight calculation
1	Σ 22 + 23 + 6 = 51
2	Σ 27 + 7 + 10 = 44
3	Σ 23 + 7 + 13 = 43
4	Σ 6 + 10 + 13 = 29

Table 3.3   Figure of Merit Calculations for Sample
Star Topology Problem

Node	Node weight calculation
1	90,867 = (44*80) + (43*1307) + (29* 1074)
2	97,187 = (51*80) + (43*1387) + (29* 1154)
3	171,997 = (51*1307) + (44*1387) + (29* 1528)
4	171,254 = (51*1074) + (44*1154) + (43* 1528)

(1b) Calculate a figure of merit for each node i using Equation 3.2.

(1c) Sort the figures of merit from low to high. *See entries in Table 3.2 above, sorted from low to high.*

(1d) Select the node with the smallest number as the network center, designated c. *See Table 3.3 above, which shows that node 1 should be selected as the network center.*

Step 2: Assign links.

(2a) Given links (i,j) with i in the tree and j outside the tree, compute: $L'(i,j) = \text{Cost}_{i,j} + \alpha \, (\text{Cost}_{i,c})$. Set the design parameter $\alpha = 1$. *See Table 3.4 for the first iteration calculations.*

(2b) Sort $L'(i,j)$ values from low to high. *Based on Table 3.4, this corresponds to the values 80, 1074, and 1307.*

Table 3.4   L'(i, j) Calculations for Sample Star Topology Problem

Node just brought	Distance from node (L'(i,j) = Cost$_{i,j}$ + α (Cost$_{i,c}$))			
into the network	1	2	3	4
1 (this is the center node, c)	—	80	1307	1074

(Note: Use values from the first line of the cost
matrix for Cost$_{i,j}$, for i = 2,3,4, since the Cost$_{i,c}$ = 0.)

Table 3.5   L'(i, j) Calculations for Second Iteration of Sample Star Topology Problem

Node just brought	Distance from node (L'(i,j) = Cost$_{i,j}$ + α (Cost$_{i,c}$))			
into the network	1	2	3	4
1 (this is the center node, c)	—	80	1307	1074
2 (this was selected above)	—	—	1467	1234

Checking (2,3) = 1387 + 1(80) = 1467
Checking (2,4) = 1154 + 1(80) = 1234
(Note: Cost$_{i,c}$ = 80, since this is the cost of link (2c)
which is now in the network)
Since both these new link costs are greater than a
direct connection to the center node, they are
rejected. Node 4 is brought into the network
through a connection through the center.

(2c) Check to see if all nodes have been connected into the network. *Only the center node is in the network at this stage of the algorithm.*

(2d) Select the lowest L'(i,j) value and remove from sorted list. Assign node j and link (i,j) as part of the network. If the constraints have been violated, remove node j and link (i, j) from the solution. Return to step (2a) above. The lowest L'(i,j) value is 80, corresponding to link (1,2). This link does not violate any constraints.

(2e) Update L'(i,j) to reflect the addition of a new link into the network. Compute the cost of bringing the remaining nodes into the network, either through the center node or by attachment to the end of the current tree built from the center. *These calculations are shown in Table 3.5.*

(2f) Sort L'(i,j) values from low to high. *Based on Table 3.5, this corresponds to 1074, 1234, 1307, and 1467.*

(2g) Check to see if all nodes have been connected into the network. *Nodes 1 and 2 have been connected to the network, but nodes 3 and 4 remain unconnected. So the algorithm proceeds to the next step.*

(2h) Select the lowest L'(i,j) value and remove from sorted list. Assign node j and link (i,j) as part of the network. *The lowest L'(i,j) value is 1074, corresponding to link (1,4). The L'(i,j) calculated for connecting node 3 to*

Table 3.6   L'(i, j) Calculations For Third Iteration of Sample Star Topology Problem

Node just brought into the network	Distance from node (L'(i,j) = Cost$_{i,j}$ + α (Cost$_{i,c}$))			
	1	2	3	4
1 (this is the center node, c)	—	80	1307	1074
2 (selected in previous step)	—	—	1467	1234
4 (selected in previous step)	—	—	2602	

Checking (3,4) = 1528 + 1(1074) = 2602
(Note: Cost$_{i,c}$ = 1074, since this is the cost of link (4,c)
which is now in the network)
Since this link cost is greater than a direct connection
to the center node, it is rejected. Node 3 is brought
into the network through a connection through the
center.

Table 3.7   Final Solution of Sample Star Topology Problem

Node just brought into the network	Distance from node (L'(i,j) = Cost$_{i,j}$ + α (Cost$_{i,c}$)			
	1	2	3	4
1 (this is the center node, c)	—	80	1307	1074
2 (selected in previous step)	—	—	1467	1234
4 (selected in previous step)	—	—	2602	
3 (final node selection)	—	—	—	—

node 2 is 1467, and the value calculated for connecting node 4 to node 2 is
1237. Since these values are higher than the cost of connecting node 4 directly
to the center, the links associated with them are not selected.

(2i) Update L'(i,j) to reflect the addition of a new link into the network.
These calculations are shown in Table 3.6.

(2j) Sort L'(i,j) values from low to high. *Based on Table 3.6, this corresponds
to 1307, 1234, and 1467.*

(2k) Check to see if all nodes have been connected into the network. *Nodes
1, 2 and 4 have been connected to the network, but node 3 remains unconnected. So the algorithm proceeds to the next step.*

(2l) Select the lowest L'(i,j) value and remove from sorted list. Assign node
j and link (i,j) as part of the network. *The lowest L'(i,j) value is 1307,
corresponding to link (1,3).*

(2m) Check to see if all nodes have been connected into the network. *All
nodes have been connected to the network, so the algorithm terminates.*

The final solution and the progression of the algorithm for this sample
problem are illustrated in Figure 3.1.

3.2.3.3   Case study

In our case study, we present a typical star topology used in WAN networks.
This topology is illustrated in Figure 3.2.

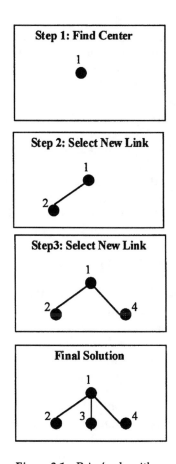

Figure 3.1 Prim's algorithm.

In this example, a customer employs a star network for their terminal-to-host based applications as well as voice/fax communications between end sites and the host location. The IBM 3174 Cluster Controllers serve to aggregate traffic from the various terminals/CRTs and coordinate communications with the front-end processor (FEP). The data communications protocol used is typically IBM's synchronous network architecture (SNA), which operates in a master-slave arrangement whereby each station takes a turn communicating with the far-end host computer by way of the cluster controller.

3.2.4 Tree Networks

3.2.4.1 Overview

Tree WAN topologies are distinguished by their use of a multipoint line that is shared by many stations. A controlling device serves as a common, central processor to which each station connects directly via a branching point.

Transmission is usually restricted so that all communications between each node go through and are controlled by a central mainframe or node. Tree WANs are a form of star topology. They are either *passive,* with multipoint circuits, or *active,* with switching nodes at the branch points. The primary difference between star and tree WANs is that in the former each station homes directly to a central processor, whereas in a tree they need not. In addition, star WANs do not contain branching points as do tree WAN topologies. This is illustrated in Figure 3.3.

In local area networks (LANs) that are configured as trees, a transmission from one station is received by all other stations, because the signal propagates the length of the medium in both directions. This phenomenon is also present in bus LAN topologies where the stations attach directly to a linear transmission medium, or bus, and communicate with the head end.[6] However, bus LANs do not employ branching points as do tree LANs. If the branches are removed from a tree LAN, the result is a bus LAN. Hence, a tree LAN topology can be thought of as a general form of a bus LAN topology.

Tree WAN networks frequently use front-end processors (FEPs) and mainframe central processing units (CPUs) as the central networking device(s). If the branching points are active, this implies that data concentrators or switches are being used to poll and to control the transmissions from mainframe to station, and from station to mainframe. Business and commercial applications that frequently use tree WANs include automatic teller machine/banking transaction networks, stock market quotation systems, and parts and inventory control networks.

Multipoint tree topologies require more complex mechanisms to control which station or device may transmit data than those required by point-to-point star networks. In point-to-point networks, no addressing is needed. In primary-secondary multipoint networks, a secondary address is needed (although a primary address is not needed because all the secondary terminals communicate only with the primary mainframe). In addition, a shared access method must be employed. This usually takes the form of cyclic polling and selection. A *poll* occurs when the central processor, or primary station, requests whether or not a terminal (i.e., a secondary station) has data to transmit. In the opposite transmission direction, a primary station may issue a *select* to transmit data back to the requesting terminal.

Delays occur in tree topologies when each secondary station must wait its turn while the other secondary stations are being served. Because of these inherent delays, tree networks are usually restricted to data transmissions, and are generally not recommended for voice, fax, or video applications.

---

[6] Head end — This is also known as the central processor.

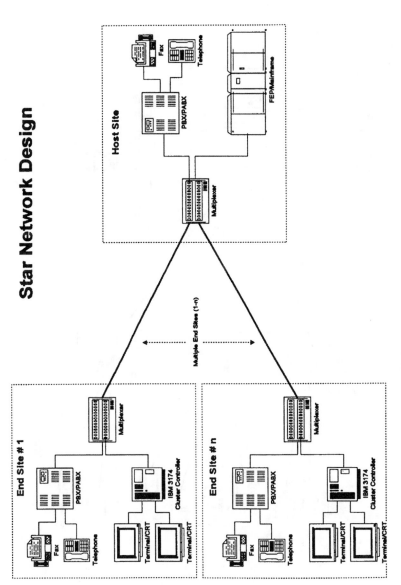

Figure 3.2 Star topology case study (Source: B. Piliouras).

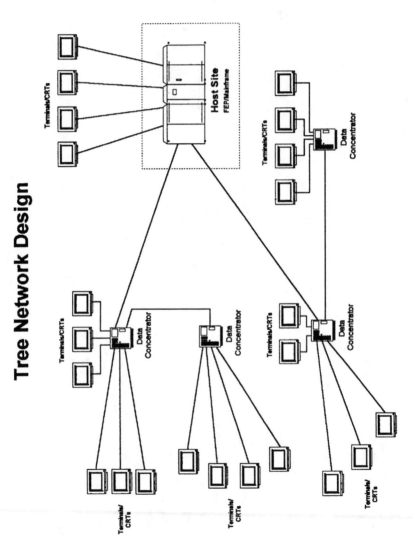

Figure 3.3   Tree network design (Source: B. Piliouras).

As compared to a star topology, the advantages of a tree topology include:

- Reduced facility costs
- Lower cost to expand
- Reduced number of hubbed facilities required

The disadvantages associated with tree topologies include:

- More complex to design
- More complex to expand
- More difficult to manage network
- Number of drops affects network performance

Recommended uses of tree topologies include:

- Large, centralized networks
- Time-sharing data applications
- Poll-select (inquiry-response) applications

3.2.4.2 Design considerations and techniques

In this section, we review several classic heuristic design techniques for designing multipoint lines. In particular, we discuss:

- Prim's algorithm (see Section 3.2.3.2)
- Kruskal's capacitated minimal spanning tree algorithm
- Esau-William's (E-W) algorithm
- Unified algorithm
- Bin packing algorithms
- Terminal assignment algorithm
- Center of mass algorithm (see Section 3.2.5.2)
- Concentrator add algorithm (see Section 3.2.5.2)
- Concentrator drop algorithm (see Section 3.2.5.2)

A capacitated minimal spanning tree[7] (CMST) topology is especially appropriate for designing a low-speed (i.e., 9.6 Kbps, 56 Kbps, or 128 Kbps) multipoint line in which the traffic flowing over the link is small compared to the link size. When traffic flows are comparable in magnitude to the smallest link in the network, it may be necessary to design several shared-access multipoint lines that are, in turn, connected to concentrators that are part of a larger backbone network. When there is enough traffic flowing

---

[7] Capacitated minimal spanning tree — This topology represents a minimal spanning tree that has been designed to conform to link capacity constraints.

through the network to fill many smaller links, this leads to a choice between several lower speed lines or a single high-speed line [Cahn98]. In the latter two cases, the optimal placement of the concentrators (or multiplexers, etc.) in the backbones can be determined using terminal assignment or concentrator location algorithms. We discuss the terminal assignment procedures later in this section, and concentrator location algorithms in Section 3.2.5.2.

The heuristic design techniques in this section are low in computational complexity, yet they provide high-quality[8] results. They are all considered a form of greedy algorithm. The first three CMST algorithms (e.g., Prim's, Kruskal's, and E-W's) listed above are special cases of the unified algorithm. CMST algorithms should be used when there are *no* constraints on the number of incoming multipoint lines that the host or concentrator can support. When the number of multipoint lines connecting to the host or concentrator must be minimized, bin-packing algorithms should be used instead of CMST algorithms. Terminal assignment problems are used to determine which terminals should be associated with a given concentrator or host location. Thus, the terminal assignment determines which terminals are grouped together on a multipoint line.

Sometimes the order in which terminals or links are considered may make a significant difference in the cost of the resulting design. However, with greedy algorithms, once a link assignment is made, it is not taken back. This is one of the ways the computational complexity of these procedures is kept under control. However, in some cases, one may wish to selectively reorder a chain of link assignments in the network. Branch and exchange algorithms are designed to do this, but they are *much more* complex and work only on limited problems. Branch and exchange algorithms work by exchanging specific link pairs to see if cost reductions can be found in the design solution. For a more detailed discussion of branch and exchange algorithms, the reader is referred to [Kersh93].

After the multipoint lines are designed, the add, drop, and center of mass algorithms can be used to determine the concentrator or backbone node to which they should connect.

Kruskal's capacitated minimal spanning tree algorithm

This algorithm begins with no links, only nodes in the network. The first step is to create a sorted list, from low to high, of all possible link costs. The algorithm then constructs a minimal spanning tree, link by link. As links are added to the network they are checked to make sure that the capacity and cycling restrictions are not violated. If the constraints would be violated by a link addition, the link is ignored and another link is examined to see if it is suitable for inclusion in the network.

---

[8] For the cases they tested, Chandy and Russell have shown that heuristic CMST techniques are usually within 5% to 10% of optimal [Ahuj82, p. 130].

1. Sort all possible links in ascending order by cost and put in an ordered list.
2. Check to see if all the nodes are connected. This will occur when (n-1) links have been selected (this is true by the definition of a minimal spanning tree).
   - If all the nodes are connected, then terminate the algorithm with the message "Solution complete."
   - If all the nodes are not connected, continue to the next step.
3. Select the link at the top of the list.
   - If no links are on the list, terminate algorithm. Check to see if all nodes are connected, and if not, terminate the algorithm with the message "Solution cannot be found."
4. Check to see if the link selected creates a cycle in the network.
   - If the link creates a cycle, remove it from the list. Return to step 2.
   - If the link does not create a cycle, check to see if it can handle the required traffic load.
     - If so, add the link to the network, and remove it from link list. Return to step 2.
     - If not, remove the link from link list and return to step 2.

An efficient implementation of this algorithm can be shown to be of O (m log n), where m is the number of edges tested and n is the number of nodes in the network [Kersh93]. A potential problem with this algorithm is that it may strand the last nodes connected to the network far from the center. The next algorithm we present — the Esau-William's algorithm — attempts to address this problem by starting off with all the nodes connected directly to the center. Links to the center are only replaced when they can be cost-justified.

We now present an example to demonstrate Kruskal's CMST algorithm. We summarize the line cost data for our sample problem in Table 3.8 and the traffic flow data in Table 3.9. If an entry in a table is missing, as indicated by "-", this means that there is no cost or traffic associated with this particular location. Assume for this problem that the maximum traffic a line can carry is 10 units. We want to find the lowest cost multiline network to connect terminals 2, 3, 4, and 5 to the host 1.

Table 3.8 Cost Matrix for Sample CMST Problem

	Host 1	Node 2	Node 3	Node 4	Node 5
Host 1	—	6	3	4	5
Node 2	6	—	3	5	7
Node 3	3	3	—	3	5
Node 4	4	5	3	—	3
Node 5	5	7	5	3	—

Table 3.9 Traffic Matrix for
Sample CMST Problem

	Traffic units generated
Host 1	—
Node 2	5
Node 3	4
Node 4	3
Node 5	5

Table 3.10   **Sorted Link Costs**
for Sample CMST Problem

Link	Cost
1-3	3
2-3	3
4-3	3
4-5	3
1-4	4
2-4	5
3-5	5
1-5	5
1-2	6
2-5	7

Solution of sample problem using Kruskal's CMST:

1. Sort all possible links in ascending order of cost and put them in a link list (Table 3.10).
2. Check to see if all the nodes are connected. *This will occur when (n − 1 = 4) links have been selected. No links have been added to the solution, therefore, we proceed to the next step.*
3. Select the link at the top of the list. *This is link (1-3). Note that when ties occur in the sorted link costs, it does not matter which link is selected, even though this may change the final solution.*
4. Check to see if the link selected creates a cycle in the network. *Link (1,3) does not create a cycle. It does not violate a capacity constraint, since only four (4) units of traffic are carried between nodes 1 and 3. So we add the link to the network.*
5. Check to see if all the nodes are connected. *Since only one of 4 required links have been selected, we proceed to the next step.*
6. Select the link at the top of the list. *This is link (2-3).*
7. Check to see if the link selected creates a cycle in the network. *Link (2,3) does not create a cycle. It does not violate a capacity constraint, since only nine (9) units of traffic would be carried between nodes 2,3, and 1. So we add the link to the network.*

8. Check to see if all the nodes are connected. *Since only two of four required links have been selected, we proceed to the next step.*
9. Select the link at the top of the list. *This is link (4-3).*
10. Check to see if the link selected creates a cycle in the network. *Link (4,3) does not create a cycle. This link does violate the capacity constraint, since twelve (12) units of traffic would be carried between nodes 1, 2, 3, and 4. So we do not add the link to the network.*
11. Check to see if all the nodes are connected. *Since only two of four required links have been selected, we proceed to the next step.*
12. Select the link at the top of the list. *This is link (4-5).*
13. Check to see if the link selected creates a cycle in the network. *Link (4-5) does not create a cycle. It does not violate a capacity constraint, since only eight (8) units of traffic would be carried between nodes 4 and 5. So we add the link to the network.*
14. Check to see if all the nodes are connected. *Since only three of four required links have been selected, we proceed to the next step.*
15. Select the link at the top of the list. *This is link (1-4).*
16. Check to see if the link selected creates a cycle in the network. *Link (1,4) does not create a cycle. It does not violate a capacity constraint, since only eight (8) units of traffic would be carried between nodes 4,5, and 1. So we add the link to the network.*
17. Check to see if all the nodes are connected. *Since four of four required links have been selected, we terminate the algorithm.*

The results of this algorithm are illustrated in Figure 3.4.

Esau-William's (E-W) capacitated minimal spanning tree algorithm

This algorithm starts by connecting all terminals to the center node or concentrator. It then attempts to replace the links, one by one, with cheaper ones. When a new link is added, the corresponding connection to the host/concentrator is removed. The cycling and traffic constraints are checked as the links are selected, and no link replacements are allowed that would cause a violation of the constraints.

1. Connect all the nodes to the center.
2. Calculate a trade-off value, d_{ij}, representing the difference in cost C_{ij} between the connection (i, j) and the cost C_{i1} to connect terminal i to center node 1.

$$d_{ij} = C_{ij} - C_{i1}$$

(calculate for all i, j, where 1 is the concentrator location)   (3.4)

3. Sort the links in increasing order of d_{ij}.

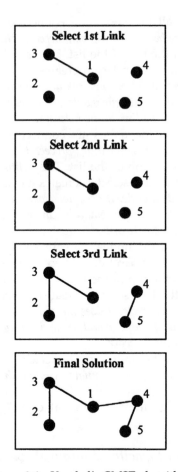

Figure 3.4   Kruskal's CMST algorithm.

4. Remove all links with an associated d_{ij} that is equal to or greater than zero (0). If no links have a negative d_{ij} value, then the algorithm terminates and the network remains as it is.
5. Select the link with the lowest (i.e., the most negative) d_{ij} value. Check to see if the link selected creates a cycle in the network.
   - If the link creates a cycle, set the d_{ij} value = $+\infty$ (this is a very large positive number) Return to step 4.
   - If the link does not create a cycle, check to see if it can handle the required traffic load.
     - If the link has sufficient capacity to carry the required traffic, replace link (i, l) with link (i, j). Recalculate affected d_{ij} values to reflect the fact that node i is now connected to node j, and not the center. Thus, replace C_{i1} with C_{ij} in the d_{ij} calculations. Return to step 2.

- If the link lacks sufficient capacity to carry the required traffic, do not replace the link. Set the link d_{ij} value = $+ \infty$ and return to step 2.

An efficient implementation of this algorithm can be shown to be of O ($n^2 \log n$), where n is the number of nodes in the network [Kersh93].

We now present an example to demonstrate E-W's CMST algorithm. We use the same cost and traffic data we used in our previous example (see Tables 3.8 and 3.9). Assume, as before, that the maximum traffic constraint is 10 units. We want to find the lowest cost multiline network to connect terminals 2, 3, 4, and 5 to the host 1.

Solution of Sample Problem Using E-W's CMST:

1. Connect all the nodes to the center. This is shown in the first diagram of Figure 3.5.
2. Calculate a trade-off value, d_{ij}, representing the difference in cost C_{ij} between the connection (i, j) and cost C_{i1} to connect i to center node 1. The results of these calculations are summarized in Table 3.11 below.
3. Sort the links in increasing order of d_{ij} (Table 3.12).
4. Remove all links with an associated d_{ij} that is equal to or greater than zero (0). This results in Table 3.13 below. Since there are still links with a negative d_{ij} value, the algorithm proceeds.
5. Select the link with the lowest (i.e., the most negative) d_{ij} value. Check to see if the link selected creates a cycle in the network. *This is link (2,3) which replaces link (2,1). The (2,3) link does not create a cycle. The combined traffic flows on links (2,3) and (3,1) will equal 9 units, which does not exceed the traffic constraint of 10 units. So, the link is accepted as a replacement for (2,1). Set $d_{23} = C_{23} - C_{23} = 0$. Updating the other d_{ij}'s affected by the link replacement: Set $d_{24} = C_{24} - C_{23} = (5 - 3) = 2$. Set $d_{25} = C_{25} - C_{23} = (7 - 3) = 4$. Since these values are positive, we do not need to consider these d_{ij}'s further as potential link replacements.*
6. Returning to Step 4, we remove all links with an associated d_{ij} that is equal to or greater than zero. *Since there are still negative d_{ij} values, we proceed with the algorithm.* (Table 3.14)
7. Select the link with the lowest (i.e., the most negative) d_{ij} value. Check to see if the link selected creates a cycle in the network. *As shown in Table 3.14, link (5,4) should be chosen to replace link (5,1). The (5,4) link does not create a cycle. The combined traffic flows on links (5,4) and (4,1) will equal 8 units, which does not exceed the traffic constraint of 10 units. So, the link is accepted as a replacement for (5,1). Set $d_{54} = C_{54} - C_{54} = 0$. Updating the other d_{ij}'s affected by the link replacement: Set $d_{53} = C_{53} - C_{54} = (5 - 3) = 2$. Set $d_{52} = C_{52} - C_{54} = (7 - 3) = 4$. Set $d_{45} = C_{45} - C_{45} = (3 - 3) = 0$. Since these values are greater or equal to zero, we do not need to consider these d_{ij}'s further as potential link replacements.*

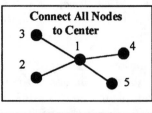

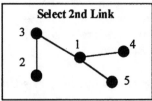

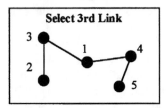

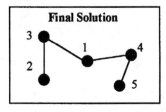

Figure 3.5   E-W's CMST algorithm.

Table 3.11 d_{ij} Calculations for
E-W's CMST

d_{ij}	d_{ij} calculations
d_{23}	$C_{23}\text{-}C_{21} = -3$
d_{32}	$C_{32}\text{-}C_{31} = 0$
d_{24}	$C_{24}\text{-}C_{21} = -1$
d_{42}	$C_{42}\text{-}C_{41} = 1$
d_{25}	$C_{25}\text{-}C_{21} = 1$
d_{52}	$C_{52}\text{-}C_{51} = 2$
d_{34}	$C_{34}\text{-}C_{31} = 0$
d_{43}	$C_{43}\text{-}C_{41} = -1$
d_{35}	$C_{35}\text{-}C_{31} = 2$
d_{53}	$C_{53}\text{-}C_{51} = 0$
d_{45}	$C_{45}\text{-}C_{41} = -1$
d_{54}	$C_{54}\text{-}C_{51} = -2$

Table 3.12   Sorted d_{ij}
Values for E-W's CMST

d_{ij}	d_{ij} calculations
d_{23}	$C_{23}\text{-}C_{21} = -3$
d_{54}	$C_{54}\text{-}C_{51} = -2$
d_{24}	$C_{24}\text{-}C_{21} = -1$
d_{43}	$C_{43}\text{-}C_{41} = -1$
d_{45}	$C_{45}\text{-}C_{41} = -1$
d_{32}	$C_{32}\text{-}C_{31} = 0$
d_{34}	$C_{34}\text{-}C_{31} = 0$
d_{53}	$C_{53}\text{-}C_{51} = 0$
d_{42}	$C_{42}\text{-}C_{41} = 1$
d_{25}	$C_{25}\text{-}C_{21} = 1$
d_{35}	$C_{35}\text{-}C_{31} = 2$
d_{52}	$C_{52}\text{-}C_{51} = 2$

Table 3.13   Negative d_{ij}
Values for First Link
Selection

d_{ij}	d_{ij} calculations
d_{23}	$C_{23}\text{-}C_{21} = -3$
d_{54}	$C_{54}\text{-}C_{51} = -2$
d_{24}	$C_{24}\text{-}C_{21} = -1$
d_{43}	$C_{43}\text{-}C_{41} = -1$
d_{45}	$C_{45}\text{-}C_{41} = -1$

Table 3.14   Negative d_{ij}
Values for Second Link
Selection

d_{ij}	d_{ij} calculations
d_{54}	$C_{54}\text{-}C_{51} = -2$
d_{43}	$C_{43}\text{-}C_{41} = -1$
d_{45}	$C_{45}\text{-}C_{41} = -1$

Table 3.15   Negative d_{ij}
Values for Third Link
Selection

D_{ij}	d_{ij} calculations
d_{43}	$C_{43}\text{-}C_{41} = -1$

8. Returning to step 4, we remove all links with an associated d_{ij} that is equal to or greater than zero (0). This results in Table 3.15. Since there are links with a negative d_{ij} value, the algorithm proceeds.

9. Select the link with the lowest (i.e., the most negative) d_{ij} value. Check to see if the link selected creates a cycle in the network. *This is link (4,3). This link would exceed the traffic constraint of 10 units, since nodes 2, 3, 4, and 5 would create a combined traffic load of 17 units. So, the link is rejected. Set $d_{43} = +\infty$.*

10. Returning to Step 4, we remove all links with an associated d_{ij} that is equal to or greater than zero (0). Since there are no links with a negative d_{ij} value, the algorithm terminates.

The solution using E-W's algorithm is illustrated in Figure 3.5. Although the results for this sample problem are the same as those obtained using Kruskal's algorithm, in general, they may not be. The *only* time you can be certain the results will be identical (in cost) is when the capacity constraint is not active and does not affect the solution. This means that no link was rejected during the course of the algorithm due to insufficient capacity.

In general, the E-W algorithm produces very good designs for multipoint lines. It also does a good job configuring access routers on a shared T1 line. The algorithm performs well (i.e., it produces low cost designs) when the traffic requirements vary considerably at each node. It also usually does better configuring a network with a smaller number of sites on the line than it does configuring a network with a large number of sites on the line [Cahn98, p. 170].

Unified algorithm

Kershenbaum and Chou have shown that Kruskal's and E-W's algorithm, and other CMST algorithms, are special cases of what they call the unified algorithm. Not surprisingly, the Unified algorithm solution procedure is very similar to the E-W algorithm we have just presented. However, the Unified algorithm incorporates a weighting factor w that is defined for each terminal location. Variations on the definition of w yield specific heuristic CMST algorithms. For example:

- If the weight $w = 0$, this reduces to Kruskal's algorithm.
- If the weight $w = C_{i1}$ = terminal distance from center, this reduces to E-W's algorithm.
- If the weight $w = 0$ at center and $w = -\infty$, this reduces to Prim's algorithm.

Other values for w are possible and may be useful depending on the type of design that is desired. It can be shown that an efficient programming implementation of the unified algorithm is of $O(n^2)$, where n is the number of nodes in the network [Kersh93].

We present the unified algorithm below. We begin by defining w_i as the weighting factor associated with terminal I and C_{ij} as the link cost connecting terminals i and j.

1. Initialize w_i for $i = 1$ to N where N is the number of terminals. Evaluate $d_{ij} = C_{ij} - w_i$ for all links (i,j) when C_{ij} exists and no constraints are violated with connection i-j.
2. Sort the d_{ij} and determine the minimum d_{ij} for all links (i,j), such that i does not equal j. If d_{ij} is equal to positive infinity, terminate the algorithm.
3. Evaluate the constraints under connection (i,j). If any constraints are violated, set d_{ij} equal to positive infinity and return to step 2. Otherwise proceed to the next step.
4. Add link (i,j) and relabel one of the terminals i or j to correspond to each other. Reevaluate the constraints. Obtain new values for w_i and d_{ij} (for all i such that w_i has changed) and return to step 2.

Bin packing algorithms

If we need to minimize the number of lines connected to the backbone node or central facility, then CMST algorithms are not suitable. Minimizing the number of lines may be necessary if there are many groups of terminals in the same city with similar link costs. We present three simple and effective bin packing algorithms that are of computational complexity, when implemented efficiently, of $O(N \log N + NB)$, where N is the number of nodes in the network and B is the number of bins (or lines) to which the nodes are assigned. These algorithms are:

- First fit decreasing
- Best fit decreasing
- Worst fit decreasing

These algorithms assume that the line capacities and traffic requirements at each site are known.

First fit decreasing bin packing algorithm:

1. Sort terminal traffic requirements from high to low. Set the highest traffic requirement equal to t_{MAX}.
2. Check highest traffic requirement, t_{MAX}, against the line capacity. If the line capacity is less than the largest traffic requirement, terminate the algorithm with the message, "No solution possible due to insufficient line capacity." If the line capacity is sufficient, proceed to the next step.
3. Check to see if all the traffic flows have been assigned to a link. If not, continue to the next step. Otherwise, terminate the algorithm with the message, "Final solution found."
4. Put the first terminal requirement (i.e., the current t_{MAX}) on the *first* line with capacity. If a line does not have capacity for the requirement, install a new line and place the traffic requirement on it. Remove the current t_{MAX} from the sorted list. Return to step 3.

Table 3.16 Sample Data For
Bin Packing Algorithms

Terminal	Traffic requirement
1	17
2	10
3	9
4	15
5	6
6	8

Table 3.17 Sample Data For
Bin Packing Algorithms

Terminal	Traffic requirement
1	17
4	15
2	10
3	9
6	8
5	6

We now illustrate the first fit decreasing bin packing algorithm with an example. The data for the sample problem are summarized in Table 3.16. Assume that the maximum line capacity for the problem is 30 traffic units.

Solution to sample problem using first fit decreasing bin packing algorithm

1. Sort terminal traffic requirements from high to low (see Table 3.17).
2. Check highest requirement against line capacity. *The line capacity is sufficient, and therefore we proceed to the next step.*
3. Check to see if all traffic requirements have been assigned. *They have not, so we proceed to the next step.*
4. Put first terminal requirement on the *FIRST* line with capacity. *We select terminal 1 for placement on a new line A. Remove terminal 1 from Table 3.17. We proceed with the algorithm.*
5. Check to see if all traffic requirements have been assigned. *They have not, so we proceed to the next step.*
6. Put first terminal requirement on the *FIRST* line with capacity. *We select terminal 4 for placement on a new line B, since placement of terminal 2 on line A would exceed the maximum traffic constraint. Remove terminal 4 from Table 3.17. We proceed with the algorithm.*
7. Check to see if all traffic requirements have been assigned. *They have not, so we proceed to the next step.*

8. Put first terminal requirement on the *FIRST* line with capacity. *We select terminal 2 for placement on line A. Remove terminal 2 from Table 3.17. We proceed with the algorithm.*

9. Check to see if all traffic requirements have been assigned. *They have not, so we proceed to the next step.*

10. Put first terminal requirement on the *FIRST* line with capacity. *We select terminal 3 for placement on line B. Remove terminal 3 from Table 3.17. We proceed with the algorithm.*

11. Check to see if all traffic requirements have been assigned. *They have not, so we proceed to the next step.*

12. Put first terminal requirement on the *FIRST* line with capacity. *We select terminal 6 for placement on a new line C, since it would exceed the maximum traffic constraint if placed on either line A or B. Remove terminal 6 from Table 3.17. We proceed with the algorithm.*

13. Check to see if all traffic requirements have been assigned. *They have not, so we proceed to the next step.*

14. Put first terminal requirement on the *FIRST* line with capacity. *We select terminal 5 for placement on line B. Remove terminal 5 from Table 3.17. We proceed with the algorithm.*

15. Check to see if all traffic requirements have been assigned. *They have, so we terminate the algorithm, having found the final solution.*

The final solution using this algorithm places terminals 1 and 2 on line A, terminals 3, 4, and 5 on line B, and terminal 6 on line C.

Best-fit decreasing bin packing algorithm

1. Sort terminal traffic requirements from high to low. Set the highest traffic requirement equal to t_{MAX}.

2. Check highest requirement t_{MAX} against the line capacity. If the line capacity is less than the largest traffic requirement, terminate the algorithm with the message, "No solution possible due to insufficient line capacity." If the line capacity is sufficient, proceed to the next step.

3. Check to see if all the traffic flows have been assigned to links. If not, continue to the next step. Otherwise, terminate the algorithm with the message, "Final solution found."

4. Put current t_{MAX} requirement on the first feasible line with the *least* capacity. If no line has sufficient capacity for the requirement, install a new line and place the traffic requirement on it. Remove the current traffic requirement t_{MAX} from the sorted list. Return to step 3.

We now illustrate the best-fit decreasing bin packing algorithm with an example. We use the same data as before (see Table 3.16). We continue to assume that the maximum line capacity for the problem is 30 traffic units.

Table 3.18   Sorted Link
Requirements For Best Fit
Bin Packing Problem

Terminal	Traffic requirement
1	17
4	15
2	10
3	9
6	8
5	6

Solution to sample problem using best fit decreasing bin packing algorithm

1. Sort terminal traffic requirements from high to low. *See Table 3.18.*
2. Check highest traffic requirement t_{MAX} against line capacity. *The line capacity is sufficient, and therefore we proceed to the next step.*
3. Check to see if all traffic requirements have been assigned. *They have not, so we proceed to the next step.*
4. Put first terminal requirement on the line with the least capacity. *We select terminal 1 for placement on a new line A. Remove terminal 1 from Table 3.18. We proceed with the algorithm.*
5. Check to see if all traffic requirements have been assigned. *They have not, so we proceed to the next step.*
6. Put first terminal requirement on the line with the least capacity. *We select terminal 4 for placement on a new line B, since placement of terminal 4 on line A would exceed the maximum traffic constraint. Remove terminal 4 from Table 3.18. We proceed with the algorithm.*
7. Check to see if all traffic requirements have been assigned. *They have not, so we proceed to the next step.*
8. Put first terminal requirement on the line with the least capacity. *We select terminal 2 for placement on line A. Remove terminal 2 from Table 3.18. We proceed with the algorithm.*
9. Check to see if all traffic requirements have been assigned. *They have not, so we proceed to the next step.*
10. Put first terminal requirement on the line with the least capacity. *We select terminal 3 for placement on line B. Remove terminal 3 from Table 3.18. We proceed with the algorithm.*
11. Check to see if all traffic requirements have been assigned. *They have not, so we proceed to the next step.*
12. Put first terminal requirement on the line with the least capacity. *We select terminal 6 for placement on a new line C, since it would exceed the maximum traffic constraint if placed on either line A or B. Remove terminal 6 from Table 3.18. We proceed with the algorithm.*

13. Check to see if all traffic requirements have been assigned. *They have not, so we proceed to the next step.*
14. Put first terminal requirement on the line with the least capacity. *We select terminal 5 for placement on line B. Remove terminal 5 from Table 3.18. We proceed with the algorithm.*
15. Check to see if all traffic requirements have been assigned. *They have, so we terminate the algorithm, having found the final solution.*

In the final solution, terminals 1 and 2 are on line A, terminals 3, 4, and 5 are on line B, and terminal 6 is on line C. This is the same solution obtained using the First Fit algorithm. In general, one would not necessarily expect the results to be the same.

Worst-fit decreasing bin packing algorithm

1. Sort terminal traffic requirements from high to low. Set the highest traffic requirement equal to t_{MAX}.
2. Check the highest requirement t_{MAX} against the line capacity. If the line capacity is less than the largest traffic requirement, t_{MAX}, terminate the algorithm with the message, "No solution possible due to insufficient line capacity." If the line capacity is sufficient, proceed to the next step.
3. Check to see if all traffic flows have been assigned to links. If not, continue to the next step. Otherwise, terminate the algorithm with the message, "Final solution found."
4. Put first terminal requirement (i.e., the current t_{MAX}) on the first feasible line with the *most* capacity. If no line has sufficient capacity for the requirement, install a new line and place the traffic requirement on it. Remove the traffic requirement from the sorted list. Return to step 3.

We now illustrate the worst fit decreasing bin packing algorithm using the same data as before. We continue to assume that the maximum line capacity for the problem is 30 traffic units.

Solution to sample problem using worst fit decreasing bin packing algorithm

1. Sort terminal traffic requirements from high to low. *The results are shown in Table 3.18.*
2. Check highest requirement against line capacity. *The line capacity is sufficient, and therefore we proceed to the next step.*
3. Check to see if all traffic requirements have been assigned. *They have not, so we proceed to the next step.*

4. Put first terminal requirement on the first feasible line with the MOST capacity. *We select terminal 1 for placement on a new line A. Remove terminal 1 from Table 3.18. We proceed with the algorithm.*

5. Check to see if all traffic requirements have been assigned. *They have not, so we proceed to the next step.*

6. Put first terminal requirement on the line with the MOST capacity. *We select terminal 4 for placement on a new line B, since placement of terminal 2 on line A would exceed the maximum traffic constraint. Remove terminal 4 from Table 3.18. We proceed with the algorithm.*

7. Check to see if all traffic requirements have been assigned. *They have not, so we proceed to the next step.*

8. Put first terminal requirement on the line with the MOST capacity. *We select terminal 2 for placement on line B. Remove terminal 2 from Table 3.18. We proceed with the algorithm.*

9. Check to see if all traffic requirements have been assigned. *They have not, so we proceed to the next step.*

10. Put first terminal requirement on the line with the MOST capacity. *We select terminal 3 for placement on line A. Remove terminal 3 from Table 3.18. We proceed with the algorithm.*

11. Check to see if all traffic requirements have been assigned. *They have not, so we proceed to the next step.*

12. Put first terminal requirement on the line with the MOST capacity. *We select terminal 6 for placement on a new line C, since it would exceed the maximum traffic constraint if placed on either line A or B. Remove terminal 6 from Table 3.18. We proceed with the algorithm.*

13. Check to see if all traffic requirements have been assigned. *They have not, so we proceed to the next step.*

14. Put first terminal requirement on the line with the MOST capacity. *We select terminal 5 for placement on line C. Remove terminal 5 from Table 3.18. We proceed with the algorithm.*

15. Check to see if all traffic requirements have been assigned. *They have, so we terminate the algorithm, having found the final solution.*

The final solution using this algorithm places terminals 1 and 3 on line A, terminals 2 and 4 on line B, and terminals 5 and 6 on line C.

Terminal assignment problem

The terminal assignment problem attempts to find the optimal connection of a terminal to a center or concentrator location. We briefly summarize the terminal assignment algorithm below. The algorithm we present below is nearly identical to the E-W algorithm. The only difference between this algorithm and the CMST version of E-W presented earlier is in the calculation of the trade-off function. Recall that the trade-off function is used to evaluate whether or not to replace a connection to the center node with the current link under consideration.

1. Calculate for each terminal the value associated with the trade-off function:

$$t_i = c_{i1} - \alpha \, c_{i2} \tag{3.5}$$

   where t_i is the trade-off cost; c_{i1} and c_{i2} are the cost of connecting terminal to its nearest and second nearest neighbor respectively; and α is a parameter between 0 and 1 reflecting our preference toward "critical" terminals. As the α value approaches 1, there is more of a penalty for failing to connect the terminal to its second nearest neighbor.
2. Find the minimum trade-off value and assign the associated terminal to the cheapest *feasible* concentrator. If a terminal has lost its last feasible neighbor, stop the algorithm. If only one feasible neighbor remains, the terminal must be assigned to that concentrator.
3. Check and update trade-offs (setting them to $+\infty$ as assignments are made) and remaining concentrator capacity affected by the last link assignment.
4. Continue until all t_i are greater than or equal to zero. At this point, terminate the algorithm.

3.2.4.3   Case study

In Figure 3.3, we presented an example of a tree WAN. This network has several CRTs running IBM 3270 terminal emulation that are connected to 3174 data concentrators (which are also known as cluster controllers). These data concentrators either connect to other data concentrators or directly to the front end processor (FEP), which communicates with the mainframe computer. Note that in many tree networks, the terminals connect directly to the FEP — typically a 3745 communications controller — or to a data concentrator. Many tree networks are actually a combination of one or more star networks interconnected with one or more tree networks. Since the costs of the links are usually mileage sensitive, it is sometimes economical to group various terminals into data concentrators via analog lines with modems, and then to connect the concentrators to the FEP using higher bandwidth digital facilities (such as 56 Kbps DDS or Nx64 Kbps Fractional T-1 links). The reduction in networking costs must be weighed against the decrease in reliability resulting from fewer links in the network. In a tree topology, if one of the higher bandwidth links fails, then multiple terminals are affected. In contrast, star networks are more reliable because each end terminal has a separate link to the central processor. However, these additional links add to the facility costs. The network designer must carefully weigh which terminals require direct links and which terminals should be aggregated via data concentrators.

3.2.5   Backbone networks

3.2.5.1   Overview

A backbone network is comprised of high-capacity links and nodes (typically, routers or multiplexers) that consolidate traffic from smaller access networks. Backbone networks allow a diverse traffic mix to be carried on a single network infrastructure, irrespective of protocol. Backbone networks are attractive because they simplify the network and offer cost economies of scale.[9] For these reasons, many large U.S. and multinational corporations implement private backbone networks to support integrated and secure voice, data, and video applications across widely distributed locations.

Typically, backbone networks are hierarchically structured in three major ways, depending on price and performance requirements:

- Second-tier private access network(s) connecting into larger, private backbone networks
- Second-tier private access network(s) connecting to local public network backbone networks
- Second-tier private access network(s) connecting to large public backbone networks supplied by IEC carriers and service providers

Private backbone networks are commonly configured with T1 and T3 circuits. The traditional alternative to a private backbone, the public X.25 network, has a maximum bandwidth capacity of 56 Kbps, and thus is not suitable for high-speed internetworking (such as would be required for LAN-to-WAN connectivity). Of the new fast packet, high bandwidth options (which include frame relay, SMDS, and ATM), frame relay is the most popular. Frame relay supports both private and public backbone networking. SMDS and ATM are public backbone options.

As companies merge to form larger corporate entities, many telecom managers are faced with the arduous task of integrating numerous external subsystems into their existing corporate network infrastructure. As such, vast quantities of traffic flow and requirements data must be analyzed before a backbone network can be cost justified, designed, engineered, implemented, and maintained.

With the enormous influx of new telecom products from local, long-distance, and international carriers, network integrators, and value-added resellers (VARs), the network designer/manager can easily become inundated by the wide variety of networking solutions and alternatives. On the user side, software applications are growing in complexity and increasing the demands on the network. At the same time, management applies constant

---

[9] Backbone networks offer economies of scale because higher speeds links carry more traffic and cost less than an equivalent number of lower speed lines. For example, one "T1 is cheaper than 24 DS0; 1 T3 is cheaper than 28 DS1; 1 E3-c is cheaper than 3 E1s" [Mino95].

pressure to increase network use and user productivity while maintaining or cutting costs.

The telecom manager, therefore, must pursue many options to juggle the needs of users and upper management. In this climate, many large corporations place their networks up for bid every three to seven years in the hopes of lowering costs and keeping pace with changing user demands. These bids often take the form of requests for information (RFI), requests for quote (RFQ), and requests for proposal (RFP) from telecom providers. When evaluating bids, the telecom manager should be guided by the business and technical requirements and should avoid adopting new products and services for the sake of status or using the latest technology.

We now summarize some of the major advantages of private backbone topologies. Private backbone topologies are:

- Used to integrate various subnetworks
- Relatively resistant to node and link failures
- Secure and private means of data transport
- Effective for certain time-sensitive applications
- Effective for distributed voice and data applications

Backbone networks have disadvantages especially as compared to local access networks. They:

- Are harder and more complex to design
- Are more difficult to manage
- Need more redundant sites to be supported
- With increased redundancy needed for network reliability, there is the additional expense of idle bandwidth.
- Can be costly if distributed rerouting and time-sensitive applications must be supported

Backbone networks are recommended to support:

- Large, multiapplication networks
- Host-to-host connectivity
- Certain distributed voice and data applications
- Multiple host connectivity for disaster recovery
- Network availability for critical applications

3.2.5.2 Design considerations and techniques
Two major topological design problems must be solved when laying out a backbone network:

- Node placement
- Backbone connectivity (see Section 3.2.6.2)

Potential backbone nodes should be chosen at centers of major traffic volumes in preference to small or remote sites. This reduces the potential for traffic bottlenecks, and it also increases the chances that the node will be located at a site where a full range of local and IEC services are available for supporting the backbone requirements.

Since backbone networks operate at high speeds, carrying lots of traffic between many nodes, the effects of network failures are magnified. For this reason, reliability is much more important when designing a backbone network than it is when designing a local access network. The performance and reliability demanded of backbone networks should be reflected in the selection and design of the network nodes and links.

With these practical guidelines in mind, four algorithms are useful for analytically deriving recommendations for locating potential backbone nodes. They are:

- Center of mass (COM) algorithm
- Add algorithm
- Drop algorithm
- Mentor algorithm (see Section 3.2.6.2)

These algorithms are broadly applicable to many network design problems. We will review how they can be used in the context of identifying concentrator locations in centralized networks and backbone nodes in distributed networks.

We now introduce the center of mass (COM) algorithm. This algorithm attempts to find natural traffic clusters based on estimated traffic between source and destination nodes. The COM algorithm is used when there are no candidate sites for the concentrator or backbone nodes. The node placements recommended by the COM algorithm must be checked to see if they are feasible.

Center of mass algorithm

The algorithm assumes that the following information is given:

- w_i traffic-based weights (calculated by totaling the traffic to and from each node).
- Desired maximum total weight W_M for a cluster (calculated by adding the w_i for each node used to construct a cluster center). This represents an upper limit on W that cannot be exceeded.
- Desired minimum total weight W_m for a cluster (calculated by adding the w_i for each node used to construct a cluster center). This represents a lower limit on the cluster size.
- Desired maximum total distance D_M between two clusters considered for a merge. This represents an upper limit on D_M that cannot be exceeded.

- (x_i, y_i) coordinates for candidate i sites. These coordinates are used to calculate the distance between two sites, using a standard distance formula. The distance calculated is used as a proxy to estimate the cost of the link to connect the nodes. Therefore, in the discussion that follows, cost and distance are used synonymously.
- Desired number of final clusters, C.

If during the course of the algorithm there are conflicts enforcing the restrictions on W_M, W_m, and D, a trade-off function must be defined to resolve the differences. The problem must then be solved again, using the trade-off function.

1. Start with each (x_i, y_i) in a cluster by itself.
2. The cost to connect nodes i and j is assumed to be directly proportional to the distance between the two. The distance between each pair of nodes is calculated using a standard distance formula shown below:

$$Cost_{i,j} = \sqrt{\left[\left(x_i - x_j\right)^2 + \left(y_i - y_j\right)^2\right]} \qquad (3.6)$$

3. Sort the costs computed using Equation 3.6 from low to high for each node pair.
4. Find the two closest nodes as candidates for merging.
   - If there are no merge candidates, check to see if the desired number of clusters, C, has been found. If not, terminate the algorithm with the message, "COM cannot find a solution." If the target for C and all other constraints are satisfied, terminate the algorithm with the message, "Final Solution Complete."
   - If the constraints are violated, reject the cluster merge and remove it from the candidate list. Return to the beginning of step 4.
   - If the constraints are not violated, merge the two clusters (i.e., i and j) that are closest to each other to form a new cluster k. The new cluster k is chosen as the center of mass based on the traffic flowing at i and j.

The x coordinate of the new cluster k is:

$$x_k = \frac{\left(w_i * x_i\right) + \left(w_j * x_j\right)}{w_i + w_j} \qquad (3.7)$$

The y coordinate of the new cluster k is:

$$y_k = \frac{\left(w_i * y_i\right) + \left(w_j * y_j\right)}{w_i + w_j} \qquad (3.8)$$

Table 3.19 Sample COM
Problem Data

Node	X_i	Y_i	Node weight
1	31	19	1
2	45	13	1
3	59	92	1
4	22	64	1
5	86	55	1
6	95	78	1
7	98	63	1
8	39	44	1
9	27	38	1
10	48	85	1

5. Remove clusters i and j from further consideration. Add cluster k to the list of clusters to be merged. Return to step 2 and calculate the distance between the existing nodes and the new cluster k.

The task of finding the two closest centers to merge is a major contributor to the algorithm's complexity. It can be shown that an efficient programming implementation of this algorithm has an overall complexity of $O(n^2)$, where n is the number of nodes considered [Kersh93].

Sample COM Problem
We now present an example to illustrate the COM algorithm. The data for the sample problem are given below:

- See Table 3.19 for (x_i, y_i) coordinates and w_i traffic-based weights.
- Desired maximum total weight W_M for a cluster = 4.
- Desired minimum total weight W_m for a cluster = no minimum.
- Desired maximum total distance D_M = 43 units. Any two clusters further apart than D_M cannot be merged.
- Desired number of final clusters C = 3.

1. Start with each (x_i, y_i) in a cluster by itself. *This means that initially each node is treated as a separate node center.*
2. Calculate the cost to connect nodes i and j. *The positional node coordinates are shown in Table 3.19. Using these coordinates and the distance formula, then cost to connect nodes i and j can be estimated.*
3. Sort the costs computed using Equation 3.6. *This is shown in Table 3.20 below, as are the results of the cluster merges.*
4. Remove clusters i and j from further consideration and cluster k is added to the clusters to be merged. Return to step 2 and calculate costs for distance between existing nodes and the new k cluster. *This is shown in Table 3.20. At the end of iteration 1, four (4) new nodes were created to reflect the merges of nodes A [3,10], B [8,9], C [5,7], and D [1,2].*

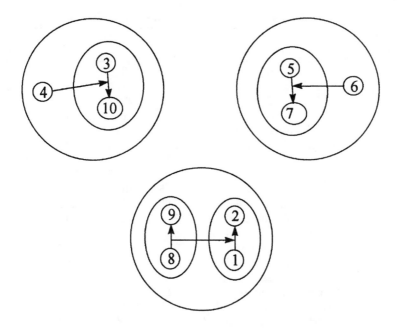

Figure 3.6 Sample COM problem — final solution (Source: B. Piliouras).

Note that due to the constraints, nodes 4 and 6 are not merged with other clusters.

5. Calculate the cost to connect nodes i and j, and sort. *The updated cost data is computed in Table 3.21 and shows the costs to merge the remaining nodes A, B, C, D, 4, and 6. During the second iteration, the following merges were made: E [C, 6], F [8, 9, 1, 2], and G [3, 10, 4]. The final result is that 3 clusters were created, as was specified in the original design objective. The final solution is shown in Figure 3.6. Note that the final solution provided by the COM algorithm must be checked for feasibility before it is accepted.*

Concentrator/backbone placement:

We introduce two new algorithms — the Add and the Drop algorithms — at this time. Both these algorithms are used to select the best sites for concentrator/backbone nodes. These algorithms assume that the following information is available:

- Set of terminal locations i
- Set of potential concentrator/backbone locations j
- Cost matrix specifying c_{ij} to connect terminal i to concentrator/backbone node j for all i and j
- Cost matrix specifying d_j the cost of placing a concentrator/backbone node at location j
- Maximum number of terminals that can be supported by each concentrator/backbone node j as MAX_j

Table 3.20   Sample COM Calculations — First Iteration

Node pairs (original)	New node (merge-1)	X	Y	Cost	Weight
3,10	A	53.5	88.5	13	2
8,9	B	33	41	13.4	2
5,7	C	92	59	14.4	2
1,2	D	38	16	15.2	2
6,7	Reject: 7 already used	96.5	70.5	15.3	2
1,9	Reject: 9 already used	29	28.5	19.4	2
5,6	Reject: 5 already used	90.5	66.5	24.7	2
1,8	Reject: 8 already used	35	31.5	26.2	2
4,8	Reject: 8 already used	30.5	54	26.2	2
4,9	Reject: 9 already used	24.5	51	26.5	2
2,9	Reject: 9 already used	36	25.5	30.8	2
2,8	Reject: 8 already used	42	28.5	31.6	2
4,10	Reject: 10 already used	35	74.5	33.4	2
3,6	Reject: 3 already used	77	85	38.6	2
8,10	Reject: 10 already used	43.5	64.5	42	2
3,5	Reject: cost limit exceeded	72.5	73.5	45.8	2
1,4	Reject: cost limit exceeded	26.5	41.5	45.9	2
3,4	Reject: cost limit exceeded	40.5	78	46.4	2
6,10	Reject: cost limit exceeded	71.5	81.5	47.5	2
5,8	Reject: cost limit exceeded	62.5	49.5	48.3	2
5,10	Reject: cost limit exceeded	67	70	48.4	2
3,7	Reject: cost limit exceeded	78.5	77.5	48.6	2
9,10	Reject: cost limit exceeded	37.5	61.5	51.5	2
3,8	Reject: cost limit exceeded	49	68	52	2
7,10	Reject: cost limit exceeded	73	74	54.6	2
2,4	Reject: cost limit exceeded	33.5	38.5	56	2
2,5	Reject: cost limit exceeded	65.5	34	58.7	2
5,9	Reject: cost limit exceeded	56.5	46.5	61.4	2
7,8	Reject: cost limit exceeded	68.5	53.5	62	2
3,9	Reject: cost limit exceeded	43	65	62.8	2
4,5	Reject: cost limit exceeded	54	59.5	64.6	2
6,8	Reject: cost limit exceeded	67	61	65.5	2
1,5	Reject: cost limit exceeded	58.5	37	65.7	2
1,10	Reject: cost limit exceeded	39.5	52	68.2	2
2,10	Reject: cost limit exceeded	46.5	49	72.1	2
2,7	Reject: cost limit exceeded	71.5	38	72.9	2
4,6	Reject: cost limit exceeded	58.5	71	74.3	2
7,9	Reject: cost limit exceeded	62.5	50.5	75.3	2
4,7	Reject: cost limit exceeded	60	63.5	76	2
1,3	Reject: cost limit exceeded	45	55.5	78.2	2
6,9	Reject: cost limit exceeded	61	58	78.9	2
1,7	Reject: cost limit exceeded	64.5	41	80.2	2
2,3	Reject: cost limit exceeded	52	52.5	80.2	2
2,6	Reject: cost limit exceeded	70	45.5	82	2
1,6	Reject: cost limit exceeded	63	48.5	87	2

Table 3.21 Sample COM Calculations — Second Iteration

New & original node pairs	New node (Merge-2)	X	Y	Cost	Weight
C,6	E	62.3	65.3	19.2	3
B,D	F	35.5	28.5	25.5	4
B,4	Reject: B already used	29.3	48.7	35	3
A,4	G	43	80.3	39.9	3
A,6	Reject: A already used	67.3	85	42.8	3
A,C	Reject: cost limit exceeded	72.7	73.8	48.5	4
A,B	Reject: cost limit exceeded	43.3	64.8	51.7	4
B,C	Reject: cost limit exceeded	62.5	50	61.7	4
C,D	Reject: cost limit exceeded	65	37.5	69	4
C,4	Reject: cost limit exceeded	68.7	60.1	70.2	3
C,6	Reject: cost limit exceeded	93	65.3	72.2	3

Add algorithm

The add algorithm starts by assuming that no concentrator/backbone nodes are in the solution and that all the terminals are connected directly to some central facility or node. It then proceeds to add concentrator/backbone nodes one at a time to lower the network costs. As each new node is added, some terminals are moved from the center or a previous node to a new node in order to obtain cost savings. The algorithm continues to add concentrator/backbone nodes until no more cost savings can be achieved.

Because the Add algorithm is a greedy algorithm, once a backbone/concentrator node is placed in the network it is not reevaluated nor is it replaced, even if in subsequent steps all the terminals associated with the node are moved to another backbone/concentrator node. Thus, it is possible at the end of the algorithm to have added nodes that have no connections. This should be checked, and if this situation exists, these nodes should be removed. The overall worst case computational complexity of an efficient programming implementation of this algorithm is estimated at O (T^*B^2), where T is the number of terminals and B is the number of potential backbone/concentrator sites [Kersh93, p. 228].

We now review the major steps of the Add algorithm:

1. Connect all terminals to the central site. Do not include any backbone/concentrator sites in the initial solution. This is illustrated in Figure 3.7.
2. Calculate the cost savings associated with connecting terminal i to backbone j. This is computed as the difference in cost between connecting each terminal i directly to a potential backbone location j and a central site, minus the cost of the backbone node j. This can be expressed mathematically as:

$$\text{Background / concentrator savings} = S_j = \sum_{i=1}^{n} \left(c_{i,0} - c_{i,j} \right) - d_j \qquad (3.9)$$

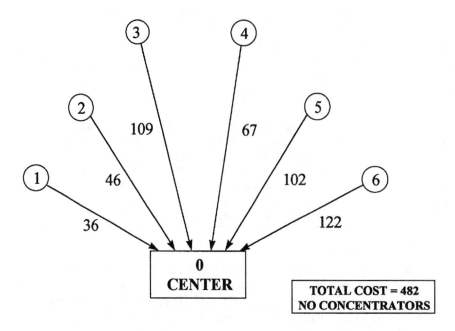

Figure 3.7 Initial Solution for add algorithm (Source: B. Piliouras).

where $c_{i,0}$ is the cost to connect a terminal directly to a central node; $c_{i,j}$ is the cost to connect terminal i to backbone j; and d_j is the cost of the backbone node. This figure of merit must be calculated for each potential backbone node j with respect to all the terminals i. Sort the S_j values from high to low.

3. If all the S_j values are negative, no savings can be obtained by adding a backbone node. Therefore, the algorithm terminates with the message, "Final solution obtained." Otherwise, pick the node associated with the largest positive S_j and add it to the solution. Move the terminals associated with the largest S_j from the center or the previous connection to the new backbone node.

4. Return to step 2, recalculating the S_j for each remaining unassigned backbone/concentrator node.

Example of Add algorithm

In Table 3.22, we present data for a sample problem that we will use to illustrate the add algorithm. This table provides data on the set of terminal locations i, the set of potential concentrator/backbone locations j, and the cost matrix specifying c_{ij} to connect terminal i to concentrator/backbone node j. Assume that the d_j cost of placing a concentrator/backbone node in the network is equal to 50 units and is the same for all nodes. Assume as well that the maximum number of terminals that can be supported by any concentrator/backbone node is equal to 4 units. There are six terminals, one center location, and six potential backbone sites in this sample problem. Note that in this example, there is no cost to connect terminals to co-located backbone nodes.

1. Connect all terminals to the central site. Do not include any backbone/concentrator sites in the initial solution. *This is illustrated in Figure 3.7.*
2. Calculate the cost savings associated with connecting terminal i to backbone j. This is computed as the difference in cost between connecting each terminal i directly to a potential backbone location j and a central site, minus the cost of the backbone node j. *For the sample problem, this involves calculating the possible cost savings associated with connecting Terminals 1-6 to Concentrators C1-C6. The Concentrator producing the largest savings is picked and added to the network design solution. This is shown in the first iteration calculations shown below.*

Table 3.22 Cost Data for Sample Concentrator/Backbone Location Problem

Terminal	Center 0	C1	C2	C3	C4	C5	C6
1	36	0	15	78	45	65	87
2	46	15	0	80	55	58	82
3	109	78	80	0	46	45	38
4	67	45	55	46	0	64	74
5	102	65	58	45	64	0	24
6	122	87	82	38	74	24	0
Total Cost	482						

3. Find the node associated with the highest savings S_j and add it to the solution. *The largest savings is obtained with the addition of either C3 or C6, which both yield a potential savings of 221 units. These results are shown in the first iteration calculations that follow. In cases of ties, either node can be selected for inclusion in the solution. We select node C3 for addition at site 3, and link terminals 3, 4, 5, 6 to it. This is shown in Figure 3.8.*
4. Return to step 2, calculating the S_j for each remaining unassigned backbone/concentrator node. *This is shown in the second iteration calculations below. In iteration 2, we check to see if any additional savings are possible with the concentrators not used in iteration 1 (e.g., C1, C2, C4, C5 and C6). We also verify that the original connections to concentrator C3 are still the cheapest. For example, we compare the savings with C3 vs. the savings with C1, etc.*
5. Find the node associated with the highest savings S_j and add it to the solution. *The largest savings is obtained with the addition of C1, as shown*

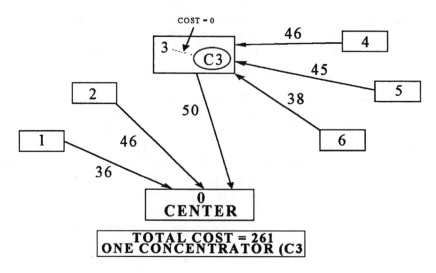

Figure 3.8 Solution obtained using add algorithm after first iteration (Source: B. Piliouras).

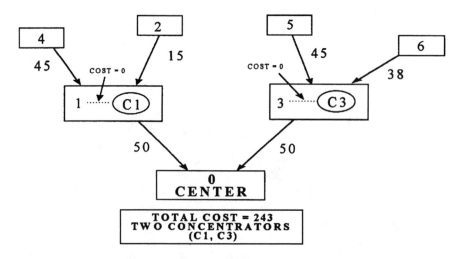

Figure 3.9 Final solution obtained using add algorithm (Source: B. Piliouras).

in the savings calculations for Iteration 1. We select node C1 for addition at site #3, and home terminals 1, 2, and 4 to it. Note that in this iteration, terminal 4 has been moved from its prior assignment at C3 to C1. This is shown in Figure 3.9.

6. Since all the remaining S_j values are negative, the algorithm is terminated with the message, "Final solution obtained."

Savings Calculations for Iteration 1 of the Add Algorithm

CHECK C1

Cost Alternatives	Cost Savings (to Center vs. to C1)
C10-C11	36-0 = 36
C20-C21	46-15 = 31
C30-C31	109-78 = 31
C40-C41	67-45 = 22
C50-C51	102-65 = 37
C60-C61	122-87 = 35
Savings (4 largest savings)	139
Concentrator C1 Cost	–50
Total Savings using C1	**89**

CHECK C4

Cost Alternatives	Cost Savings (to Center vs. to C4)
C10-C14	36-45 = –9
C20-C24	46-55 = –9
C30-C34	109-46 = 63
C40-C44	67-0 = 67
C50-C54	102-64 = 38
C60-C64	122-77 = 48
Savings (4 largest savings)	216
Concentrator C4 Cost	–50
Total Savings using C4	**166**

CHECK C2

Cost Alternatives	Cost Savings (to Center vs. to C2)
C10-C12	36-15 = 21
C20-C22	46-0 = 46
C30-C32	109-80 = 29
C40-C42	67-55 = 12
C50-C52	102-58 = 44
C60-C62	122-82 = 40
Savings (4 largest savings)	159
Concentrator C2 Cost	–50
Total Savings using C2	**109**

CHECK C5

Cost Alternatives	Cost Savings (to Center vs. to C5)
C10-C15	36-65 = –29
C20-C25	46-58 = –12
C30-C35	109-45 = 64
C40-C45	67-64 = 3
C50-C55	102-0 = 102
C60-C65	122-24 = 98
Savings (4 largest savings)	267
Concentrator C5 Cost	–50
Total Savings using C5	**217**

CHECK C3

Cost Alternatives	Cost Savings (to Center vs. to C3)
C10-C13	36-78 = –42
C20-C23	46-80 = –34
C30-C33	109-0 = 109
C40-C43	67-46 = 21
C50-C53	102-45 = 57
C60-C63	122-38 = 84
Savings (4 largest savings)	271
Concentrator C3 Cost	–50
Total Savings using C3 (Best Savings; tied with C6)	**221**

CHECK C6

Cost Alternatives	Cost Savings (to Center vs. to C6)
C10-C16	36-87 = –51
C20-C26	46-82 = –36
C30-C36	109-38 = 71
C40-C46	67-74 = –7
C50-C56	102-24 = 78
C60-C66	122-0 = 122
Savings (4 largest savings)	271
Concentrator C6 Cost	–50
Total Savings using C6 (Best Savings; tied with C3)	**221**

Savings Calculations for Iteration 2 of the Add Algorithm

CHECK C1

Cost Alternatives	Cost Savings (to Center vs. to C1)
C10-C11	36-0 = 36
C20-C21	46-15 = 31
C30-C31	109-78 = 31
C40-C41	67-45 = 22
C50-C51	102-65 = 37
C60-C61	122-87 = 35

Cost Alternatives	Cost Savings (to previous C3 vs. to C1)
C33-C31	0-78 = –78
C43-C41	(Cheaper using C1) 46-45 = 1
C53-C51	45-65 = –20
C63-C61	38-87 = –49

89	Savings (C11, C21, C41)
–50	Concentrator C1 Cost
–21	Remove savings from Concentrator 3, Iteration #1 (C40-C43)
18	**Total Savings using C1**

Savings Calculations for Iteration 2 of the Add Algorithm (continued)

CHECK C2

Cost Alternatives	Cost Savings (to Center vs. to C2)	Cost Savings Cost Alternatives(to previous C3 vs. to C2)
C10-C12	36-15 = 21	
C20-C22	46-0 = 46	C33-C320-80 = –80
C30-C32	109-80 = 29	C43-C4246-55 = –9
C40-C42	67-55 = 12	C53-C5245-58 = –13
C50-C52	102-58 = 44	C63-C6238-82 = –44
C60-C62	122-82 = 40	
	67	Savings (C12, C22)
	–50	Concentrator C1 Cost
	17	**Total Savings using C2**

CHECK C4

Cost Alternatives	Cost Savings (to Center vs. to C4)	Cost Savings Cost Alternatives(to previous C3 vs. to C4)
C10-C14	36-45 = –9	
C20-C24	46-55 = –9	C33-C340-46 = –46
C30-C34	109-46 = 63	C43-C44(Cheaper using C4) 46-0 = 46
C40-C44	67-0 = 67	C53-C5445-64 = –19
C50-C54	102-64 = 38	C63-C6438-74 = –36
C60-C64	122-77 = 48	
	67	Savings (C44)
	–50	Concentrator C1 Cost
	–21	Remove savings from Concentrator 3, Iteration #1 (C40-C43)
	-4	**Total Savings using C4**

CHECK C5

Cost Alternatives	Cost Savings (to Center vs. to C5)	Cost Savings Cost Alternatives (to previous C3 vs.to C5)
C10-C15	36-65 = –29	
C20-C25	46-58 = –12	C33-C350-45 = –45
C30-C35	109-45 = 64	C43-C4546-64 = –18
C40-C45	67-64 = 3	C53-C55(Cheaper using C5) 45-0 = 45
C50-C55	102-0 = 102	C63-C65(Cheaper using C5) 38-24 = 14
C60-C65	122-24 = 98	
	200	Savings (C55, C65)
	–50	Concentrator C1 Cost
	–141	Remove savings from Concentrator 3, Iteration #1 (C50-C53 and C60-C63)
	9	**Total Savings using C5**

CHECK C6

Cost Alternatives	Cost Savings (to Center vs. to C6)	Cost Savings Cost Alternatives(to previous C3 vs. to C6)
C10-C16	36-87 = –51	
C20-C26	46-82 = –36	C33-C360-38 = –38
C30-C36	109-38 = 71	C43-C4646-74 = –28
C40-C46	67-74 = –7	C53-C56(Cheaper using C6) 45-24 = 21
C50-C56	102-24 = 78	C63-C66(Cheaper using C6) 38-0 = 38
C60-C66	122-0 = 122	
	200	Savings (C56, C66)
	–50	Concentrator C1 Cost
	–141	Remove savings from Con. 3, Iteration #1 (C50-C53 and C60-C63)
	9	**Total Savings using C6**

Drop algorithm

The Drop algorithm starts by assuming that all the concentrator/backbone nodes are in the solution and that all terminals are connected directly to the nearest backbone node. It then proceeds to delete concentrator/backbone nodes one at a time to lower the network costs. As a node is removed, some terminals are moved to the center or to other nodes in order to obtain the savings. The algorithm continues to delete concentrator/backbone nodes until no more cost savings can be found.

Like the Add algorithm, the Drop algorithm is a greedy algorithm. Despite apparent similarities between the two algorithms, there are also substantial differences. Thus, one should not expect that they will produce the same results, although it is possible that they might. The overall worst case computational complexity of an efficient programming implementation of this algorithm is estimated at $O(T^*B^3)$, where T is the number of terminals and B is the number of potential backbone/concentrator sites. This is significantly worse than the computational complexity of the Add algorithm. Thus, for large problems, the Add algorithm may be preferable to the Drop algorithm, particularly since the Add algorithm produces solutions that are similar in quality to those produced by the Drop algorithm [Kersh93, p. 233].

We now review the major steps of the Drop algorithm:

1. Connect all terminals to the nearest backbone or concentrator site. Do not include any backbone/concentrator sites in the initial solution.
2. Calculate the cost savings associated with connecting terminal i to backbone j. This is computed as the difference in cost between connecting each terminal i directly to a potential backbone location j and a central site, and the cost of the backbone node j. This can be expressed mathematically as:

$$\text{Backbone / concentrator savings} = S_j = -\sum_{i=1}^{n}\left(c_{i,0}-c_{i,j}\right)-d_j \qquad (3.10)$$

   where $c_{i,0}$ is the cost to connect a terminal directly to a central node; $c_{i,j}$ is the cost to connect terminal i to backbone j; and d_j is the cost of the backbone node. This figure of merit must be calculated for each potential backbone node j with respect to all the terminals i. Sort the S_j values from high to low.
3. If all the S_j values are negative, no savings can be obtained by adding a backbone node. Therefore, the algorithm terminates with the message, "Final solution obtained." Otherwise, pick the node associated with the largest positive S_j, and drop it from the network. Move the terminals associated with the largest S_j from the previous connection to the center or to a new backbone node.
4. Return to step 2, recalculating the S_j for each backbone/concentrator node remaining in the solution.

Table 3.23   Sample Cost Data for Drop Algorithm

T_i	C_0	C_1	C_2	C_3
1	2	1	2	4
2	1	0	1	2
3	4	1	2	2
4	1	2	1	2
5	2	3	2	0
6	4	4	3	2

Example of Drop algorithm

The set of terminal locations i, the set of potential concentrator/backbone locations j, and the cost matrix specifying the cost c_{ij} to connect terminal i to concentrator/backbone node j for a new sample problem is given in Table 3.23. Assume that the d_j cost of placing a concentrator/backbone node is equal to two units (and includes the cost of the node and the cost to connect the node to the center) and is the same value for all j. Assume that the maximum number of terminals allowed on each concentrator/backbone node, MAX_j, is equal to three units and is the same value for all nodes. There are six terminals, a C_0 center location, and three potential backbone sites in this sample problem.

1. Connect all terminals to the nearest backbone site. Include all backbone/concentrator sites in the initial solution. This solution can easily be obtained by selecting the lowest row value in the cost matrix for each terminal. *The results of the initial solution are illustrated in Figure 3.10 b. The cost of the initial solution is calculated as the sum of the line costs to connect to concentrator plus the number of concentrators multiplied by the cost of each concentrator. For this problem, the initial solution cost = 1 + 0 + 1 + 1 + 2 + (3 * 2) = 11. Sites with an overlapping square and circle have a backbone/concentrator and terminal co-located at the same site.*

2. Calculate the cost savings associated with disconnecting terminal i from backbone j. *For the sample problem, this involves calculating the possible cost savings associated with connecting terminals 1 to 6 to the center as opposed to concentrators 1, 2, or 3. The concentrator whose removal produces the largest savings is picked and deleted from the network design solution. This is shown in the first iteration calculations shown below. Recall that only three terminals, at most, can be assigned to a concentrator, based on a prespecified constraint. When a negative savings is computed (as indicated in the entries in Table 3.24), this means it is cheaper to connect the terminal to the center directly than it is to maintain the current connection to the concentrator. Conversely, positive entries indicate that it is cheaper to connect the terminal to the concentrator being considered than it is to connect the terminal directly to the center.*

3. Find the node associated with the highest savings S_j and add it to the solution. *The largest savings is obtained with the removal of C3, which yields a potential savings of two units. We select node C2 for removal, and home terminal 4 to the center. This is shown in Figure 3.11.*

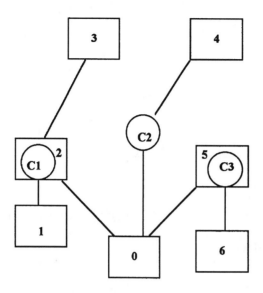

Figure 3.10 Initial solution to sample problem using Drop algorithm.

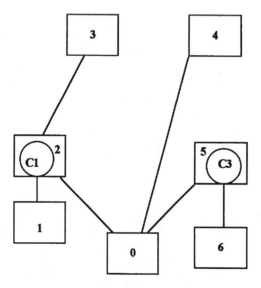

Figure 3.11 Solution to sample problem Using Drop algorithm.

4. Return to step 2, calculating the S_j for each remaining backbone/concentrator node. *As shown in the first iteration calculation, no additional savings are possible by deleting concentrator 1 or 3. Since all the remaining S_j values are negative, the algorithm is terminated with the message "Final solution obtained." The solution remains the same as it was at the end of the first iteration.*

Table 3.24 First Iteration Calculations for Drop Algorithm

Checking C1:	Checking C2:	Checking C3:
$c_{10} - c_{11} = 2 - 1 = 1$ (T1)	$c_{40} - c_{42} = 1 - 1 = 0$ (T4)	$c_{50} - c_{53} = 2 - 0 = 2$ (T5)
$c_{20} - c_{21} = 1 - 0 = 1$ (T2)		$c_{60} - c_{63} = 4 - 2 = 2$ (T6)
$c_{30} - c_{31} = 4 - 1 = 3$ (T3)		
Removing saves: –5	Removing saves: 0	Removing saves: –4
Total savings after concentrator cost considered: $(-5+2) = -3$	Total savings after concentrator cost considered: $(0+2) = 2$	Total savings after concentrator cost considered: $(-4+2) = -2$

3.2.5.3 Case Study

As illustrated in the following diagram, large backbone networks are sometimes comprised of an inner backbone, referred to as a "core backbone," and multiple interconnecting semiautonomous remote subsystems. These subsystems may be comprised of one or several regions or districts, where each region controls several smaller end sites. Many topologies and designs can be used to interconnect the core backbone and the remote subsystems. Which design is best depends on the needs of the users and the allocated telecom budget.

When the subsystems are semiautonomous, both intraregion and interregion communication is needed. Examples of intraregional communications include point-to-point voice and client/server or peer-to-peer distributed data applications where sites share information within the district area. In comparison, interregional communications require data exchange across regions. This type of exchange might occur between a sales office and a district office, and from the district office to the core backbone. Examples of interregional applications include compressed broadcast video, point-to-point voice, and mainframe data applications. As LAN file servers replace mainframes, the corporate backbone infrastructure must provide increasing support for client/server applications.

In our case study, as shown in Figure 3.12, multiple high-speed private digital links are used to interconnect the company's core backbone. This network is comprised of corporate headquarter facilities, secondary main processing centers, and designated backup and disaster recovery location(s). In this scenario, the interconnecting links are usually dedicated circuits employing a range of speeds and mediums, including T-1 (1.544Mbps), DS-3 (44.736Mbps), or OC-n[10] SONET digital facilities. Note that within the core backbone network, a separate site has been designated as the corporate backup node or disaster recovery site. This backup site may be an office of the corporation or it may be located within a company that provides disaster

---

[10] OC-n – This is a SONET-related term for the optical carrier level N. This is the optical signal that results from an STS-N signal conversion. The STS-1 is the basic SONET signal that operates at a transmission rate of 51.84 Mbps. An STS-N signal has a transmission rate of N times the basic STS-1 transmission rate.

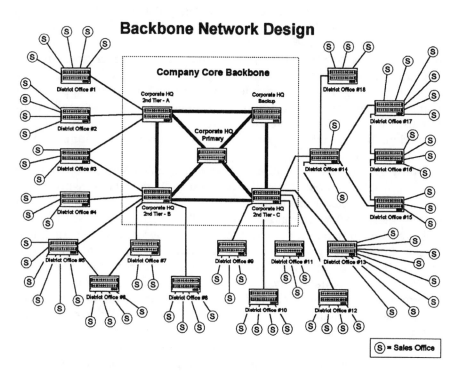

Figure 3.12 Backbone case study (Source: B. Piliouras).

recovery services. To ensure true redundancy in the event of a major network failure, some companies install duplicate equipment (for instance, voice and data switches, multiplexers, routers/bridges, mainframes, and file servers). In general, the telecom costs rise quickly when network failures cannot be tolerated and redundancy must be built into the network.

Many company backbones are a compilation of various separate network topologies overlaid to produce an overall asymmetrical corporate network. In our example, some district offices have only a single path to the core backbone, while other districts employ redundant rings, multidrop trees, or star topologies. Locations employing single paths are also called "single-threaded" sites. In our case study, this includes district offices 1, 2, 4, 8-12, and 18. These sites are relatively inexpensive to interconnect, but do not have an alternate path back to the core backbone network. For mission critical districts and sales offices, redundant paths may be needed, as is illustrated for district offices 3, 5-7, 13-14, and 15-17.

As can be seen in Figure 3.12, redundancy may take many forms and depends on the types of applications supported, the urgency of the trans-missions, and the associated networking costs. In the case of district office groups 5-7 and 14-17, for example, the company has installed TCP/IP routers using the open shortest path first (OSPF) routing protocol. Therefore, if the path is disrupted from district office 6, through district office 7, to the core backbone, traffic could be rerouted through district office 5. Likewise, if the links from district offices 5 and 7 to the core backbone become inoperative, traffic can be rerouted in the opposite direction. However, these reroute procedures introduce additional node processing and facility propagation delays and are usually recommended only for time-insensitive applications, such as large file transfers or noninteractive client/server data applications. As hardware and software vendors introduce products to compensate for the various delays, this type of network configuration can support more time-sensitive applications.

Network redundancy can take other forms, as shown in Figure 3.12 at district office 3. Suppose that at this location we have delay-sensitive voice, video, or data applications that are critical to the company's operations. At this site, there are two alternate paths providing redundant access to the core backbone and to two 2nd-tier backbone nodes. Therefore, if a single-link or backbone node fails, transmissions can be routed to the alternate facility or node. It should be stressed, however, that redundancy does not always equate to network infallibility. For instance, if the 2nd-tier–B node were out of service due to maintenance or failure, and the link from district office 3 to the 2nd-tier–A node were to fail, then a reroute would prove futile. The redundant link would switch over and become the primary link, and retransmission would be attempted with the failed 2nd-tier–B node.

The reader can easily deduce that when a given link fails and data is rerouted onto an active circuit carrying other user traffic, the use of the active link will increase. If full redundancy is required, the alternate link will need enough capacity to accommodate both the primary traffic and the rerouted traffic. Full link and equipment redundancy can be expensive and is usually reserved only for those sites that are considered mission critical by corporate headquarters.

A good compromise solution is illustrated in Figure 3.12 at district office 13. Instead of leasing a separate path to one of the 2nd-tier nodes, the link from district office 13 to district office 14 can be used if necessary to reroute traffic. This scenario might occur when the district offices are closer to each other than to any one of the core backbone nodes, and the facilities are leased on a cost-per-mile basis. Thus, a partial backup path can be provided for some of the more critical applications at a relatively small incremental cost.

In addition to the facility (or line) redundancy, separate hardware ports are needed if physical diversity is required. As we discuss in later sections of this chapter, networks using X.25 packet, frame relay, and asynchronous transfer mode (ATM) can logically map multiple virtual circuits onto the same physical port. This approach can lead to substantial cost savings since the amount of hardware and the number of facilities ports can be reduced. These transport methods can be deployed on a private basis or procured from telecom providers through various public and semiprivate product offerings.

3.2.6   Mesh Networks

3.2.6.1   Overview

A fully meshed topology allows each node to have a direct path to every other node in the network. This type of topology provides a high level of redundancy and reliability. Although a fully meshed topology usually facilitates support for all network protocols, it is not tenable for large packet-switched networks.[11] However, by combining fully meshed and star approaches into a partially meshed environment, telecom managers can improve network reliability without encountering the performance and network management problems usually associated with a fully meshed approach. A partially meshed topology reduces the number of routes within a region that have direct connections to other nodes in the region. Since all nodes are not connected to all other nodes, a nonmeshed node must send traffic through an intermediary node acting as a collection point. There are many forms of partially meshed topologies.

Although a partial mesh diminishes the amount of equipment and carrier facilities required in the design, thus reducing networking costs, it may also introduce intolerable network delays. For example, voice/fax, video, and some data applications do not fair well with a partial mesh topology, due to the network latency introduced by circuit back-hauling, and nodal processing delays. To compensate for these delays, some legacy data protocols require local and remote "spoofing" to ensure that application timeouts do not occur. However, in general, partially meshed approaches provide the best overall balance for regional topologies in terms of the number of circuits, redundancy, and performance.

Many insurance companies, banking and financial institutions, and wholesale/retail operations employ mesh networks because their business transactions and order entry operations depend on guaranteed connectivity to their processing centers. This is especially critical in situations where the value of the product or commodity varies widely with respect to time. As an example, the loss incurred in selling stock too late in a volatile market can more than compensate for the additional costs of building redundancy into the network.

Some of the advantages associated with mesh topologies include:

- Resistance to nodal and link failures
- Cost effective method for distributed information computing
- Relatively easy to expand and modify

---

[11] Packet-switched networks are not suited to full mesh topologies due to the large number of virtual circuits required (one for every connection between nodes), problems associated with the large number of packet/broadcast replications required, and the resulting configuration complexity in the absence of multicast support in nonbroadcast environments.

The disadvantages of mesh topologies relate to the fact that they are:

- More complex to design
- More difficult to manage
- Require more node site support
- Redundancy at the expense of idle, wasted bandwidth

Mesh topologies are recommended for:

- Large, multiapplication networks
- Host-to-host connectivity
- Distributed application support
- User-initiated switching
- Redundant network support of mission critical applications

3.2.6.2   Design considerations and techniques

As an alternative to maintaining a private mesh network, some companies use public switched networks. This involves attaching CPE[12] devices or nodes to the shared carrier network, usually by means of high bandwidth T-1 (1.544Mbps) or DS-3 (44.736Mbps) digital facilities that provide separate voice/fax and data channels. The digital facilities are usually demultiplexed at the carrier's point of presence (POP), where they are interfaced to separate voice, data, and video networks owned and operated by the carrier. After traversing different networks, the channels are remultiplexed and transmitted to the company's remote location over T-1 copper, DS-3 coaxial, or OC-n SONET fiber local loops. At the company's end site, the high bandwidth local loops are demultiplexed back into the original voice, data, and video signals.

Using a public carrier reduces the number of I/O CPE ports and local access circuits needed. It may also be a cost-effective way to transmit each user application over the most appropriate transport network. For example, delay sensitive and highly interactive voice, fax, and legacy data applications can be sent over dedicated circuit-switched connections with minimal delay. Conversely, less time sensitive client/server data applications can be sent over packet- or frame-switched networks where the facilities (and costs) are shared among several users. A public carrier network provides the benefits of a mesh topology at a cost that is often much less than the cost of a private network. The costs of a private mesh are often substantial due to the sheer quantity of circuits that must be dedicated to interconnect the various nodes.

When designing a network, the node traffic should be carefully analyzed to determine which nodes and applications require a full or partial mesh topology, and which ones might benefit from other networking architectures

---

[12] CPE — This stands for customer premises equipment. This is the terminal equipment that is connected to the public switched network. The CPE can be provided by the common carrier or some other supplier.

such as a star or tree topology. For most organizations, the cost of making a network completely fault tolerant is prohibitive. Determining the appropriate level of fault tolerance to be built into the network is neither a trivial matter nor an insignificant undertaking. As the complexity of the network increases (as it does with a mesh topology), it becomes increasingly important to use modeling and simulation tools to help design the network. The network decisions that must be made when planning a large mesh topology are too complex and numerous to manage without a systematic, analytical approach.

At this time, we introduce the Mentor algorithm, which was developed specifically for mesh topologies. Mentor is a heuristic, hierarchical design algorithm. As a first step, it selects the backbone sites to be used in the mesh topology. The algorithm then proceeds to design an "indirect" routing tree to handle traffic between nodes that is insufficient in volume to justify direct links. After an indirect routing tree is designed, the algorithm installs direct links between sites that have traffic above a prespecified threshold. Mentor is designed specifically to favor highly utilized, direct links to take advantage of the economies of scale offered by high-speed links. In addition to the traffic requirements, the designs produced by Mentor are determined by various preset design parameters, which we discuss in detail in the following paragraphs. These design parameters strongly influence the final design produced by Mentor. By making changes to the design parameters, it is possible to produce very different network topologies. Because the algorithm is so fast — it is of $O(n^2)$ — it is possible to produce lots of network designs for analysis in a very short period of time. Mentor is available in the WinMIND™ design tool.

In the last section, we presented several procedures — the COM, Add, and Drop algorithms — for selecting backbone nodes. Mentor also offers a procedure — Mentor part I — for locating backbone nodes in a mesh network. All of these algorithms are suitable for locating backbone nodes and may be used to complement each other when exploring various design strategies. After the backbone nodes have been selected, they need to be interconnected to produce a final design. The Mentor algorithm part II performs this task. An overview of the Mentor Algorithm is provided in Figures 3.13 and 3.14. For more details on the Mentor algorithm, the interested reader is referred to [Kersh91].

Mentor algorithm Part I

The purpose of this part of the algorithm is to select backbone sites for the mesh topology. The algorithm examines all the potential node sites and separates them into two categories: end nodes e_i and backbone nodes b_j. Mentor Part II ignores the end nodes when developing the backbone mesh topology. End nodes are connected to backbone nodes at a later time using the same techniques introduced earlier to solve the terminal assignment problem.

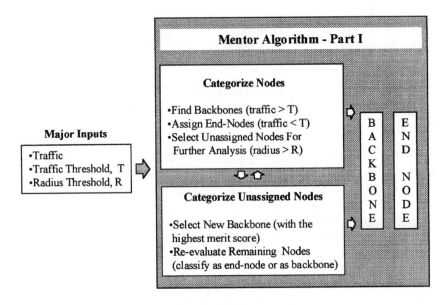

Figure 3.13   Overview of Mentor — Part I.

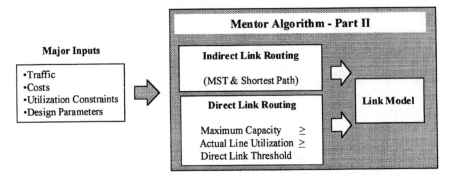

Figure 3.14   Overview of Mentor — Part II.

Mentor Part I takes as given a radius parameter r and a weight limit parameter w. The radius parameter specifies the largest acceptable distance between a node and the nearest backbone node. The weight limit specifies the traffic threshold that must be satisfied to automatically qualify a node as a backbone site.

1. This step finds all the nodes that qualify automatically to become a backbone node because of the amount of traffic flowing through them. The remaining nodes are segregated for further analysis in the next

step. Start by comparing the total traffic t_i flowing through each node i (i.e., the sum of all traffic entering and leaving the node) to the weight w:

- If $t_i \geq w$, then assign terminal i as a backbone node, designated b_j.
- If $t_i < w$, then assign terminal i as an unassigned node, designed u_i.
- Find the maximum t_i, designated t_{MAX}.

2. This step examines all the u_i nodes that did not qualify, based on the previous calculations, to become a backbone. If a node u_i is within the allowed radius of a backbone site b_j, the node is assigned to the nearest backbone and is redesignated as an end node e_i. If the node u_i is too far from any currently designated backbone node, the node is relabeled c_i to indicate that further clustering is necessary. This step begins by computing the distance d_{ij} between each node u_i and each b_j. The distance calculation provided in Equation 3.6 can be used for this purpose.

   - If $d_{ij} \geq r$, then designate terminal i as a node c_i that requires assignment to a closer backbone.
   - If $d_{ij} < r$, compare against all dij values calculated for terminal i. Find the smallest d_{ij} and assign terminal i as an end-site e_i associated with backbone b_j
   - Find the maximum d_{ij}, designated d_{MAX}.

3. This step examines all the nodes c_i that have not been assigned as either a backbone or an end node. A trade-off function, or figure of merit, m_i is calculated below for each c_i. The node with the highest m_i represents the node with the best balance between central location and sufficient traffic flow to justify designation as a new backbone node. Compute m_i for each c_i as:

$$m_i = 1/2 \left(\frac{d_{max} - d_{ij}}{d_{MAX}} + \frac{t_{ij}}{t_{MAX}} \right) \qquad (3.11)$$

Find the node with the largest m_i and designate it as a new backbone node b_j. Rename all c_i nodes as u_i and return to step 2 (where the u_i nodes will be compared against the b_j just selected to see if they now qualify as end nodes). Continue until all nodes are designated as either an end node or a backbone node, at which time the algorithm is terminated.

Mentor algorithm Part II

This part of the algorithm takes as given the following design parameters:

- A design parameter, α, which controls the topology of the indirect routing tree. This parameter is set somewhat arbitrarily to a value between 1 and 0, inclusive. When α is set to 0, a minimal spanning tree is generated from the center. When α is set to 1, a starlike configuration is built around the center. Typically, values between .2–.5 yield good mesh topologies.

- A slack parameter, S, which specifies the maximum percentage of the available bandwidth that can be used during the design process. For example, if S = .40, this means that only 40% of the available line capacity can be used. Any traffic levels up to and including this amount can be placed on the link. However, any traffic level exceeding this amount must be routed on other links.
- A direct link threshold, D, which specifies the level of traffic that is sufficient to justify a direct link between any two nodes.

1. Find the most central node. This is used to build an indirect routing tree. For each backbone node b_j, calculate a figure of merit as the sum over all nodes j of the distance — or cost, which is an approximation of distance — $dist_{ij}$ from i to j multiplied by the total traffic to and from j, w_{ij}:

$$f_j = \sum_{i=1}^{n} \left(dist_{ij} \times w_{ij} \right) \tag{3.12}$$

Find the minimum f_{ij}, designated f_{min}. The node associated with this value is chosen as the center of the indirect routing tree, c.

2. Design the indirect routing tree. Start with node c, which is the center of the network as computed in step 1. Initially c is designated as part of the indirect routing network and all other nodes are considered outside the network. For all links (i,j) with i in the tree and j outside the tree, we define L'(i,j) as:

$$L'(i,j) = (i,j) + \alpha \ (i,c) \tag{3.13}$$

Where α is a prespecified design parameter between 0 and 1, inclusive.

- Find the minimum computed value of L'(i,j). The node associated with this value is brought into the tree.
- Update the L'(i,j) values affected by the addition of this node. Find the minimum computed value of L'(i,j). The node associated with this value is now brought into the tree. Continue this process, bringing in one node at a time, until all the nodes are connected.

3. Sequence the traffic requirements for each pair of nodes i and j. Logically, we are trying to establish a sequence for loading traffic on the indirect routing tree in such a way that direct links will be encouraged. The indirect tree is designed to carry traffic that is insufficient to justify a direct link between two nodes. The indirect tree will also carry

overflow traffic from direct links that do not have sufficient capacity to carry all the traffic demand between two directly linked nodes. Overflow traffic can be routed on the indirect routing tree, up to the point that the traffic reaches the critical direct link threshold (Note: At this point, a new direct link is installed on that portion of the indirect routing tree. This is done in step 4 below). We start by identifying all the pendent node pairs.[13] These pairs are considered first-level node pairs. Find the nearest neighbors of all the other nodes. We compute the cheapest route for detouring traffic through the nearest neighbors of the two nodes under consideration. These alternate routes represent dependencies that must be sequenced *after* the node pairs. Thus, these node pairs are considered second-level node pairs, and their dependencies are considered third level node pairs, and so on. Although these dependencies *are not actually used* in routing the traffic during the course of the algorithm, we are attempting a routing that detours the traffic through the fewest number of hops, and that encourages optimal loading of traffic to justify direct links. The node pairs must be sequenced according the following rules: (1) node pairs at the highest sequence level (i.e., the pendent node level) can be considered in *any* order, (2) node pairs at a lower level can sequenced only *after all* dependencies at higher levels have first been considered. Node pairs that are separated by the most number of hops must be sequenced before node pairs separated by fewer hops. For example, all the node pairs separated by four hops should be sequenced before all the node pair dependencies separated by three hops, and so on. This sequencing is not necessarily unique, and it is possible that many valid sequences can be defined. Only one valid sequence is used in the fourth and final step of the Mentor algorithm.

4. Assign direct links. Consider each node pair in the sequence determined in step 3. If the direct link traffic threshold between the node pairs is equaled or exceeded, then assign a direct link between the two nodes. If the traffic threshold is not reached, route the traffic between the nodes on the indirect routing tree. If the traffic on the indirect routing tree reaches or exceeds the traffic threshold, install a new link on the indirect routing tree to handle the overflow. Conceptually, if an indirect link is loaded to the threshold level, the indirect link is converted into a direct link, and a new indirect link is inserted. In this way, the indirect tree is never overloaded beyond its usable link capacity. Once a direct link is assigned, no further attempt is made to put more traffic on the link, even if the link has excess capacity. After every node on the sequence list has been evaluated, the algorithm terminates.

---

[13] Pendent node — A pendent node is characterized as having only one link into the node. A pendent node pair, as the name implies, consists of two pendent nodes.

Inherent in this presentation of the Mentor algorithm are several key assumptions, which we list below:

- Traffic is bidirectional. This is realistic for voice data, and not necessarily for packet-switched data. If this assumption is not true, the algorithm can be modified to explicitly consider the load on each link in each direction during the link selection.
- Only one link capacity at a time is considered. The algorithm can be modified to account for multiple link capacities; however, this adds to the computational complexity of the procedure.
- Cost is represented by an increasing function of distance. This assumption is made when distance and cost are treated synonymously. The cost data comes into play when the indirect routing tree is being constructed. If exact tariff data are available, these data can be used in the cost matrix without changing the execution of the algorithm. The algorithm would have to be modified only if the distance approximation is altered and therefore the cost matrix used is affected.
- Cost is symmetric in that the line costs are the same irrespective of the starting and ending node. Thus, a link (i, j) will have the same cost as a link (j, i). This assumption is consistent with the use of virtual circuits in a TDM-based network. The algorithm can be modified to handle different costs in different directions. However, this adds to the computational complexity of the algorithm.
- The traffic requirement can be split over multiple routes. If this is not true, the algorithm can be modified to use a bin-packing algorithm to assign traffic to the links.
- Each link is moderately utilized. This assumption helps to account for the exceptionally low computational complexity of the Mentor algorithm. This assumption is enforced through the selection of the slack parameter, which ensures that no link is overutilized.

The version of Mentor presented here is best suited for designing TDM-based networks. If a router-based network needs to be designed, modifications may be needed to ensure that reasonable designs are produced. For example, this might involve decreasing the usable line capacity assumed by the algorithm, as represented by the slack parameter. In addition, it may be advisable to modify the construction of the indirect routing tree to encourage a topology that minimizes the number of hops.[14] The interested reader is referred to [Cahn98], which presents numerous modifications and extensions to the Mentor algorithm for various types of networks.

At this point, we elaborate briefly on the design parameters used by Mentor. The design of the indirect routing tree is controlled by an α parameter,

---

[14] Hop — The number of hops in the network refers to the number of nodes that the data must traverse in the network when going from source to destination.

which governs the trade-off between tree length and path length. When the α parameter is set to 0, a minimal spanning tree is produced. This type of structure will tend to produce lower-cost designs. When the parameter is set to 1, a star is produced. This type of structure will tend to produce smaller path delays. With everything else held constant, increasing the direct link threshold — i.e., the slack parameter — increases the available line capacity used during the design process. Since more line capacity is available, there will be a tendency to use fewer lines in the overall design. To increase the reliability of the networks generated using Mentor, try decreasing the link utilization threshold. Decreasing the link utilization will encourage more links in the final design, thus providing opportunities for more alternative routes. In addition, since the links are less heavily used with a lower link utilization threshold, the overall impact of a specific link failure will tend to be reduced correspondingly. However, if the relative traffic requirement is very small in relation to the line capacity, changing the slack parameter may have little overall impact on the final topology.

Sample problem using Mentor algorithm

The Mentor algorithm is perhaps best understood by example. We present data for a sample problem below. The assumptions we make for this problem include:

- All nodes are potential backbone nodes (this assumption means that we do not need to use Mentor — Part I to determine the backbone locations).
- Traffic can be split on the network as required to fill the links.
- All lines are full-duplex.

The design parameters we have selected are:

- α = .2
- Maximum usable capacity = 19.2 units
- Direct link traffic threshold = 80% of the maximum usable capacity = 15.36 units

Table 3.25   Traffic Data For
Sample Mentor Problem

Traffic requirements matrix					
	A	B	C	D	E
A	0	4	11	10	7
B	4	0	4	6	8
C	11	4	0	9	8
D	10	6	9	0	5
E	7	8	8	5	0

Table 3.26   Cost Data For
Sample Mentor Problem

Cost requirements matrix				
	A	B	C	D	E
A	1	7	11	11	8
B	7	1	4	2	6
C	11	4	1	7	3
D	11	2	7	1	7
E	8	6	3	7	1

Solution to sample problem using Mentor algorithm

Step 1:   Calculate the network center.
First, add the traffic flows through each node.

Node A Traffic Flow
AB + AC + AD + AE + BA + CA + DA + EA = 4 + 11 + 10 + 7 + 4 + 11 + 10 + 7 = 64

Node B Traffic Flow
BA + BC + BD + BE + AB + CB + DB + EB = 4 + 4 + 6 + 8 + 4 + 4 + 6 + 8 = 44

Node C Traffic Flow
CA + CB + CD + CE + AC + BC + DC + EC = 11 + 4 + 9 + 8 + 11 + 4 + 9 + 8 = 64

Node D Traffic Flow
DA + DB + DC + DE + AD + BD + CD + ED = 10 + 6 + 9 + 5 + 10 + 6 + 9 + 5 = 60

Node E Traffic Flow
EA + EB + EC + ED + AE + BE + CE + DE = 7 + 8 + 8 + 5 + 7 + 8 + 8 + 5 = 56 bps

Now find the figure of merit, m_i, for each node using the traffic flows calculated above multiplied by the link costs.

Node A: (44 * 7) + (64 * 11) + (60 * 11) + (56 * 8) = 2120
Node B: (64 * 7) + (64 * 4) + (60 * 2) + (56 * 6) = 1160
Node C: (64 * 11) + (44 * 4) + (60 * 7) + (56 * 3) = 1468
Node D: (64 * 11) + (44 * 2) + (64 * 7) + (56 * 7) = 1632
Node E: (64 * 8) + (44 * 6) + (64 * 3) + (60 * 7) = 1388

Since node B has the smallest figure of merit, it is selected as the network center.

Step 2:   Find the indirect routing tree which is used to carry overflow traffic
(where $\alpha = .2$) Distance to the tree = (distance$_{i,j}$) + .2 (distance$_{i,\ center}$)
Since D is the cheapest link to attach to the center node B, it is brought
into the tree first. Next, the connection costs to bring A, C, and E into
the network through attachment to D are computed. When these costs

Table 3.27 Indirect Routing Tree Calculations for Mentor

Node just brought into tree	A	B	C	D	E
B (center node)	7	—	4	2	6
D (attached to center)	11 + .2(2) = 11.4	—	7 + .2(2) = 7.4	—	7 + .2(2) = 7.4
C (attached to center)	11 + .2(4) = 11.8	—	—	—	3 + .2(4) = 3.8
E (attached through C)	8 + .2(4+3) = 9.4	—	—	—	—
A (attached to center)					

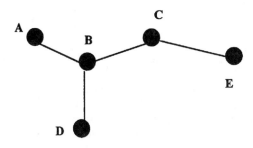

Figure 3.15 Indirect routing tree created by Mentor algorithm.

are compared to the cost of bringing a node in directly through the center node, the next node selected is C, which is homed to the center node. The costs to connect A and E into the network through an attachment to C are computed in the third line of Table 3.27. When these costs are compared to the cost of bringing these nodes in directly through the center node, the decision is made to select node E, and to home it through node C. The final node A is attached to the network at the center node, since it is cheaper to attach node a through the center than it would be attach A to the end of the current tree (which consists of links from node B, to node E, to node C).

The final indirect routing tree produced in this step is shown in Figure 3.15.

Step 3: Find the sequencing requirements.

For each potential node pair, the alternative traffic detour routes are examined. The alternative detour routes are formed by detouring each node's traffic to its nearest neighbor and connecting the nearest neighbor to the other end point of the node pair under consideration. This is done only for the purpose of sequencing the links in preparation for Step 4. This does not represent the actual traffic assignment and routing, which is done in Step 4.

For the problem at hand, we list the possible node pairs in lexicographic order:

(A,B), (A,C), (A,D), (A,E), (B,C), (B,D), (B,E), (C,D), (C,E), (D,E).

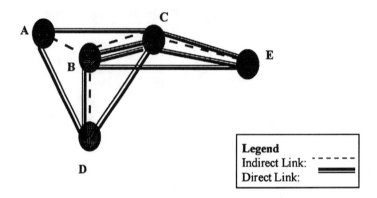

Legend
Indirect Link: `------`
Direct Link: ▬▬▬

Figure 3.16   Final design created by Mentor algorithm.

At this point, the lexicographic node pair ordering is not significant. This ordering is used solely to clarify the notation.

By computing the cost of the alternative routes for each node pair, we can logically deduce the traffic overflow dependencies, and from this a link sequence for step 4. We determine the via-node for all nonadjacent nodes, as shown below. This is done by examining each node in a node pair to find its nearest neighbors. Each nearest neighbor is a potential via-point. From this, we then list all possible routes for nonadjacent nodes. We choose the via-route with the cheapest path. We list each node pair in the lexicographic list with the associated dependencies, thereby creating a dependency matrix as shown in Table 3.28.

Sequencing the links, we must always start with the pendent node pairs. In this case, there are three pendent node pairs: (A,D), (A,E), and (D,E). Therefore, a possible sequence is:

AD, AE, DE, AC, AB, BE, CE, BC, CD, BD.

This is not the only possible valid sequence that we could have found.

Step 4:   Assign direct links. The direct link placements made by the Mentor algorithm are summarized in Table 3.29. The final design produced by Mentor, including both indirect and direct links, is provided in Figure 3.16.

Modification to Mentor algorithm using half-duplex lines
To use Mentor to design a half-duplex network, we have only to recalculate the direct link assignments made in step 4 above. In the case of half-duplex links, the direct link assignments are based on the highest traffic requirement in either direction. As given previously, the maximum usable line

Table 3.28 Dependency Matrix for Sample Mentor Problem

Node pair	Possible detour routes	No. of Hops between nodes	Cost of cheapest route
AB	Nearest neighbors	1	(Note: Since A and B are separated by only 1 hop and are nearest neighbors, an alternative path does not need to be considered.)
AC	AE to C	2	There is only one alternate path
AD	AB to D	2	There is only one alternative path
AE	AB to E	3	7 + 6 = 13 *
	EC to A		3 + 11 = 14
BC	BD to C	1	2 + 7 = 9
	CE to B		3 + 6 = 9
BD	Nearest neighbors	1	(Note: Since B and D are separated by only 1 hop and are nearest neighbors, an alternative path does not need to be considered.)
BE	BD to E	2	2 + 7 = 9
	EC to B		3 + 4 = 7 *
CD	DB to C	2	2 + 4 = 6 *
	CE to D		7 + 3 = 10
CE	Nearest neighbors	1	(Note: Since C and E are separated by only 1 hop and are nearest neighbors, an alternative path does not need to be considered.)
DE	DB to E	3	2 + 2 = 4 *
	EC to D		3 + 7 = 10

capacity = 19.2 units, and the direct link traffic threshold is 15.36 units. The resulting half-duplex line placements for this revised problem are summarized in Table 3.30.

Step 5: Assign direct links.

3.2.6.3 Case Study

In Figure 3.17, we present a typical mesh network. In this example, a company uses a two-tier network that is interconnected to two backup and disaster recovery sites. In this case study, time-division multiplexers (TDM) are used as circuit-switches to dedicate specific information channels between each node pair. Thus, all the network switching occurs within and between the PBXs, the DEC VAX computers, and the LAN/WAN routers. The reader should note that although poll-select mainframe applications can operate over mesh topologies, the applications should be routed on the most

Table 3.29 Direct Link Assignment for Sample Mentor Problem

Links	Traffic path	Traffic requirement	Load on direct links	Load on indirect links
AD	AB; BD; and AD	20	AD (19.2)	AB (.8)
				BD (.8)
AE	EC, CB, BA	14		EC (14)
				CB (14)
				AB (14.8)
DE	EC, CB	10	EC (19.2)	EC (4.8)
			CB (19.2)	CB (4.8)
				BD (10.8)
AC	AC and AB, BC	22	AC (19.2)	BC (7.6)
				AB (3.6)
AB	AB	8		AB (11.6)
BE	BE	16	BE (16)	
CE	CE	16	CE (16)	
BC	BC	8	BC (15.6)	
CD	CD	18	CD (18)	
BD	BD	12	BD (19.2)	BD (3.6)

Table 3.30 Direct Link Assignment Using Half-Duplex Links

Links	Traffic path	Traffic requirement	Load on direct links	Load on indirect links
AD	AB, BD	10		AB (10)
				BD (10)
AE	AB, BC, CE	7	AB(17)	BC (7)
				CE (7)
DE	EC, CB, BD	5		EC (12)
				CB (12)
				BD (15)
AC	AB, BC	11	BC (19.2)	BC (3.8)
				AB (11)
AB	AB	4		AB (15)
BE	BC, CE	8	CE(19.2)	BC (11.8)
				CE (.8)
CE	CE	8		CE (8.8)
BC	BC	4	BC(15.8)	BC (0)
CD	CB, BD	9	BD(19.2)	CB (9)
				BD (4.8)
BD	BD	6		BD (10.8)

direct paths that are economically feasible. This is to minimize the delays inherent in this type of distributed computing environment. The factors that determine which users are connected to what sites include the following:

- Distance to the nearest site, Site A or Site B
- Location of application(s) requested by the users

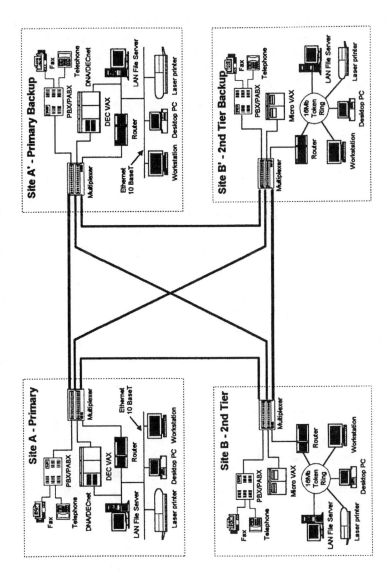

Figure 3.17 Mesh network case study (Source: B. Piliouras).

- User security clearance issue
- Protocol interoperability and conversion/proxy requirements,
- Delay and processing intolerance limits

It is easy to see why mesh networks can be very expensive. In a fully meshed network, there are $(N*(N-1))/2$ full-duplex links (where N is the number of nodes in the network). This quickly becomes a large number of links as the number of nodes added to the network increases. The input/output (I/O) hardware required grows as the square of the number of devices, since each device requires (N-1) I/O ports for network connectivity. For example, when four multiplexer devices (N = 4) are used in the network, N-1 (or 4-1 = 3) I/O port cards and cables are needed by each multiplexer to connect to the other multiplexers. As the number of links and I/O ports and associated hardware increases, so does the network cost.

3.3   Major features and functions of automated design tools

In previous sections, we summarized the major steps involved in planning and designing a network. These steps include:

1. *Requirements Analysis* — This involves collecting data on potential line types and costs, node types and costs, sources and destinations of traffic, and traffic flows between nodes.
2. *Topological Design* — This involves using various design techniques, including heuristic design algorithms, to produce a network topology specifying link and node placements.
3. *Performance Analysis* — This involves assessing the cost, reliability, and delay associated with the topological designs under consideration.

Network design tools help to automate some or all of the above design activities. A good design tool can greatly assist the process of collecting the requirements data. For instance, the major network design tools have built-in databases containing cost data for various node types, and tariff databases or cost generators for calculating circuit charges. As part of their service, the vendors of network design tools provide periodic updates to these databases so the information remains current. In addition to providing as much built-in data as possible, a good design tool should facilitate the entry and collection of organization specific data needed to plan the network. This includes providing a means to collect data on the potential node locations and traffic flows. It might also include a traffic generator. The automated tool should relieve the designer, to the extent that it is possible, of the burden of collecting and massaging a lot of requirements data. During the design phase, the network tool should provide a variety of tools, techniques, and algorithms so that many types of designs can be produced and evaluated. Automated tools are then needed to calculate the reliability, cost, and delay characteristics of each design candidate. Sensitivity analysis is also an important aspect of network design. Tools that allow the network designer to selectively modify certain aspects of the network and to recalculate the associated network performance characteristics are essential to a comprehensive network assessment. After completing the network design and analysis, reports and graphic displays documenting the design process are needed. In addition to supporting all these functions, a good network design tool should have an easy to use interactive, graphic user interface.

Some organizations rely almost exclusively on manual techniques to design a network. Although this approach certainly is flexible, it is usually wholly inadequate for a network of any substantial size. There are just too many things to consider and too many calculations to make to develop an optimal network solution without the aid of automated tools. If there are n potential node placements in the network, there are $n * (n-1)/2$ potential lines and $2^{(n*(n-1)/2)}$ possible topologies. There are simply too many possibilities to enumerate or to test by hand! Organizations that rely on manual methods may do so because they are not aware that network design tools are available, or they may lack sufficient resources (i.e., staff or dollars) to acquire a network design tool. WAN design tools tend to be very expensive. They may cost tens of thousands of dollars and, therefore, be out of the price range of smaller organizations. Service providers and outside consultants frequently make use of automated design tools, because in a competitive bid situation, the organization offering the design with the best price and performance is likely to win the contract.

By way of illustration, at this time we introduce the MIND™ family of network design tools (which includes WinMIND™, WinMIND™-International (I-MIND), and Pricer) offered by the Network Analysis Center, Inc. (NAC), of Jericho, New York. The WinMIND™ software products can be used to design, optimize, price, and analyze all types of WANs, LANs, backbone networks, and local access networks. Both domestic and international versions are available. It is one of the most comprehensive design toolkits available in the market. The software is well designed and easy to use. In addition to their software products, NAC also offers its customers training, technical support, and consulting services, some of which are free. A *few* of the specialized network design tasks that the WinMIND™ tools support are listed below.

Backbone design

WinMIND™ can be used to design optimal links, route traffic, determine backbone and access-switch sites, analyze performance, analyze switch or path failures and enhance the design so that it can survive any switch or path failure. The networks are designed on a least-cost basis using a built-in tariff database. Designs for fixed routing, frame relay, virtual circuits TDM, X.25, and ATM are supported. See Figure 3.18 for a sample screen.

Using prespecified switch sites (see Figure 3.19 for a sample screen for entering node locations) and traffic vectors, the software can determine the optimal interconnections between switches or other types of locations. The

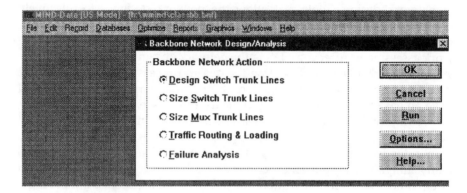

Figure 3.18   Backbone design screen (Source: NAC, Inc.).

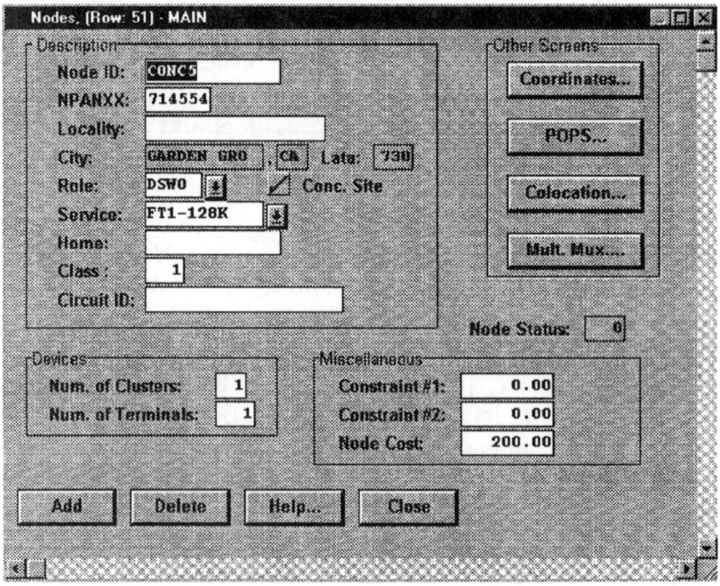

Figure 3.19   Node specification screen (Source: NAC, Inc.).

network designer can constrain the design based on the maximum number of hops, the minimum number of paths between nodes, and the maximum line loading. The software then routes the traffic, determines the topology, and sizes the links according to user-specified link speeds and tariff combinations.

Hierarchical network design

WinMIND™ can be used to perform pricing, performance analysis, traffic routing, link sizing, circuit design, optimal multiplexer placement for point-to-point or multipoint hierarchical networks, and backbone circuits for polled and SNA hierarchical topological networks. The tool determines the optimal connections between terminal and controller locations to multiplexers and/or hosts. Designs are configured on the basis of least-cost, maximum number of controllers per circuit, maximum number of terminals per circuit, maximum response time, maximum line utilization, and other factors.

Concentrator location
WinMIND™ can be used to determine the optimum number and location of backbone switches for a mesh network. The network designer can control the number and selection of the node placements (see Figure 3.19 for a sample screen). The tool can be used to identify where additional switches should be placed in order to reduce traffic congestion and to provide alternative routing capabilities.

Failure analysis
WinMIND™ can be used to assess the survivability of a private network in the event of a circuit or switch failure by analyzing the network and determining: (1) What will happen to the network if any network component fails? (2) How must the network be changed to allow all priority traffic to be rerouted in the event of failure? (3) How much will it cost to make the modifications needed to improve the network reliability?

Once these questions have been answered, the tool can be used to identify where new links should be placed to ensure network survivability. The tool begins by simulating the failure of the most critical component, then the second, and so on. When a failure disables the traffic flows on the network, the existing links are augmented to support the required traffic flows.

Network pricing
Pricer™ calculates the costs for leased communication point-to-point circuits, multipoint circuits, or entire networks. The network designer can specify choices for circuit routing, LEC bridging, maximum bridges and cascades, and POP selection.

Using a built-in tariff database, Pricer™ produces four cost reports: (1) a brief report showing total fixed and recurring costs by circuit, (2) a summary report that breaks down the total prices provided in the brief report, (3) a logical map that provides diagrams of each circuit, and (4) a detailed report that provides cost and engineering information, including bridging offices, local loop, and carrier information, for each circuit segment. See Figure 3.20 for an example of a sample cost report.

Network performance analysis
WinMIND™ evaluates the network performance in terms of the message characteristics, communication protocols, propagation delay, and host processing time. The tool can be used to model traffic flows on the network and to report circuit usage and line loading in terms of the originating, terminating, and transient traffic.

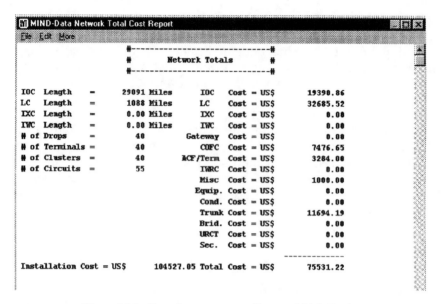

Figure 3.20 Sample cost report (Source: NAC, Inc.).

Other Special Features

WinMIND™ also includes the following standard features:

- **Network analysis center tariff services** — NAC's tariff library features over 1700 U.S. tariffs and a technical staff dedicated to maintaining its accuracy. Tariff rates include both monthly recurring and one-time installation charges. Tariff information is available on the following:
  - Analog, digital (DDS), fractional T-1, T-1 and T-3 services
  - Major interexchange carriers: AT&T, MCI, Wiltel, WorldCom, Sprint, and Cable & Wireless
  - AT&T and Sprint FTS-2000 ("A" and "B")
  - Local exchange carrier (LEC) tariffs: both for state PUC & federal FCC access rates
  - Special access and digital hubbing tariffs
  - Many of the large independent tariffs throughout the U.S. — GTE, ConTel, etc.
  - AT&T subrate digital multiplexing services: SDM5, SDM10, SDM20, and high-capacity M24
  - Special intrastate digital services: Synchronet, ADN, DACS, Mega-Link, Digipath II, Digicom I, and Optinet
  - Frame relay tariffs for many carriers
  - C- and D-type conditioning charges, and analog and digital LEC bridging

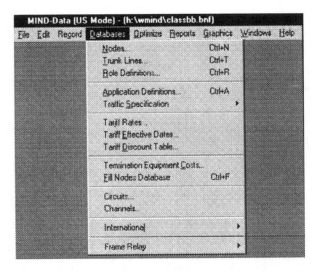

Figure 3.21 Sample tariff and data base screen (Source: NAC, Inc.).

- Discount pricing plan tables including AT&T Tariff 12
- Meet-point billing
- Custom tariffs and special discounted rates
- User-defined tariffs
- **Graphical Displays** — This includes accurate geographic maps, with user-defined color coding to differentiate different types of circuits and device types and locations. Maps can be plotted or printed. See Figure 3.22 for a sample graphical display.
- **Comprehensive on-line help**

There are other network design tools available for WAN design and analysis (in Chapter 5, we also briefly survey LAN network design tools). We list a few of them below. Note that we have not attempted to provide a comprehensive survey of all the network design tools on the market. The selection and purchase of a network design tool should be predicated on the project's objectives and requirements. This, in turn, provides the basis for developing a specification of the functional requirements and potential uses of the network design tool. For example, it may be necessary to have a tool that supports tariff databases. Some tools provide comprehensive tariff information for international and domestic networks. This functionality may be necessary if a large multinational network is being designed. If the organization is designing a large WAN that must be billed back to the users, then a tool may be needed to produce detailed cost and usage reports. In addition to considering the specific technical design requirements the tool should support, other factors that might be relevant include the availability of the technical and/or consulting support, training, frequency and cost of tariff upgrades, and the company financial status and years in business.

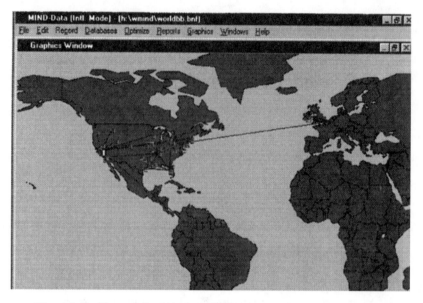

Figure 3.22 Sample graphical network display (Source: NAC, Inc.).

Network Design and Analysis Corporation (NDA), Toronto, Canada
NDA is a software company specializing in network design and pricing tools. They offer a family of tools called AUTONET, consisting of five (5) modular, integrated tools, which are listed below:

Performance-3. This tool is used for planning network bandwidth. It is designed to model the performance of multiprotocol LANs and WANs. To use the tool, you must enter the network topology, equipment specifications, and application workload profiles. From this, the tool calculates the estimated response time and network utilization. It can also be used to simulate the effects of link and node failures.

Designer. This is a distributed network design used to optimize multipoint networks. It can also be used to identify cost-effective concentrator locations, and to explore "what-if" scenarios using different carriers, services, speeds, and protocols. It can be used to develop either new or incremental designs.

MeshNet. This is a tool for optimizing the design of WAN backbones. It can be used to design least-cost mesh and hierarchical networks. It is an iterative design tool that can be used to develop new or incremental designs. It provides multiple design algorithms and performs failure analysis.

Advisor. This is a new network design and pricing tool that is completely graphical. It is used to design mesh, star, ring, and multipoint networks. The tool provides pricing for most leased line and frame relay services. It supports an MS OLE interface to MS Excel and Word, and offers point and click, cut and paste functionality for ease of use. It provides high-quality maps and reports using MS Word. It also provides Internet ready HTML reports.

Auditor. This tool is used for order processing, facility management, and billing reconciliation. It can be used to:

- Generate work orders electronically
- Track carrier confirmations and order completions
- Compare carrier bills against network inventory
- Maintain an up to date network inventory automatically
- Track inventory and equipment for LANs, WANs, and in house cables

Network Tools, Inc., San Jose, California

Network Tools develops and markets applications designed to help internetworking vendors and their customers maintain a competitive edge in the rapidly evolving technology and product marketplace. Combining core competencies in real-time network management, Internet technologies, network modeling, and solution-based design, the Network Tools application suite helps vendors simplify the development, testing, marketing, sales, and support of internetworking products. Network Tools's products also help end users evaluate competing network and product designs and integrate new products and technologies within existing systems efficiently and cost effectively. They have four major products summarized below:

Chisel. This is a multifunction tool that monitors and tests the end-to-end behavior and performance of applications under varying loads across a network. It uses a single graphical user interface that integrates the following functions:

- Baseline calculation of application performance and monitoring of the application quality of service (QOS)
- Load and stress-test of the scalability of network architectures, devices, and applications
- Verification of the effectiveness of corporate firewalls across all seven OSI layers — for both Internet and Intranet configurations

Virtual Agent. This is a dynamic network modeling and management product designed to demonstrate and validate real-time network management capabilities. It uses MIB and RMON data to build a replicated

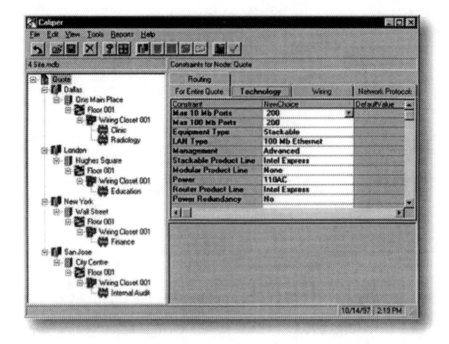

Figure 3.23 Sample caliper input data screen (Source: Network Tools, Inc.).

LAN/WAN internetwork that can be monitored and controlled by any SNMP management application.

Caliper. This is a fully automated design and configuration tool for designing multivendor, multitechnology internetworks. It requires minimal user effort and input and automatically designs low-cost LAN/WAN internetworks. It uses a single Windows® screen. It can produce multiple network designs and pricing alternatives. Two sample screens illustrating Caliper's easy to use user interface are shown in Figures 3.23 to 3.26.

Web Raider. This tool enables end users and network service providers to replicate actual HTTP application sessions and to test and document Web-server response time.

Cisco Netsys Technologies Group (formally Netsys Technologies); San Jose, California

In 1996, Cisco Systems completed its acquisition of Netsys Technologies, a provider of network infrastructure management and performance analysis software. Netsys Technologies is now part of Cisco's NTG (Netsys Technologies Group) service provider line of business. Their two major products are:

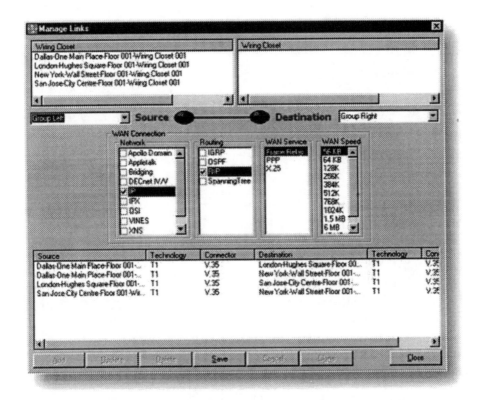

Figure 3.24 Sample caliper WAN design screen (Source: Network Tools, Inc.).

Netsys Service-Level Management Suite 4.0. This tool provides an integrated end-to-end network management solution.

Netsys Enterprise Solver 3.0. The Netsys Enterprise Solver family of network tools is used to aid network problem solving, management, and planning. It includes: the Netsys Connectivity Tools, Performance Tools, and Netsys Advisor. It provides policy-based, service-level management solutions that enable network managers to define, monitor, and assess the network connectivity, security, and performance policies.

ImageNet LTD, Israel
ImageNet Ltd. develops and markets an integrated suite of software tools for network design, analysis, and simulation. These tools help network managers and system integrators to design, build, and upgrade midsize and large complex, high-quality networks (including those that use multimedia and Internet technologies).

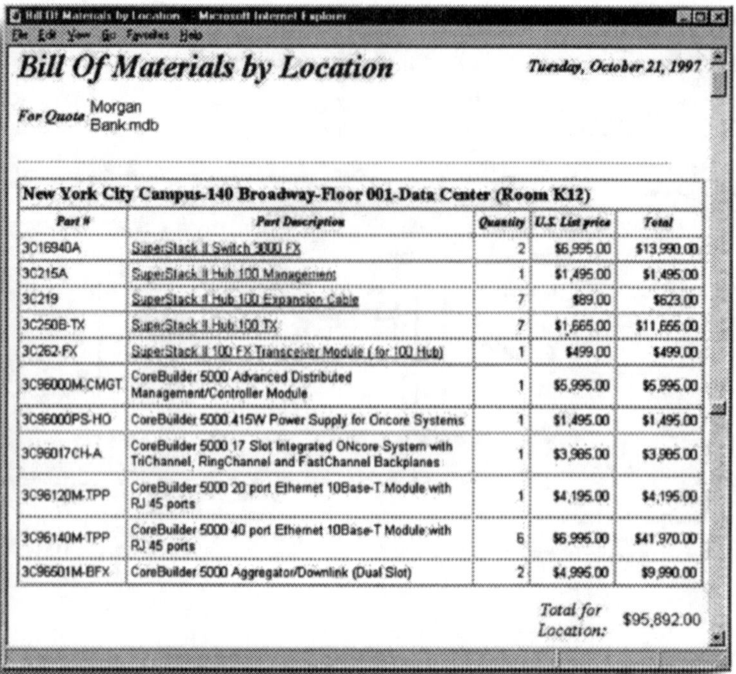

Bill Of Materials by Location Tuesday, October 21, 1997

For Quote Morgan Bank.mdb

New York City Campus-140 Broadway-Floor 001-Data Center (Room K12)				
Part #	Part Description	Quantity	U.S. List price	Total
3C16940A	SuperStack II Switch 3000 FX	2	$6,995.00	$13,990.00
3C215A	SuperStack II Hub 100 Management	1	$1,495.00	$1,495.00
3C219	SuperStack II Hub 100 Expansion Cable	7	$89.00	$623.00
3C250B-TX	SuperStack II Hub 100 TX	7	$1,665.00	$11,666.00
3C262-FX	SuperStack II 100 FX Transceiver Module (for 100 Hub)	1	$499.00	$499.00
3C96000M-CMGT	CoreBuilder 5000 Advanced Distributed Management/Controller Module	1	$5,995.00	$5,995.00
3C96000PS-HO	CoreBuilder 5000 415W Power Supply for Oncore Systems	1	$1,495.00	$1,495.00
3C96017CH-A	CoreBuilder 5000 17 Slot Integrated ONcore System with TriChannel, RingChannel and FastChannel Backplanes	1	$3,985.00	$3,985.00
3C96120M-TPP	CoreBuilder 5000 20 port Ethernet 10Base-T Module with RJ 45 ports	1	$4,195.00	$4,195.00
3C96140M-TPP	CoreBuilder 5000 40 port Ethernet 10Base-T Module with RJ 45 ports	6	$6,995.00	$41,970.00
3C96501M-BFX	CoreBuilder 5000 Aggregator/Downlink (Dual Slot)	2	$4,995.00	$9,990.00

Total for Location: $95,892.00

Figure 3.25 Sample caliper bill of materials screen (Source: Network Tools, Inc.).

CANE Suite. This is a suite of Windows® NT-based software tools that network designers, system integrators, network managers, product vendors, and LAN consultants use to design, analyze, and simulate new and existing computer networks. It supports requirements specification and conceptual design, detailed design, testing/simulation, and vendor-specific configuration and ongoing maintenance and expansion. Optional add-on modules to the basic CANE application include:

- IP Planner — This tool is used for planning Intranets and other Internet Protocol networks.
- VLAN Planner — This tool is used to map and configure virtual LANs. Typical uses of this product include designing new networks, upgrading existing networks, evaluating alternative configurations, simulating and testing "what if" scenarios, and documenting network designs.

In summary, the motivation for using a design tool is that it helps the network designer to quickly develop a number of design options under a variety of traffic and cost assumptions. The designs are then analyzed to discover which design types produce the best results for the requirements

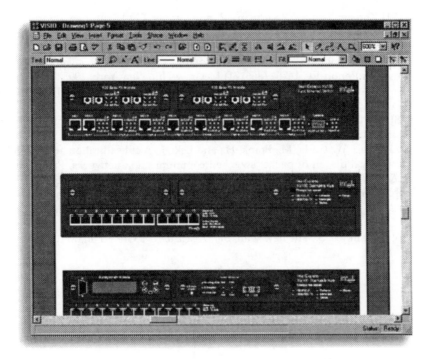

Figure 3.26 Sample caliper network diagram screen (Source: Network Tools, Inc.).

that must be supported by the network in the present and in the foreseeable future. Thus, all good design tools encourage exploration of multiple design options, with some allowance for intervention by the designer. The complexity of network design and analysis is such that it is not feasible to perform these tasks by hand for medium to large-scale networks.

Bibliography

[Ahuj82] Ahuja, V., *Design and analysis of computer communication networks*, McGraw-Hill, New York, 1982.

[Cahn94] Cahn, R., *MENTour: an algorithm for designing reliable, high-speed networks*, T.J. Watson Research Center, September 2, 1994.

[Cahn98] Cahn, R., *Wide area network design: concepts and tools for optimization*, Morgan Kaufmann, 1998.

[Davi94] Davidson, R., *Broadband networking abc's for managers*, John Wiley & Sons, New York, 1994.

[Fran72], Frank, H., and Chou, W., Topological optimization of computer networks, *Proceedings from the IEEE:* 60:1385-1397, 1972.

[Gage95] Gage, B., Choosing a public data service, *Datapro*, McGraw-Hill, Inc. May 1995.

[Hend96] Henderson, L., The price of wan connectivity, *Byte Magazine*, McGraw-Hill, May 1996.

[Kersh89] Kershenbaum, A., Interview with T. Rubinson on April 27, 1989.

[Kersh91] Kershenbaum, A., Kermani, P., and Grover, G., Mentor: an algorithm for mesh network topological optimization and routing, *IEEE Transactions on Communications*, Volume 39, No. 4, April 1991, pp. 503-513.

[Kersh93] Kershenbaum, A., *Telecommunications network design algorithms*, (IBM Research Report RC 14764/#66171) McGraw-Hill, New York, 1993.

[Lake96] Lake, L., Value-added networks (VANs): overview, *Datapro*, McGraw-Hill, November 1996.

[Mar78] Maruyama, K., Designing reliable packet switched communication networks, *Proceedings of the IEEE ICCC*, pp.493-498, 1978.

[MGE74] Chou, W., Gerla, M., Frank, H., and Eckl, J., A cut saturation algorithm for topological design of packet switched communication networks, *Proceedings of IEEE National Telecommunication Conference*, pp. 1074-1085, December 1974.

[MS86] Monma, C., and Sheng, D., Backbone network design and performance analysis: a methodology for packet switching networks, *IEEE Journal of Selected Areas of Communication*, 4:946-965, 1986.

[Sala96] Salvamone, S., The new wan, *Byte Magazine*, McGraw-Hill, May 1996.

chapter four

Value-added networks

Contents

4.1   Packet-switched networks

4.1.1   Overview

In 1976, the Consultative Committee for International Telegraphy and Tele-phony (CCITT)[1] adopted the ITU-T X.25 standard for packet switching. The X.25 standard specifies the interface between a host system and a packet-switched network via a dedicated circuit. Packet switching and its associated X.25 protocols were developed specifically to support data communications during the early days of analog networking. Analog lines are inherently noisy, creating signal distortion and transmission errors. Although a certain amount of error in voice communications is tolerable in data communications — which involve transmission of precise binary bits — it is not. X.25 was designed to compensate for these reliability problems by providing extensive error control and handling to ensure that transmissions are correctly received. As the reliability of modern communications equip-ment steadily improved, and as digital and fiber lines have replaced analog lines, the need for network level error control and handling has declined. The inherent latency and delay introduced by X.25 error checking limits the performance of packet-switched networks, especially in handling high-vol-ume data transmissions. As we discuss in Section 3.2.7.2, frame relay was developed to overcome some of the performance limitations of X.25 and has evolved as a popular successor to X.25.

Nonetheless, X.25 packet switching is available domestically and inter-nationally, providing dependable, economical networking for geographically dispersed sites. Packet switching is a flexible, scalable, and robust network-ing technology. Packet-switched networks are relatively easy to manage and they readily accommodate the addition of new WAN links as the need arises. Packet-switched networks are particularly well suited to supporting small- and medium-scale data networking applications, such as terminal to main-frame access, electronic mail transmission, and data file transfer. Packet switching is also used to support low-volume internetworking.

Packet switching is implemented in two basic ways using either data-grams or virtual circuits. Datagrams, which are employed by the Internet community, are created by "chopping" a message into a series of minimes-sages. These minimessages are, in turn, forwarded through a series of inter-mediate network nodes. The path each datagram takes through the network is independent of the path taken by the other datagrams. The datagrams are then resequenced at the final destination into their original order. Virtual circuits, in contrast, establish a logical or "virtual" end-to-end path prior to the transmission and reception of data packets. Each approach has its own inherent advantages and disadvantages. They also introduce transmission delays in different ways.

---

[1] CCITT — Since 1993, this organization has been known as the ITU-T. It is an advisory committee dedicated to the development of standards and recommendations for the tele-communications industry.

Datagrams do not require established transmission paths, since the network determines the packet routing at the time of the transmission. This is a "connectionless" method of packet transport. This method is very robust in compensating for network failures. If a path on the packet-switched backbone fails, the datagram is simply sent over an alternate path. There is no need to reestablish the connection between the sender and receiver to continue the transmission. As is evident from this discussion, no call setup or call teardown procedures are required when using datagrams. This helps to control one aspect of transmission delay.

When using datagrams, there are no guarantees that a packet will be successfully transmitted and received. Instead, datagrams offer a "best effort" packet delivery service. It is possible that packets will not be delivered successfully, or they may arrive out of sequence, since different paths may be taken by each datagram. Thus, the devices at the receiving end must resequence the packets into their original order. If the packets cannot be sequenced correctly, or they are corrupted during transport, it is usually up to the end devices to request a retransmission. As the number of packets in the transmission increases so does the chance that there will be an error in the transmission. Therefore, this method of error control is best suited to transmissions consisting of relatively few packets, since errors will not be corrected until the end of the transmission. If the transmission is very long, it will also take a long time to re-transmit the message when an error occurs.

Datagrams are similar in concept to "message switching" in that both methods use connectionless packet transport. One major difference between the two, however, is that message switching sends the entire message as one continuous segment whereas datagrams are sent as a series of small packets.

When a traditional circuit switch is overloaded, "blocking" results. Any newcomers attempting to use the overloaded network are denied service. In contrast, datagram packet switching allows packets to be rerouted around congested areas using alternate paths. The trade-off against this resiliency is the increased transmission time needed to calculate alternative routes and to perform the necessary error checking and handling. The overhead included in the datagram packets (to perform such functions as addressing, routing, and packet sequencing) also reduces the possible end-to-end line utilization. These are significant factors in determining the performance limits of packet-switched networks.

Packet addressing, resequencing, and error control functions are performed by devices called packet assemblers/dissemblers (PADs). Terminals and host devices send data in asynchronous (i.e., data are sent as individual characters) or synchronous (i.e., data are sent as blocks of characters) format. PADs convert these data streams into packets. Many PADs act as a concentrator or multiplexer, consolidating transmission streams from multiple devices and device types. Devices are typically connected to PADs in one of three ways: (1) through a direct, local connection, (2) through a leased line connection, (3) or through a public dial-up line. The PAD, in turn, can be connected to an X.25 network or some other non-X.25 device.

Virtual circuits provide another form of packet switching. In contrast to datagrams, virtual circuits are "connection" oriented. With virtual circuits, paths are set up once, using permanent virtual circuits (PVCs), or on a call-by-call basis, using switched virtual circuits (SVCs). PVCs require only one call setup (at the start of the transmission) and are permanently mapped to a particular end site. However, PVCs require predetermined mappings specifying which sites can communicate with each other, and they do not support any-to-any communication, as do SVCs.

Although SVCs experience more call delay (relating to call set-up and call teardown procedures) than PVCs, they offer more flexibility since virtually anyone can communicate with anyone else on the network. The open networking approach afforded by SVCs may, however, translate into increased CPU[2] and switch processing, and if the circuit becomes inoperative during a transmission, a new connection must be established. These factors contribute to the potential for transmission delays. However, once a path is established in a virtual circuit, each packet is transmitted over the same path. Since the same logical path is employed for the duration of the transmission, multiple packets can be sent before an acknowledgment (ACK) or negative-acknowledgment (NAK) is returned to the originating packet node. This significantly increases the efficiency of the link utilization. The number of packets that can be sent within a given ACK/NAK interval is known as the "sliding window." For X.25 networks, window sizes of 8 or 128 are common. When packets are lost or corrupted during transport, the window "slides" down to a smaller number of frames per ACK/NAK. Similarly, when the quality of transmission improves, the window slides open. The sliding window provides an effective way for the network to adapt to changing traffic loads and conditions.

4.1.2   Design considerations

A packet-switched network consists of the following components:

- Non-X.25 devices — These are the terminals and end devices on the network
- PADs — These devices are used to attach non-X.25 devices to the X.25 network via a packet switch.
- Packet switch — These devices attach to PADs, other packet switches, and to the X.25 backbone.
- Private/carrier facilities — These are the links that comprise the access network(s) and the X.25 backbone.

A packet-switched network can be implemented as a public, private, or hybrid solution, as illustrated in Figure 4.1. All you need to use a public X.25 network is a connection, usually via a packet switch. In a typical configuration,

---

[2]   CPU — This stands for central processing unit, i.e., a computer.

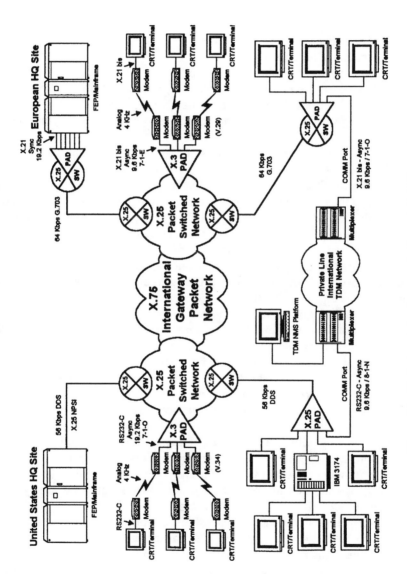

Figure 4.1   Packet-switched network (Source: B. Piliouras).

multiple end devices are clustered into a single PAD, and in turn multiple PADs are connected to a single packet switch. The packet switch can be connected to other PADs or to other packet switches. The major decisions when implementing a public X.25 network solution involve selecting the service provider and the packet switch. The packet switch should be chosen with the following in mind:

- Switch capacity
- Switching functions supported
- Type of access (public or private)
- Modularity required (memory, CPU, number of ports, etc.)
- SNMP support (for network management purposes)
- Interoperability with existing network equipment (routers, bridges, etc.)

Most packet switches are configured to support frame relay, T1/E3 speeds, and multiple protocols. This means it is fairly easy to make the transition from X.25 to frame relay. In addition, most packet switches are modular so they can be easily expanded to support growing traffic demands. This facilitates network scalability and helps to preserve the organization's investment in equipment. When selecting a packet switch, a word of caution is in order. Just because two packet switches comply with X.25 standards does not mean that they are compatible and can be used in the same network. X.25 specifies the standards for interfacing to an X.25 network, and it does not apply to equipment interfaces. Most packet switches use proprietary protocols and cannot be interconnected.

There are several compelling reasons for using a public-switched X.25 network. First, X.25 may provide a cost-effective network solution, particularly if the sites to be connected are far apart. However, when transmitting high traffic volume over short geographic distances, it may be cheaper to use a private line WAN, due to the pricing strategies of the service providers. Second, this type of network is easily scalable and provides consistent, reliable performance. Third, it also requires minimal equipment and facilities to implement. Finally, X.25 provides a way to integrate many diverse protocols onto a single network infrastructure. This ability relates to the fact that X.25 networks operate at the OSI Layer 3 or network layer. Because X.25 networks operate at the OSI network boundary level, they are generally compatible with all network architectures.

When a company owns a private X.25 network, they must be responsible for maintaining and operating the X.25 backbone through the use of leased line facilities. This type of network is usually more costly than a public X.25 network. According to [Dint94], in order to be truly cost effective, a private X.25 network should be used at or near capacity. Toward this end, some organizations resell their excess network capacity, possibly during off-peak hours, to other organizations to help pay for the costs of the network. The need for network security, scalability, and control are the reasons most often cited for implementing a private X.25 network.

Hybrid solutions, consisting of both private and public X.25 facilities are possible, and may be justifiable, depending upon circumstances.

4.1.3   Case study

As shown in Figure 4.1, in our case study, X.25 PADs are used as port aggregators to reduce the number of network circuits and ports on the FEP/mainframes. In some cases, an X.25 PAD can take the place of a 3174 cluster controller in an SNA environment.

In general, charges for X.25 packet services typically include an access charge per location, and a usage charge based on a certain price per kilobyte (i.e., 1000 bytes). Many packet services use "postalized" rates whereby the infamous mileage-sensitive circuit charge is factored into the overall costs. To improve the cost effectiveness of the network, some organizations opt to use a small portion of the bandwidth to transport data from other networks. In our case study, the company has elected to use a low-cost X.25 link to connect to their higher-cost international private line multiplexer network. The X.25 line is used to support various network management functions.

As illustrated in Figure 4.1, many different types of packet service are available. They include the following:

Dedicated access — This provides a direct connection to serving points of presence through a dedicated private line arrangement. Some of the available access options include X.25 direct, Bisync, SDLC, SNA, and Tymnet protocol connections. Access speeds are typically in the range of 9600 bps to 64 Kbps.

Switched access — This provides dial access, which in turn offers low-cost access to the X.25 packet network. Dial access provides secure remote access via the public switched telephone network (PSTN). The access options include public asynchronous dial-up and X.25 dial-up. The speeds supported by switched access usually range from 300 bps to 19.2 Kbps.

X.75 gateway services — Connectivity can be extended through the network to international postal telegraph and telephone (PTT) and other third-party packet networks via an extensive array of X.75 gateway arrangements. This is an important aspect of international packet services as these X.75 gateways provide companies with a cost-effective way to extend their global reach.

X.25 to frame relay gateway — This type of connectivity provides an economical way to bring small remote offices onto a backbone WAN that has been implemented with frame relay services. In countries where frame relay is not yet offered, X.25 can be used to access other networks that use frame relay.

Protocol and rate conversion — This involves a protocol and/or rate conversion function within the network. The following protocol conversion services are generally available:

- Async to X.25
- Async to SNA
- Async to 2780/3780 HASP
- Async to TCP/IP
- Async to UTS (universal terminal service — Unisys)
- Async to 3270 Bisync/Host
- X.25 to 3270 Bisync
- X.25 to 3270 SNA (QLLC)
- X.25 to SDLC
- X.25 to TCP/IP
- 3270 Bisync to 3270 SNA

A major advantage of packet-switched networks is that rate conversions are possible on each end of the transmission, allowing different source and destination access speeds to be used. As shown in Figure 4.1, some sites use 9.6 Kbps lines while other sites use 19.2 Kbps lines. Modems provide a cost-effective way to use an analog local loop[3] to connect remote locations into the network.

Many types of modulation techniques are available around the world. In the United States, the most typical modem in use is a V.34 modem, which is capable of supporting transmission speeds up to 28.8 Kbps. In some cases, the network will not support 28.8 Kbps transmission and may only accommodate speeds up to 19.2 Kbps. This may be due to noise on the local analog circuit or to limitations on the port speeds available on the X.25 PAD/switch. Analog loops are inherently noisy, and signal distortion may result in less than optimal modulation performance. Typically, the originating modem attempts to connect with the destination modem at the highest possible speed. If the receiving modem is capable of the same modulation scheme (e.g., V.34), and the local circuit is free of noise and distortion, data will transmit up to the maximum rate of 28.8 Kbps. When the receiving modem uses a lower-speed modulation technique (e.g., V.29), the originating modem will usually lower its speed to match the receiving modem. In Figure 4.1, we have shown a connection to a European site using V.29 dial-up modem technology. This modem supports transmission speeds up to 9.6 Kbps.

4.1.4   *Advantages, disadvantages, and recommended uses*

In this section we summarize major advantages, disadvantages, and recommended uses of X.25 packet-switched networks.

---

[3] Local loop — This is the part of the communications circuit that connects the subscriber's equipment to the equipment of the local exchange.

The major advantages of X.25 packet-switched networks include:

- Easy to manage and scale in size
- Provides a way to share a link with multiple devices
- Available on a worldwide basis
- Supports some low-volume LAN-to-LAN internetworking
- Network level (OSI layer 3) operation supports data transmission across multiple protocols and network architectures
- For networks using datagrams:
  - Ability to bypass network congestion (which increases performance)
  - Ability to bypass network failures (which increases reliability)
- For networks using virtual circuits:
  - End-to-end transmission guaranteed

Some of the major disadvantages of packet-switched networks include:

- Delays and long response times due to "store and forward" transmission method
- Address, routing, and sequence packet overhead requirements
- Not good for time-sensitive applications
- Usually only allows data transport up to 64 Kbps
- For networks using datagrams:
  - "Best effort" transmission (no guarantees that data will be received)
- For networks using virtual circuits:
  - No ability to bypass network congestion (which decreases performance)
  - No ability to bypass network failures (which decreases reliability)

The recommended uses of packet-switched networks include:

- Terminal-to-host and host-to-host connectivity
- Distributed, international network topologies
- Low- and medium-speed data applications
- Low-cost networks (particularly those offering "postalized" rates)

4.2 Frame relay networks

4.2.1 Overview

X.25 packet services were developed to operate on high BER[4] analog copper circuits. To compensate for high BERs, X.25 packet networks use a substantial

---

[4] BER — This stands for bit error rate. It is a measure of the line transmission quality and specifies the probability that a given bit will have an error in transmission. The BER rate of a typical analog circuit is on the order of 1 in 10^6, while the BER rate of a typical fiber circuit is on the order of 1 in 10^9.

amount of overhead to ensure reliable transmission. With the widespread deployment of highly reliable fiber networks, this processing overhead is increasingly unnecessary. Frame relay was developed to capitalize on these improvements in network technology. Frame relay operates as a layer 2 data link protocol interface between end users and networking equipment and assumes that upper-layer protocols, such as TCP/IP or SNA, will handle end-to-end error correction. If transmission errors are detected, frame relay discards the corrupted packets (or as they are known in frame relay networks, "frames").[5] It is then up to the customer premise equipment (CPE) and the higher-layer protocols to request retransmission of the lost frames. This form of single error checking lowers network overhead and increases response time dramatically compared to X.25.

Frame relay is based on the integrated services digital network (ISDN) standards. Frame relay protocols have been established by ANSI (T1.606, T1.617, and T1.618), the CCITT (I.233 and I.370), and various CPE vendors. For example, the Local Management Interface (LMI) was developed by the "gang of four"[6] to extend standard frame relay features and to help ensure some degree of compatibility between various frame relay products and vendor solutions. LMI extensions can be "common" or "optional" and include virtual circuit status messages (common); global addressing (optional); multicasting (optional); and simple flow control (optional). Other protocols — such as I.441 and Q.922 — support frame relay and can be referenced in the ISDN signaling standards.

Frame relay is often called a "stripped down" version of X.25 packet-switched technology. In some respects, frame relay is similar to X.25 and synchronous data link control (SDLC) protocols — particularly since it is based on a form of packet switching — but it offers much better overall performance. The main difference between frame relay and X.25 packet networks is the way that the address and control bits in the packet are employed. Frame relay addressing is based on either a two- or a four-octet address field specifying a permanent virtual connection (PVC). The two-byte default address field contains a ten (10)-bit data link connection identifier (DLCI), which, in theory, represents an address range from 0 to 1023. The four-byte field, when used with an address field extension bit (EA), provides up to twenty-four (24) address bits. However, in practice, they are not all available for addressing purposes, since numerous DLCIs are reserved for control, maintenance, and housekeeping network tasks. Many companies

---

[5] Frame relay packets — In frame relay, the packets transmitted over the network are called frames. A frame is a block of data consisting of a flag sequence, an address field, a control field (which is not used by frame relay, but is defined for compatibility with other protocols), a FECN bit, a BECN bit, an EA expansion bit, a DE (delete eligibility) bit, a frame relay information field containing user data, and a frame checking sequence field used to detect transmission errors. The data link layer (layer 2) creates frame relay packets using this standard format.

[6] Gang of four — This refers to a group of four telecom vendors: Cisco Systems, Northern Telecom, StrataCom, and Digital Equipment Corporation.

provision frame relay technology over private or shared user-group networks using customized DLCI and PVC numbering plans.

Frame relay uses PVCs to establish end-user connectivity. A frame relay permanent virtual connection is similar to an X.25 permanent virtual circuit and is identified by a logical channel number (LCN). Several hundred PVCs can be defined at a single access point that connects the customer premises equipment to the frame relay network. Thus, frame relay provides a convenient, single physical interface to support multiple data streams. Since applications can be mapped directly to a specific DLCI using static virtual circuit mapping or data encapsulation methods, the corresponding data streams can be segregated according to protocol type. This may be useful in improving the performance of time-sensitive applications or "chatty" protocols. The number of frame relay permanent virtual connections used in any given network is highly dependent on the protocols in use and the traffic load that needs to be supported. In general, the number of DLCIs needed on each line depends on several interrelated factors, including line speed, static routes, routed protocols in use, amount of broadcast traffic, the size of the routing protocol overhead, and SAP[7] messaging.

The bandwidth provided by the frame relay service is based on the committed information rate (CIR). The CIR is measured in bits per second. By definition, the CIR is the maximum *guaranteed* traffic level that the network will allow into the packet-switching environment through a specific DLCI. In practice, the actual usage can be up to the actual physical capacity of the connecting line. If a transmission requires more bandwidth than the CIR and there is sufficient capacity on the line, the transmission will likely go through. However, if the line becomes congested, frame relay will respond by discarding frames. The committed burst (Bc) size is the number of bits that the frame relay network is committed to accept and transmit at the CIR. A related metric, the committed burst excess (Be) size sets the upper bit limit for a given DLCI. Hence, Be represents the number of bits that the frame relay network will attempt to transmit after Bc bits are accommodated.

Traffic increases can lead to queuing delays at the nodes and congestion in the network. The CCITT I.370 recommendation for frame relay congestion control lists several objectives that need to be balanced while maintaining flow control in the network. These include minimizing frame discard; lessening QOS[8] variance during congestion; maintaining agreed upon QOS; and limiting the spread of congestion to other aspects of the network.

Frame relay uses both congestion avoidance (which uses explicit messaging) and congestion recovery (which uses implicit messaging) methods to maintain control over traffic flows in the network. Two bits in the address field of the frame relay frame header are used to send an explicit signal to

---

[7] SAP — This stands for service advertising protocol. This protocol is used to broadcast messages to ascertain the connectivity between various end users and processing nodes.

[8] QOS — The Quality of Service metric is used by the service provider or carrier to define the guaranteed level of the transmission facilities.

the user that congestion is present in the network. One of these is the backward explicit congestion notification (BECN) bit. The BECN bit signals when the network is congested in the opposite direction of the transmission stream. The forward explicit congestion notification (FECN) bit is used to notify the user when the network is congested in the direction of the transmission stream. A single discard eligibility (DE) bit is used to indicate which data frame(s) the network considers "discard eligible" and which ones are not. When the network becomes congested, the frame(s) marked DE are the first to be discarded. Until recently, these congestion control bits were not used by device manufacturers to signal congestion problems in the network to the end user. Although frame relay is designed to provide various kinds of congestion warnings, these warnings are not always fully used or implemented in network products and services.

In summary, frame relay was developed specifically to handle bursty[9] data network applications, and is commonly used to consolidate data transmissions from and to terminal-to-host and/or LAN-to-LAN applications onto a single network infrastructure. Generally, frame relay is not mileage sensitive, so it can be a cost-effective solution for connecting widely dispersed sites. Thus, organizations use frame relay to reduce network complexity and costs. Frame relay can be used in private or public network solutions.

4.2.2 Design Considerations

Network managers need to consider the cost, reliability, and performance characteristics of frame relay to decide if it is suited to their networking applications. Frame relay is excellent at handling bursty data communications and internetworking functions. It provides bandwidth on demand (although the bandwidth allocation is not guaranteed beyond the CIR), and supports multiple data sessions over a single access line. Frame Relay is generally capable of operating at speeds up to T1 (1.544 Mbps) or E1 (2.048 Mbps), and some service providers are offering frame relay at speeds up to 6 Mbps. Since the vast majority of networking in place today uses T1 line speeds or lower, the bandwidth frame relay provides is more than adequate for the networking needs of most companies.

However, frame relay does exhibit transmission delays inherent to any packet service. While frame relay provides higher access speeds than X.25 and accommodates various data protocols, it is not meant to support delay-sensitive applications such as interactive voice and video. Therefore, some time-sensitive applications may be better suited to a private line network, since private lines offer a guaranteed quality of service (QOS) that frame relay does not. It is likely that with future advancements in the technology

---

[9] Bursty — Bursty traffic is characterized by intermittent high-volume traffic spikes interspersed with moderate levels of traffic.

there will be fewer limitations on the type of traffic that can be carried successfully on a frame relay network.

A number of issues must be addressed before implementing a frame relay network. First, frame relay is an interface standard and not an architecture. Thus, a common network architecture must be in place to support end-to-end connectivity. TCP/IP is commonly used as a network architecture that supports frame relay. Second, the network design parameters need to be estimated. This involves analyzing traffic patterns and requirements in order to select the appropriate frame relay service parameters (i.e., the CIR, Be, and Bc levels of service). Third, the impacts of frame relay on the (existing and planned) network components need to be determined. Routers, bridges, multiplexers, FEPs, packet switches, and other devices that do not support frame relay must be upgraded to support it. This involves installing FRADs[10] on the network equipment. FRADs implement frame relay interface (FRI) standards that establish how to connect a device to a frame relay network. Most routers are configured with a built-in FRAD; however, most bridges are not. Thus, the costs to upgrade to frame relay depend on the type of equipment being used in the network.

There are four major ways that frame relay networks can be implemented: as a public network, a private network, a hybrid, or as a managed service. Frame relay is most commonly implemented as a public network solution. With public frame relay service, each network location gets a connection into the frame relay public network service through a port connection. The port connection is the node's gateway into the public frame-relay service. Permanent virtual circuits (PVCs) are established between these port connections (we note at this juncture that frame relay offers both PVCs and SVCs. However, we have concentrated our discussion on PVCs because they are used far more often, particularly since not all the telecom providers offer SVC service). One big advantage of frame relay over private leased lines is that one connection into the network can support many PVCs to many other locations. This is illustrated in Figure 4.2. Compared with a private line mesh topology with dedicated point-to-point connections, frame relay offers opportunities for substantial equipment savings. In addition, since frame relay pricing is generally not distance related, it is usually cheaper to connect widely separated sites with public frame relay than it is to use private leased lines that are priced by distance. Finally, public frame relay networks are usually easily modified and reconfigured by the service provider, providing more flexibility for change and growth requirements than dedicated, private line WANs.

---

[10] FRAD — This stands for frame relay assembler disassembler. This device is similar in concept to the PAD used in packet switching. It enables asynchronous devices such as personal computers to send data and communicate over the frame relay network. FRADs are usually implemented as an integral part of the network device. In addition to frame relay, many FRADs support X.25, ISDN, and other protocols in the same device.

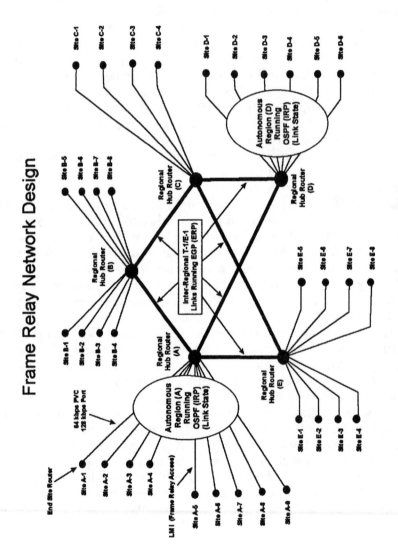

Figure 4.2   Frame relay network (Source: B. Piliouras).

In a private frame relay implementation, an organization leases private lines from a carrier. These private lines interconnect the various network nodes to corporate backbone switches. One of the primary reasons for using frame relay in a private network is that it is capable of supporting diverse traffic types over a single backbone (however, it is possible that organizations with low-volume traffic may find it more cost effective to use a TDM private-line network using T1 or E1 lines that offer guaranteed levels of performance. In fact, in this situation, the most common solution is a private, leased line network. As the number of sites in the network grows, the rationale for using a private frame relay network over a private leased line network becomes more compelling.

The costs for frame relay are usually based on the committed information rate (CIR) of each PVC, port costs for the backbone network, and the access link costs to connect the various sites to the frame relay network. Typically, there are fixed monthly fees (however, some service providers offer other pricing options, including capped usage-based fees). Although fixed fees are appealing, because they make it easier to budget and plan network expenses, they also present special challenges to the network manager. As telecom managers struggle to balance their communications budgets every year, it becomes increasingly important to establish an effective balance between network optimization and system overload. One of the dangers of using frame relay in a private network is that it may make it harder to manage these pressures, since the frame relay pricing provides a powerful incentive to use the network facilities to the maximum extent possible.

A hybrid solution interconnects frame relay and other kinds of networks into a larger network structure. In general, a hybrid network is harder to manage because of the diverse traffic and equipment that it must support. Since different portions of the network may use different technologies and protocols, there is also more potential for difficulties in the end-to-end trans-mission. However, the business and technical requirements may be such that a hybrid network solution is necessary.

In an effort to control the complexity and cost of increasingly larger and more sophisticated networks, many organizations are outsourcing their frame relay networks entirely. The service provider is responsible for all aspects of the local access provisioning, the customer premise equipment, disaster recovery arrangements, and network management. This is one of the easiest ways to implement a frame relay network.

4.2.3   Case Study

As illustrated in Figure 4.2, hierarchical meshed frame relay networks can be used to localize traffic by region. This segmentation is accomplished by placing regional hub routers at network mesh points. The regional hub routers also reduce the number of DLCIs needed at each physical interface to the network. This configuration is used to help manage network complexity. This configuration is not always suitable if there is a lot of traffic on the network, since it may result in too much packet broadcast and replication, thus overloading the DLCIs.

Packet broadcasts are intrinsic to frame relay routing protocols. OSPF and RIP are two commonly used frame relay routing protocols. OPSF, derived from OSI's intermediate system to intermediate system (IS-IS) routing protocol, is a link state intradomain hierarchical routing protocol developed for large, heterogeneous IP internetworks. Link state protocols typically employ a "shortest path first" algorithm that *floods* all network nodes, but only broadcasts routing information about the state of its links. The earlier routing information protocol (RIP), in contrast, is a distance vector routing protocol, which uses "Bellman-Ford"–type shortest-path algorithms to determine packet routing. OSPF broadcasts a partial routing table to all nodes, while RIP sends out an entire routing table, but only to its neighbors in a localized region of the network. Both schemes can produce a significant amount of broadcast traffic in a large network. However, of the two protocols, RIP generates considerably more broadcast traffic.

While differing in their methods of updating routing information, both OSPF and RIP are considered interior routing protocols (IRPs) or intradomain routing protocols. In contrast, exterior routing protocols (ERPs) provide interdomain routing to support communications between core backbone routers. Like OSPF and RIP, EGP is a dynamic routing protocol that adjusts in real time to changing network conditions.

Gateway routers are expensive; however, when used in conjunction with both intra- and interrouting protocols, they provide a relatively straightforward and easy way to interconnect large networks. In the case study design, gateway routers are used to separate autonomous systems (AS) or regions, running the OSPF interior routing protocol (IRP) from the meshed T-1/E-1 backbone running the EGP exterior routing protocol (ERP).

Frame relay can be used to implement star, partially meshed, or fully meshed topologies for regional and interregional networks. Star topologies are frequently used to minimize the number of DLCIs required in the network, thus reducing network costs. However, a "single-threaded" star topology does not offer the fault tolerance needed by many networking solutions. In a fully meshed frame relay network, every routing node is logically linked via an assigned DLCI to every other node in the network. This topology may not be practical for large frame relay internetworks because of its potential costs (which are related to the number of DCLIs needed) and performance problems (which are related to the packet broadcast scheme that frame relay uses, which can cause severe congestion in larger networks). A partial mesh frame relay network is a common compromise solution, since it combines the cost savings of a star topology with the redundancy of a fully meshed topology.

4.2.4 Advantages, disadvantages, and recommended uses of frame relay

In this section we summarize major advantages, disadvantages, and recommended uses of frame relay networks.

The major advantages of frame relay networks include:

- Good support for bursty data traffic
- Easy to manage and scale in size
- Allows resource sharing
- Improved service provided by network rerouting capabilities
- Widely available and rapidly growing use on a worldwide basis
- Supports medium volume internetworking of LAN-to-LAN applications
- Provides faster network transport (using OSI Layer 2) than X.25 packet technology
- May reduce costs and complexity as compared to a comparable private line, leased network

The disadvantages of frame relay networks relate to the following:

- Not designed to support interactive voice and video applications
- Congestion control mechanisms are not fully implemented by providers of frame relay products and services
- Wide fluctuations in QOS make it more difficult to justify end-user chargebacks
- Does not support connectionless-oriented applications
- Inadequate bandwidth support for large, steady data file transfers

Frame relay is recommended for use in:

- Small- to large-scale data applications
- Autonomous and interregional networking
- Peer-to-peer, terminal-to-host, and host-to-host connectivity requirements
- Distributed, international network topologies with wide connectivity
- Low-, medium-, and medium-high-speed data applications
- Low- to medium-cost networks (particularly when "postalized" rates are available)

4.3 SMDS networks

4.3.1 Overview

Switch multimegabit data service (SMDS) is a connectionless, cell- and packet-switched public data transport service specifically designed for high-speed data communications, such as LAN interconnectivity. SMDS typically supports speeds from 1.544 Mbps to 45 Mbps, although sometimes slower service is available. SMDS is a standard specifying carrier service and device interface and is not a technology per se (and therefore the technologies and architectures associated with SMDS are transparent to the user). SMDS was developed by Bellcore in the late 1980s to provide the RBOCs[11] with a means for providing high-speed data switching to their customers. SMDS is structured exactly like the telephone network, with connectivity through the local exchange[12] and interexchange carriers. SMDS uses a universal addressing scheme that is analogous to telephone numbers. Any SMDS device can contact any other SMDS device on the SMDS network by placing a call using the correct addressing number. SMDS is offered by the RBHCs,[13] Southern New England Telephone (SNET), EMI Communications, and MCI.

SMDS is very closely aligned with the IEEE 802.6 metropolitan area network (MAN) standard. The MAN standard is designed to operate with three main transmission mediums listed below:

- *CCITT G.703* — 34.368 Mbps and 139.264 Mbps over metallic circuits
- *ANSI DS-3* — 44.736 Mbps over coaxial cable or fiber optics
- *ANSI SONET (CCITT SDH)* — OC-3 (155.52 Mbps) and above over single-mode fiber

SMDS relies on a robust underlying layer 1 infrastructure (e.g., SONET or SDH), the customer premise equipment and upper-layer protocols to manage end-to-end communication and error handling. SMDS does not provide error recovery at the network level and it does not keep track of lost cells when congestion or other problems occur in the network (as ATM does with the CLP[14] bit and frame relay does with the BECN and FECN bits). Although SMDS is based on the IEEE 802.6 MAN standard, it does not have the distance limitations of a MAN (which is limited to an area approximately 50 kilometers in diameter).

A distributed-queue dual-bus (DQDB) architecture is defined in the IEEE 802.6 MAN standard. The DQDB is the medium access control (MAC)[15] technique used by IEEE 802.6 and, as such, functions at the lower portion of layer 2 in the OSI protocol stack (i.e., below IEEE 802.2 logical link control

---

[11] RBOC — This stands for regional Bell operating company and refers to the local or regional telco in one of seven regions in the United States. The RBOCs are also known as the "Baby Bells."

[12] Local exchange — The local exchange is the central office where the subscriber's lines terminate. The exchange accesses other exchanges and national telephone networks.

[13] RBHC — The seven (7) regional Bell holding companies were formed after the AT&T divestiture to provide both regulated and unregulated communications services.

[14] CLP — This stands for cell loss priority bit. ATM uses this bit to signal when congestion is present in the network and cells are being dropped.

[15] MAC — This stands for media access control. The MAC method enables network devices to access the physical network media and to transmit information. It corresponds to the OSI Layer 2.

[LLC]). The DQDB bus architecture supports both packet-switched and circuit-switched services and consists of a protocol syntax and a distributed queuing algorithm. Essentially, the DQDB standard specifies a slotted ring and queuing system for controlling the order in which devices may transmit data. The DQDB protocol uses a reservation scheme based on a round robin, first-come first-served basis. Once a transmitting station has finished waiting its turn in queue, it can use an "empty" slot to transmit data.

The DQDB bus slots are used by SMDS to transmit data in 53-byte fixed-length cells. Each cell consists of a 52-byte data segment and a 1-byte access control field. The 52-byte data segment, in turn, is comprised of the following:

- Network control information (32 bits)
- Segment type (2 bits)
- Message ID (14 bits)
- Segmentation unit (352 bits)
- Payload length (6 bits)
- Payload cyclic redundancy check (CRC) (10 bits)

SMDS defines two different slot types: queue arbitrated (QA) and prearbitrated (PA) slots. QA slots are employed to transport packet-switched data, while PA slots are used to carry circuit-switched data. DQDB uses the countdown (CD) parameter to ensure quick access under light loads and predictable queuing under heavy loads. This ensures optimal handling of both bursty interactive traffic (packet switched) and sustained file transfers (circuit switched).

The global CCITT E.164 address resides in the SMDS header and is used to transport data to one or several recipients at a time. The latter form of transmission is commonly referred to as "multicasting." Multicasting allows a single transmission to be sent to multiple destinations on the basis of a distribution list containing group addresses (which are established at the time of network subscription). Multicasting can substantially reduce the processing load on the local premise device processing because the internal SMDS network switches perform the packet duplication and distribution functions. For connectionless broadcasts outside of the LATA, an additional interconnect carrier interface (ICI) is required that defines how LECs and interexchange carriers (IXCs) interface to maintain end-to-end connectivity.

The SMDS E.164 addressing scheme allows users on a single SMDS access line to establish a switched connection to any other SMDS location with a valid E.164 address. This allows communications between intracompany and intercompany sites. As compared to private line networks, SMDS reduces the number of local loops, end site router ports, and channel service unit/data service units (CSU/DSUs) needed in the network. However, due to the public nature of E.164 addressing, ample security measures need to be implemented to prevent misrouted data and unauthorized access. To this end, SMDS provides network security through the use of source address authentication and destination address screening.

The *protocol* used between the SMDS network and the customer premises equipment (CPE) is called the SMDS interface protocol (SIP) while the *interface* is referred to as the subscriber network interface (SNI). In essence, the SNI provides access to the SMDS network. Interfaces have been developed to connect SMDS to frame relay, ATM, FDDI, HDLC, token ring, and Ethernet.

In contrast, frame relay uses a user network interface (UNI) to access permanent virtual circuits or, to a lesser extent, switched virtual circuits) to establish connections. The connectionless service provided by SMDS through the use of the SNI offers higher throughputs and generally lower end-to-end network latency than frame relay. SMDS uses a sustained information rate (SIR) — which is analogous to frame relay's committed information rate (CIR) — as a measure of service quality. However, the SIR refers to the combined service limit on the access line, while the CIR relates to the bandwidth limitations on individual PVCs within the access facility.

The SIP layers are partitioned logically into three layers (L3, L2, and L1) with the corresponding protocol data units (PDUs)[16] listed below:

- *SIP L3 PDU*: Encapsulates SMDS service data units (SDUs), SMDS destination address, and overhead into a high-level data link control (HDLC) frame format. SMDS SDUs can contain up to 9188 bytes of user information. HDLC frames include link flow control, error messages, and link management.
- *SIP L2 PDU*: Segments L3 PDU data into equal length 44-byte cells and adds nine (9) bytes of additional header and trailer overhead for a total of 53 bytes per SDMS cell.
- *SIP L1 PDU*: Converts (maps) L2 PDU fixed-length cells into DS-1 (1.544 Mbps) or DS-3 (44.736 Mbps) signal formats using the physical layer convergence protocol (PLCP).

We present a case study to illustrate SMDS's use of PDUs. In Figure 4.3, for Customer A, Site 3, the SMDS router outputs L3 PDU frames into an SMDS CSU/DSU that functions as both an L2 PDU and an L1 PDU. The SMDS CSU/DSU outputs either a DS-1 (digital signal level 1) or a DS-3 (digital signal level 3) signal. With respect to the customer's premise, notice that the subscriber network interface (SNI) can be used to access either a local exchange carrier (LEC) or an interexchange carrier (IEC). Intra-LATA SMDS transmission exists within the IEC SMDS DQDB looped bus backbone. For inter-LATA transport, the LEC and IXC need to implement an interconnect carrier interface (ICI) to maintain end-to-end connectivity.

Since asynchronous transfer mode (ATM) technology supports SMDS traffic through the use of AAL-3/4 segmentation, ATM networks can also interface with SMDS networks. However, this assumes that the SMDS

---

[16] PDU — A protocol data unit refers to the data transmitted between OSI layers. Each PDU contains the data to be sent, plus headers and trailers appended by each OSI layer as the transmission is processed.

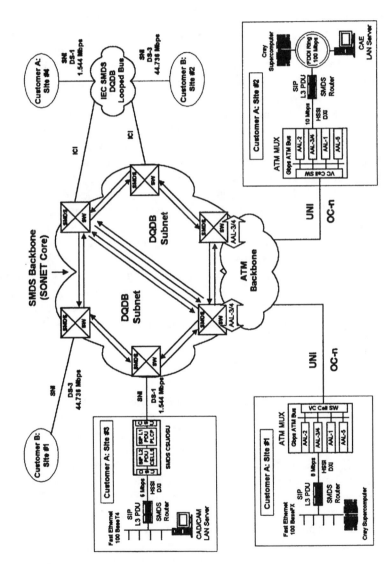

Figure 4.3   SMDS network design (Source: B. Piliouras).

deployed through ATM is compatible with the other switch provider's rendition of SMDS service.

SMDS DQDB subnetworks are typically interconnected through the use of bridges, routers, or gateways. In this configuration, the principal DQDB components are the head station, two unidirectional buses, an end station, and multiple access units. The head station generates frame synchronization on the forward bus, while the end station generates the frame on the opposite bus. The access units, which attach to both buses, *write* on the bus and *read* from the bus. Note that with DQDB architectures, data passes *by* the node, not *through* the node as with ring networks. Therefore, failed DQDB nodes can be removed from the bus without affecting other portions of the network. For example, if a node or line segment fails, the nodes on either side of the failure can take on the functions of the end station so that the network can remain operational. Within the SMDS network, the DQDB dual buses can also be partitioned into multiple subnetworks to provide network redundancy and automatic network restoration capabilities.

SMDS data units carry both source and destination address information within the SIP L3 PDU that is consistent with the conventional 10-digit North American numbering plan (NANP). To ensure that an address has been legitimately assigned by the SNI from which it originated, the SMDS network validates all source addresses. This protects network users from "address spoofing".[17] Address screening is also used to examine SMDS destination and source addresses that are entering or leaving the network. In addition to providing network security, address screening is also used to allow a subscriber to establish a "private virtual network" that will accept only those addresses that are part of a predefined customer's network.

4.3.2   Design considerations

Unlike ATM and frame relay, SMDS is not available as a private networking solution. It is strictly a public switched service. However, as such, SMDS is easy to implement, since the telco or service provider supports the maintenance and operation of the network. SMDS is implemented as a datagram service and is intended for data communications exclusively. (We note, however, that SMDS is not suited to all types of data communications. For example, it does not handle SNA-data applications efficiently, since it converts SNA-generated 8-bit character streams into 53-byte cells.) The inherent performance limitations of a datagram approach make SMDS unsuitable for time-sensitive applications such as voice and video.

SMDS is not a widely used technology. In a recent survey, the SMDS market accounted for .3% of the total data communications market and is only 3% of the size of the frame relay market [Data97]. Industry analysts

---

[17] Spoofing — This occurs when a packet claims to be from an address that is different from which it actually originated.

predict that SMDS will likely be replaced in the future by frame relay (which will be used to support low-end network applications), and by ATM (which will be used to support high-end network applications).

The flexibility and ease of SMDS in implementing point-to-point and many-to-many connectivity is one of the major reasons why it is being used today. Implementing similar levels of connectivity in frame relay involves complicated provisioning of permanent virtual circuits (PVCs). SMDS, on the other hand, is a connectionless service. Each SMDS device is assigned a unique address analogous to a phone number, which is used to make a connection in a manner similar to placing a phone call. Thus, any location with SMDS-compatible equipment can communicate with any other SMDS-compatible device. SMDS provides call screening, call verification, and call blocking functions on the basis of its universal public addressing scheme. SMDS also allows unique user groups to be defined to create, in effect, a "private" SMDS network that only the members within the defined user group can access.

SMDS is offered almost exclusively by the local telcos. This means that if the networking requirements cross inter-LATA boundaries, there may be difficulty obtaining SMDS service, since MCI is the only IEX that offers SMDS. This is one of the major reasons why SMDS is usually limited to MAN applications. The limited availability of SMDS outside of regional boundaries, and not technical reasons, creates a practical constraint on the use of SMDS in some WAN applications. However, MCI continues to broaden its offering of SMDS services both domestically and internationally, and this constraint may lessen in the future.

When implementing a SMDS network, it is important to consider the following factors:

- Distance and positioning of nodes on the DQDB bus — This is one area that must be carefully planned to avoid performance problems due to queuing delays.
- Bandwidth requirements — LAN-to-LAN and LAN-to-WAN applications benefit most from the bandwidth and performance offered by SMDS.
- Future migration paths to B-ISDN — ATM and other technologies are likely to supplant SMDS in the future and strategies to transition to this type of environment should be developed).
- Pricing — SMDS pricing is based on the SMDS port connection and bandwidth allocation. The cost effectiveness of SMDS in any given situation should be weighed against other available networking options.

4.3.3   Advantages, disadvantages, and recommended uses

In this section we summarize major advantages, disadvantages, and recommended uses of SMDS networks.

Some of the advantages of SMDS networks include:

- Effective solution for LAN-to-LAN connectivity
- Easy migration to B-ISDN services (such as ATM)
- Connectionless service (thus reducing delay and increasing through-put as compared to frame relay, which is connection-oriented)
- Easy to design and manage
- Easy to add new SMDS locations to the network
- Consolidates multiple "circuits" onto single access facility
- High network survivability and large bandwidth potential
- Easy connection procedures due to open network architecture and flexible public E.164 addressing
- Security protection from "address spoofing"
- Source address authentication, destination address screening, and verification functions can be used to establish private user groups

Some disadvantages associated with SMDS include:

- Does not provide error control for packet data
- Medium is shared with other users (i.e., it is offered only as a public offering)
- Relatively high overhead (which is over 8%, and sometimes much higher) compared to T-1 private lines (which have an overhead rate of about 1%)
- Uneven positioning and long distances between nodes on the DQDB bus can create network delays
- Viewed as a short-term solution for broadband services

Recommended uses for SMDS networks include:

- Backbone infrastructure for public LECs and IXCs
- Connectionless, bandwidth intensive LAN-to-LAN communications such as:
    - Computer aided design/computer aided manufacturing (CAD/CAM)
    - Computer aided engineering (CAE)
    - Magnetic resonance imaging (MRI)
    - Desktop publishing
    - Disaster recovery operations
    - Distributed client/server applications (database access, file transfers, etc.)
- Support of data requirements that mandate high network reliability and performance
- Large scale internetworking requiring any-to-any connectivity

4.4 ISDN networks

4.4.1 Overview

The integrated services digital network (ISDN) supports voice, data, text, graphics, music, video, and other communications traffic over twisted-pair[18] telephone wire. Although there are many ISDN network specifications and standards, the evolution of ISDN and Broadband ISDN (B-ISDN)[19] in particular, is still in its infancy.

ISDN was originally conceived by the telcos as a way to develop cost-effective voice communication services for their customers. However, ISDN is also well positioned to meet the growing needs of digital data communications. Although a key benefit of ISDN is the ability to carry simultaneous voice and data applications over the same digital transmission links, ISDN provides many other attractive services, including "intelligent" options and features, and network management and maintenance functions. B-ISDN was developed in the mid-1990s to enable the public networks to provide switched, semipermanent, and permanent broadband connections for point-to-point and point-to-multipoint applications. B-ISDN is considered to be the second-generation version of ISDN. The B-ISDN standard defines a cell relay technology that is designed to work in conjunction with ATM and SONET. Channel speeds of 155 Mbps and 622 Mbps are available through B-ISDN and ATM services. The developers of ISDN and B-ISDN have attempted to develop an infrastructure for worldwide networking solutions based on comprehensive standards and services. The integration provided by a common infrastructure facilitates the provisioning of services and a more comprehensive approach to network management.

The CCITT is an international organization responsible for the development of ISDN standards. The ISDN standards are known as the I-series recommendations. By convention, all CCITT standards are referred to as recommendations. The signaling aspects of ISDN, including the link access protocol D (LAP-D) specification, are defined in the CCITT Q-series recommendations. The B-ISDN standard is specified in ITU-T Recommendation I.121. For a complete review of ISDN and B-ISDN standards, the reader is referred to [Summ95].

Some of the more commonly available ISDN services include:

- 64 Kbps (1B+D)
- 128 Kbps (2B+D)
- 384 Kbps (H0)
- 1472 Kbps (H10: 23B+D)
- 1536 Kbps (H11: 24B)
- 1920 Kbps (H12: 30B+D)

---

[18] Twisted pair — This consists of two insulated wires twisted together without an outer covering. This is the standard wiring used in telephone installations.

[19] B-ISDN — This refers to services provided by telephone carriers offering ISDN in place of ATM.

Bearer or B channels, as listed above, operate at 64 Kbps and are used to carry digitized voice, data, or video traffic over circuit-switched, packet-switched, or semipermanent connection facilities. The D channels carry signaling information that is used to control the circuit-switched B channels. They can also carry low-speed packet-switched data. H channels are aggregates of B channels.

The ISDN architecture delineates two service classes: BRI and PRI. Basic rate interface (BRI) circuits employ a 2B+D structure, which consists of two full-duplex 64 Kbps B channels and one full-duplex 16 Kbps D channel. Overhead bits are also carried on the channel and are used to perform framing, synchronization, and management functions. BRI facilities are the ones most often used for corporate networking applications.

When the traffic requirements exceed 64 Kbps per channel, multiple channels are grouped together or *bundled* to provide the necessary bandwidth. These bundled channels are typically digital 1.544 Mbps (T-1) circuits in the U.S. and in a few other countries, such as Canada and Japan. In the rest of the world, 2.048 Mbps (E-1) digital circuits are used. Access facilities operating at these speeds are known as primary rate interface (PRI) circuits and are formally offered in H channel structures beginning at 384 Kbps (H0).

Many times PRI circuits include a separate 64 Kbps D channel for each PRI facility. When there are multiple PRI circuits to the same customer location, it is possible to control all transmissions through a single D signaling channel. In this case, all the PRI channels can be used to transmit data, except for the one PRI reserved for signal control purposes. Signaling overhead is greatly reduced when only one D channel is used to control multiple PRI circuits. However, there is the risk that if the D channel fails, no transmissions can be sent.

The actual call processing is performed by a Signaling System 7 (SS 7) network that controls the call setup and call teardown procedures for each ISDN call. This is the same SS 7 signaling network used by the United States local exchange carriers (LECs) and the interexchange carriers (IXCs) to place standard voice calls. In other parts of the world, the common channel signaling system 7 (CCSS 7) signaling technology infrastructure is used. While many of the basic concepts are the same, there are differences between the two signaling methods that necessitate signal conversion equipment. Therefore, many popular switch manufacturers support both signaling formats so their equipment can be used to handle both domestic and international ISDN calls.

4.4.2   Design considerations

Although ISDN was slow to take off in the United States marketplace, its popularity has been growing in recent years. One of the reasons for this is the increased availability of ISDN products and services. Until fairly recently, ISDN was not widely available throughout the United States, in part because telephone carriers have been slow to upgrade their equipment to support

ISDN. However, even when ISDN service is available, there is often a long lead time of weeks or months to install the lines after they are ordered.

The limited availability of ISDN contributed to the reluctance of vendors to offer ISDN-ready products. As ISDN has become generally available, so have ISDN-compatible products. The development of IDSN device interface standards has also encouraged the development of new ISDN product offerings. ISDN protocols governing communication between devices in the network are based on the notion of "functional devices" and "reference points." ISDN equipment standards define several types of device classes. In these "functional device standards," each device type is characterized by certain functional or logical features that may or may not actually be present in the device. Device-to-device interfaces are defined through "reference point standards" that are based, in turn, on the functional device types. Reference points allow different ISDN functional devices to operate under different protocols on the ISDN network. The ISDN standards are designed to promote modular product development so that equipment manufacturers can make hardware and software improvements to their product lines without compromising ISDN compatibility. From the customer's perspective, this means that ISDN equipment can be purchased from a variety of vendors, thus offering more freedom of choice in the selection of ISDN networking solutions. As illustrated in Figure 4.4, several ISDN functional groups and reference points are currently defined, including:

- **TE1** *(subscriber Terminal Equipment type 1)*: Devices that support the standard ISDN interface such as digital telephones and digital (type IV) facsimile equipment.
- **TE2** *(subscriber Terminal Equipment type 2)*: Devices that support non-ISDN equipment such as video CODECs, terminals, and mainframe computers.
- **TA** *(Terminal Adapter)*: Devices that convert non-ISDN equipment into ISDN formatted signals. The ISDN TA can often be either a stand-alone device or combined with a TE2 type device.
- **NT1** *(Network Termination 1)*: Devices that include functions associated with the physical and electrical aspects (OSI layer 1) of ISDN termination on a customer's premise.
- **NT2** *(Network Termination 2)*: Device with switching and concentration functions up to OSI layer 3 (Network level). Some examples include digital PBX/PABXs, LAN routers, and terminal controllers.
- **NT1/2** *(Combination Network Termination 1 and 2)*: Not shown in diagram. Device that combines NTI and NT2 functions into a single piece of telecom equipment. Often used by foreign postal telegraph and telephone (PTT) ISDN providers.
- **R** *(Reference Point R — Rate)*: Provides a non-ISDN interface between non-ISDN compatible equipment and a Terminal Adapter (TA).
- **S** *(Reference Point S — System)*: Separates individual ISDN user equipment from network-related telecom functions.

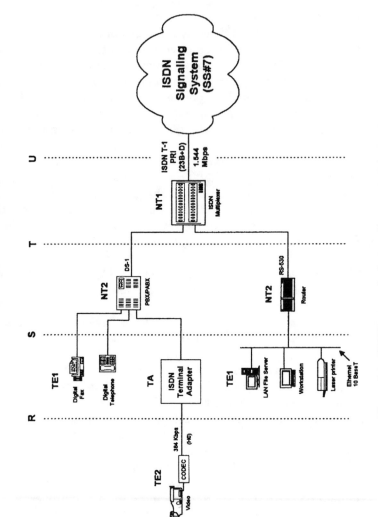

Figure 4.4 ISDN network design (Source: B. Piliouras).

- **T** *(Reference Point T — Terminal)*: Consistent with ISDN network termination, it separates end user equipment (up to NT2) from network provider equipment (NT1).
- **U** *(Reference Point U — User to Network Interface)*: Reference point between NT1 devices and line-termination equipment in the carrier network. The U reference point is typically employed in North America because the carrier network does not usually provide NT1 functionality.

Although ISDN continues to be more expensive than regular phone service, its costs have declined as service providers have developed a larger customer base. This, too, has encouraged ISDN use. A primary driver behind the market acceptance of ISDN is the suitability of ISDN for certain niche applications. In particular, ISDN provides a very cost effective and convenient networking solution for the following types of applications:

- Telecommuting
- Internet connectivity to ISP[20] provider
- Long distance videoconferencing
- Dial-up lines for backup and disaster recovery
- LAN-to-LAN connectivity (especially for small businesses and organizations)
- Branch-office connectivity between remote sites having low-volume traffic (particularly for banking and automated teller machine applications)

Not surprisingly, ordering ISDN service is more complicated than ordering regular telephone service because of the various service and circuit options available. The basic steps involved in implementing ISDN involve the following:

1. Determine the network requirements. This is true for all network design problems!
2. Determine the appropriate service. In addition to bearer services, mentioned earlier for data transport, there are supplementary ISDN services and teleservices. Supplementary services provide options for voice communication and include such things as caller Id and call waiting. Teleservices are a sort of catchall service category for various enhanced offerings, such as teleconferencing, videotext, and message handling.
3. Determine the type and number of circuits needed. B, D, and H channels are available with various options for ISDN service.

---

[20] ISP — This stands for internet service provider. This is a company providing public access to the Internet, usually through a leased line ISDN connection via a modem.

4. Order the lines and equipment. As indicated previously, there can be long lead times before the ISDN lines are installed. This means that the ISDN service needs to be planned well in advance. The network designer must also make sure that the equipment selected matches the ISDN services ordered, and vice versa, since not all equipment is compatible with ISDN service. For example, most modems do not support speeds above 28.9 Kbps. ISDN transmissions, on the other hand, begin at 56 Kbps and can exceed 1 Mbps. In addition, not all ISDN-compliant equipment supports all ISDN functions.

5. Install the equipment. The supplier of the ISDN equipment and services must be selected. ISDN service providers, including Pacific Bell, for example, are working with many of the major equipment vendors to develop value-added reseller (VAR) relationships, authorizing the VARS to sell and install ISDN lines. This will increase the number of service options available to the organization. A complaint businesses have had in the past about ISDN is that it is difficult to incorporate different products and vendors into an integrated ISDN environment. The ISDN Forum (comprised of representatives from 3COM, Ascend Communications, Lucent, and U.S. Robotics) and various standard bodies are working to alleviate this problem.

4.4.3   Advantages, disadvantages, and recommended uses

In this section we summarize major advantages, disadvantages, and recommended uses of ISDN networks and standards.

Some of the immediate benefits that can be achieved using ISDN networks include:

- Charges are based on the duration of the call and not according to a predetermined rate
- Easy to design, manage, and implement an ISDN solution
- Consolidates multiple "circuits" onto single access facility
- Improves end-to-end service quality vs. private-line (OSI layer 1), frame-switching (OSI layer 2), and packet-switching (OSI layer 3) network infrastructures

Some of the disadvantages associated with ISDN include:

- Relatively high overhead (more than 4%) compared to T-1 private lines (about 1%)
- Standards are continually evolving
- Vendor compatibility varies widely depending on ISDN function and reference point
- Uncertain overall implementation by numerous worldwide carriers and equipment providers
- Uncertain future migration path to Broadband ISDN (B-ISDN)

Recommended ISDN uses include:

- Future backbone infrastructure for public LECs, IXCs, and PTTs
- Support for a variety of end-user telecom requirements, including voice, data, text, graphics, music, video, and other communications traffic
- Telecom applications where the end user pays only for the duration of the call
- Special communications needs that require high network reliability and performance
- Large-scale internetworking requiring any-to-any "connectivity"
- Low- to high-cost networks

4.5   ATM networks

4.5.1   Overview

Until the mid-1990s, most public and private networks relied upon time division multiplexing (TDM) data transport techniques, using DS0, DS1, DS2, and DS3 circuits. Until recently, this was adequate for most voice and data applications. However, as data communications have grown and new data applications have emerged, there has been an increasing need for more speed and greater bandwidth. Current packet-switching technology is simply not adequate in speed or sophistication to handle such emerging broadband applications as multimedia, high-speed digital transports of medical images, television service, and color facsimile. The telco's/PTTs responded to this need by developing ATM (asynchronous transfer mode). The telco's/PTTs designed ATM to support a variety of tailored services for their customers, based on a new form of cell relay switching technology.

ATM transports data in small fixed-length 53-bytes cells analogous to packets. From queuing analysis, we know that segmenting data into smaller units reduces the overall transmission times in the network. The ATM cell size was designed from a queuing perspective specifically to optimize the transport of many types of voice and data applications with minimum delay. In addition, ATM makes use of advances in fiber optic and switching technology so that high bandwidth applications can be transported at very high speeds.

One of the major benefits of ATM is that it provides integrated transport on a single network for many types of data, while optimizing the *simultaneous* transmission of each data type at a guaranteed quality of service level. Like X.25 and frame relay, ATM provides multiple logical connections over a single physical interface. However, ATM can transport data on the order of several hundred megabits per second (Mbps), whereas X.25 and frame relay can only accommodate speeds up to 64 Kbps and 2 Mbps, respectively. In short, ATM combines the flexibility of packet and frame switching with the (desirable) steady transmission delays of circuit switching at new, higher levels of service. In addition, ATM is backward compatible with packet-switching, frame relay, and SMDS switching and can support these protocols on the same network.

The basic concept of ATM is analogous to an endlessly circulating set of boxcars. Although each boxcar travels synchronously with respect to the other boxcars, the rate at which the contents of the boxcars are loaded or removed is asynchronous. When a data cell is ready for transport, it is loaded into a particular slot (or boxcar). In ATM, each slot can accommodate at most one cell. Other data cells waiting for transport are queued until a slot is available. When there are no data cells awaiting transport, there may be empty slots (i.e., boxcars) circulating. The transmission rate is, in essence, the rate at which the slots, or boxcars, are moving. The queuing method ATM uses for data awaiting transport guarantees that transmission delays and QOS can be maintained consistently throughout the network

ATM is an outcome of the broadband integrated services digital network (B-ISDN) standards. B-ISDN is based on common channel signaling (CCS).[21] In the United States, CCS commonly takes the form of signaling system 7 (SS7), whereas throughout Europe and other parts of the world common channel signaling system #7 (CCSS7) is used. Since differences exist between the two CCS methods, international signaling may require conversion in each direction before B-ISDN transmission can occur. Broadband-ISDN is designed to support switched, semi-permanent, and permanent broadband connections for point-to-point and point-to-multipoint networks. Channels operating at speeds of 155 Mbps and 622 Mbps are possible under B-ISDN. B-ISDN is based on cell switching and is intended to work with ATM. B-ISDN uses SONET as the physical networking conduit for ATM transmissions.

ATM standards define three operating layers that correspond roughly to OSI layer 1 (physical) and layer 2 (data link). The ATM physical layer defines the physical means of transporting ATM cells. Although ATM cells may be transported over many different types of media, their transport is optimized by synchronous optical NETwork (SONET) fiber optic technology. SONET, a high-speed transport medium based on single-mode and multimode fiber optic technology, provides a very reliable layer one physical backbone infrastructure. Furthermore, because SONET can operate at gigabits per second (Gbps) data speeds, larger bandwidth applications are possible over ATM networks that employ SONET technology.

The second layer, the ATM layer, defines the switching and routing functions of the network. Located within the ATM five-byte header are various UNI[22] fields, which include the virtual channel identifier (VCI) and

---

[21] CCS — This is a signaling protocol developed by the ITU-T for high-speed digital networks.
[22] UNI — This stands for user network interface. This is the point at which the user connects to an ATM network.

the virtual path identifier (VPI). The VCI serves a function similar to that of an X.25 virtual circuit. The VCI field is used to define a logical connection between two ATM switches. Thus, the VCI serves as an access point to the "virtual circuit" connection established over one or more virtual paths (VPs). The VPI field, eight bits (UNI) or twelve bits (NNI)[23] in length, is used to aggregate VCIs corresponding to multiple virtual channels (VCs). When a VPI value is assigned, a virtual path is established, and when a VPI value is removed, a virtual path is terminated. As their names imply, both VPIs and VCIs are virtual and coexist within the same ATM interface, and therefore switches can act upon one or the other or on both the VPI and VCI fields. For example, in a straight-through connection, the entire VPI is mapped across the ATM input and output ports, and there is no need to process the VCI addresses. However, in other types of connections, a VCI can enter the ATM switch via one VPI and can exit by way of another VPI. Thus, a virtual circuit is fully identified using *both* the VCI and VPI fields (since two different virtual paths may have the same VCI value).

The cell loss priority (CLP) bit is used to control congestion in ATM networks. This bit indicates whether a cell should be discarded. ATM's method of congestion control is similar in some respects to frame relay's use of the discard eligible (DE) bit. ATM attempts to balance the discard of incoming data against a preestablished traffic threshold while steadily "leaking" data into the network. A "leaky bucket" algorithm is often used to estimate and control how much congestion is present in the network based on the number of cells traversing the network with the CLP bit turned on. ATM also uses operations, administration, and maintenance (OAM) cells to carry network management information between the ATM switches.

ATM's third layer, the ATM adaptation layer, provides the mechanism for handling different types of simultaneous user applications. To accommodate dissimilar transmission requirements, ATM separates the incoming data into the following categories:

- *AAL-1*: Constant bit rate (CBR) connection-oriented transport for time-sensitive-type applications including voice, full-motion video, and TDM data.
- *AAL-2*: Variable bit rate (VBR) connection-oriented mechanism for time-sensitive applications, such as compressed H.261 video, which do not require permanent connectivity.
- *AAL-3/4*: Unspecified bit rate (UBR) connectionless-oriented transport for LAN-to-LAN data such as switched multimegabit digital service (SMDS).

[23] NNI — This stands for network-network interface. This is the interface between two devices in an ATM network. There is more discussion on NNIs later in this section.

- *AAL-5*: Available bit rate (ABR) connection-oriented method for bursty data traffic such as frame relay.

These groups are processed through a single network entry point referred to as the user-to-network interface (UNI). The network-to-network interface (NNI) controls the internetworking between the various ATM switches. The private NNI (PNNI) protocol is designed specifically to support private ATM networks. ATM's traffic grouping promotes an integrated approach to network management, eliminating the need for separate local access facilities for different types of transmissions.

The AALs or ATM adaptation layers reside "logically" above the ATM 53-byte common switching fabric and below the transport data layer. The AAL functions are organized into two logical sublayers: the convergence sublayer (CS) and the segmentation and reassembly (SAR) sublayer. The CS resides logically below the transport protocols, such as TDM voice and frame relay, and above the SAR. CS components include the common part (CP) and either a service specific part (SSP) or a service specific convergence procedure (SSCP). The CS components are associated with the AALs and are used to create protocol data units (PDUs).[24] The PDUs consist of the transported data, a beginning header, an end tag, and a field length trailer. The SAR functions are responsible for segmenting (going down the OSI stack from the CS) and reassembling (going up the OSI stack from the ATM switching fabric) the payload (i.e., the data to be transmitted) of the ATM cells. The payloads are 48 bytes in length and, when coupled with a 5-byte header, make up the ATM 53-byte cell relay structure. One of the downsides of ATM, therefore, is the excessive amount of cell overhead ($5/53 \cong 9.4\%$) that it requires.

4.5.2 Design considerations and techniques

ATM is an emerging technology that integrates voice, data, video, and imaging traffic across a WAN backbone at gigabit speeds. However, the standards and implementation of ATM are not firmly established, and many issues must be resolved before ATM becomes a practical solution for most companies.

For instance, although ATM supports complex and sophisticated networking requirements, it is not supported by the major network management protocols in use today. SNMP, the most popular network management protocol in use today, was never intended to handle the volume of data needed to manage an ATM network and is wholly inadequate for this purpose. Thus, it is a major organizational challenge to migrate from an existing SNMP management model to an ATM-based model. This challenge is compounded by the fact that the ATM standards for network management are

---

[24] PDUs — A protocol data unit is an OSI term for the data to be transmitted, and the headers and trailers that are appended to the data by each OSI layer for processing purposes.

still under development. The most current efforts to develop network management standards for ATM networks are being driven by the ATM forum, a vendor consortium developing ATM specifications. To date, the ATM forum has recommended:

- A five-layer ATM management model for managing ATM networks and services
- An OAM (operations, administrative, and maintenance) layer management standard
- The interim local management interface (ILMI) standard to manage configuration, alarm, and control information for an ATM interface using SNMP over AAL.

In comparison to other alternatives, ATM service and equipment is expensive. Although on a price-performance basis ATM is very attractive, few organizations have sufficient bandwidth requirement to justify the cost of an ATM network. In the near future, it is unlikely that there will be a mass movement to adopt enterprise ATM networking. What is more likely is that organizations will employ ATM as part of a hybrid network solution that works side by side with more traditional networking options. Because of its versatility, ATM can be used in a variety of combined narrowband/broadband, public/private, and LAN-to-LAN/LAN-to-WAN/WAN-to-WAN network solutions.

The potential of ATM is best achieved when the organization as a whole is configured to support high bandwidth applications. This means that under optimal conditions the equipment generating the high bandwidth traffic should produce ATM cells. Implementing this level of ATM support on an enterprise basis has profound impacts on the fiber, equipment, and facilities that can be used in the network. Although ATM works on a variety of media, fiber optic cabling is needed to achieve its full bandwidth potential. However, one must have right of way or access to install fiber optic cabling, and this is often impractical over long distances. This means that a public ATM network solution is often the only viable solution for many types of WANs. It is also not easy to plan a comprehensive enterprise strategy to migrate to ATM given the fact that the ATM standards and products are still emerging. Vendors have developed ATM products based on their "best guess" estimates of where the market demands will lead. However, there is no guarantee that their product strategies will coincide with the organization's actual needs.

ATM is a high-performance, state-of-the-art technology that provides a common, simplified network architecture that is both scalable and flexible. These are sufficiently compelling reasons why some organizations have been early adopters of ATM technology. For example, United States government agencies (such as NASA and the Department of Energy) and various corporations (such as McDonald's, Amoco, and Chrysler) have developed ATM networks to support their multimedia and high bandwidth data applications. These organizations view ATM as a mission critical component of their overall organizational success [McDy95], [Dobr96].

There are a number of ways that ATM is being implemented in the marketplace:

- *ATM LAN:* The most common solution today for internetworking LANs it to use intelligent hubs operating on FDDI.[25] This type of solution supports communication speeds up to 100 Mbps. An ATM star-configured LAN, in contrast, supports speeds up to 155 Mbps, using standard twisted-pair copper wiring. Thus, ATM provides better support for high bandwidth applications than traditional FDDI LAN hubs. Many LAN switch-and-hub vendors are reporting record sales of ATM equipment, and this appears to be an increasingly popular network solution [Dobr96, p. 32].
- *ATM LAN-to-LAN:* It is possible to interconnect LANs into an ATM switch. ATM switches are more intelligent than LAN hubs and operate at faster speeds, allowing data to pass from one port to any other without blocking. In this solution, ATM-based routers and hubs are used to form a LAN backbone.
- *ATM private WAN:* In this solution, customer premise-based ATM switching and multiplexing devices are interconnected via leased lines. This type of network might be used in preference to a more traditional frame relay or packet-switched network to reduce the number of lines needed, to increase the number of applications supported by the network, and/or to increase network throughput. In general, ATM is not widely deployed in private WANs.
- *ATM public network:* In this solution, devices are connected into a single ATM backbone, which is, in turn, connected to a public carrier. The CPE is connected to an LEC or service provider, most typically through a microwave or fiber optic connection. Hand in hand with the selection of a public cell relay service is the selection of a provider for the access portion of the network that connects into the public ATM backbone. There are a number of access network providers, including: (1) local exchange carriers (LECs) or regional bell operating company (RBOCc), (2) dedicated LEC/RBOCX provided facilities to an IXC's POP, where the IXC's service is used, (3) or alternative access provider (AAP). A public ATM network is the easiest to implement and involves the fewest decisions of the choices listed here.

---

[25] FDDI — This stands for fiber distributed data interface. This is a LAN standard specifying LAN-to-LAN backbone transmission at 100 Mbps over fiber optic cabling.

Once the decision is made to deploy ATM technology, the network must be designed. This involves evaluating choices and requirements relating to the following:

- Design and selection of the network topology
- Identification and selection of traffic to be supported on the network
- Identification and selection of protocols to be supported by the network
- Design and selection of access network type (e.g., IEC fast packet service or dedicated line)
- Selection of public and/or private facilities (including cabling and switching needed)

ATM design tools have lagged behind the technology, making it difficult to design an optimal ATM network. One of the few design algorithms to emerge for designing ATM networks is the Mentour algorithm, developed by Dr. Robert Cahn at the IBM T. J. Watson Research Center. The Mentour algorithm is based on the Mentor algorithm. The Mentor algorithm, as presented in Section 3.2, produces treelike networks as the link speeds used in the design increase relative to the traffic requirement. This topology is too unreliable for high bandwidth ATM applications. Adaptations are made in Mentour to produce a more reliable, two-connected backbone network. Mentour is a fast computational procedure of $O(n^2 + b^k)$, where n is the number of nodes in the network and b ($3 \leq k \leq 4$) is the number of backbone sites.

Like Mentor, Mentour is a two-part algorithm. In the first part of the algorithm, Mentour determines which sites should be (lower-volume) end sites and which sites should be (higher-volume) backbone sites. The end sites are connected to the backbone nodes using a multipoint line or terminal assignment algorithm. In the second part of the algorithm, Mentor assigns links between the backbone nodes. It begins by developing an indirect routing tree to carry lower-volume traffic. Conceptually, this step is similar to the process used by Mentor. However, instead of using a CMST algorithm to develop the indirect routing tree as Mentor does, Mentour uses a heuristic furthest insertion point algorithm to solve the traveling salesman problem.[26] After the backbone network is designed, direct links are inserted to handle particularly high-volume traffic flows. As does Mentor, Mentour assumes that certain design parameters are taken as givens (i.e., a traffic threshold for determining backbone nodes, a direct link threshold, and a line slack parameter).

---

[26] Traveling salesman problem — In this problem, we attempt to find the least-cost route for a salesman who must visit each of n cities once, starting and ending at the same designated city. The number of possible tours for a traveling salesman problem with n cities is: (N-1)! For example, for N=5, (5*4*3*2*1 = 120) possible tours are possible.

Mentour algorithm Part I

The purpose of this part of the algorithm is to select backbone sites. The algorithm examines all the potential node sites and separates them into two categories: end nodes e_i and backbone nodes b_j. Mentour Part II ignores the end nodes when developing the backbone mesh topology. End nodes are connected to backbone nodes using the same techniques introduced earlier in Section 3.2.4.2 to solve the terminal assignment problem. This part of the algorithm is identical to Mentor Part I.

Mentour algorithm Part II

1. Create a traveling salesman tour for the backbone nodes using the furthest point insertion algorithm. This step ultimately creates a double-connected backbone loop for indirect traffic routing.
   - Begin by finding the two nodes i and j that are farthest apart. This can be based on known link costs or by using a standard distance formula.
   - Create a degenerate cycle between the two nodes i and j. This involves placing two parallel links between the nodes.
   - Find the node that is as far as possible from both i and j. Designate this node as x. Insert x into the loop by removing one link between i and j, and insert two new loops from i to x, and from j to x. Perform the insertion so as to minimize the link costs.
   - Continue to add nodes, one at a time, into the indirect routing tree. The node that is added at each step is the farthest node remaining from the nodes already in the loop. Stop when all the nodes have been included in the loop.
2. Assign direct links. This step begins by sorting the possible backbone links in decreasing order of link cost. Each node pair in the list is checked, starting at the top of the list, to see if the traffic between them is sufficient to justify a direct link. If the prespecified traffic threshold is not reached, route the traffic between the nodes on the indirect routing tree. If the traffic on the indirect routing tree reaches or exceeds the traffic threshold, install a new link on the indirect routing tree to handle the overflow. Conceptually, if an indirect link is loaded to the threshold level, the indirect is converted into a direct link, and a new indirect link is inserted. In this way, the indirect tree is never overloaded beyond its usable link capacity. Once a direct link is assigned, no further attempt is made to put more traffic on the link, even if the link has excess capacity. After every node on the sequence list has been evaluated, the algorithm terminates.
3. Size links to accommodate the traffic assignments made in Step 2. The algorithm is finished.

We now provide an example to illustrate the Mentour algorithm.

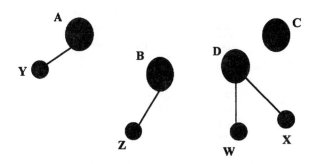

Figure 4.5 Mentour, Part 1, Solution.

Sample Mentour problem

Suppose we are given the traffic and cost data contained in Tables 4.1, 4.3, and 4.4. Assume that the threshold to qualify as a backbone node is 10 units of traffic. That is, any node with a traffic flow greater than 10 units will be designated as a backbone node, and any node with a traffic flow less than 10 units will be designated as an end node. Assume that the direct link threshold is 50 units, and the slack parameter is 40% (i.e., 60% of the available line capacity can be used). Assume that each link has a capacity of 100 traffic units.

Sample Mentour problem Part I

The algorithm examines all the potential node sites and separates them into two categories: end nodes e_i and backbone nodes b_j. The traffic flow through each node is calculated and checked against the backbone traffic threshold. The results of these calculations are shown in Table 4.2.

Using the cost data provided in Table 4.4, end nodes are linked to the nearest backbone node, as shown in Figure 4.5. The lowest value in each row of Table 4.4 represents the nearest backbone for a given end node. In the case of ties, it does not matter which backbone node is selected as the homing point for the end node. As shown in Figure 4.5, the results of this assignment are: end-nodes w and x are assigned to backbone node D, end node y is assigned to backbone node A, and end node z is assigned to backbone node B.

Sample Mentour algorithm Part II

1. Create a traveling salesman tour for the backbone nodes using the furthest point insertion algorithm. This step ultimately creates a double connected backbone loop for indirect traffic routing.
   To create the indirect routing tour, we find the two nodes that are furthest apart and connect them in a degenerate cycle. These are nodes A and C. We then find the node that is furthest from both A and C. This is node B. We then include B into the loop by replacing one (A,C) link and inserting two new links (A,B) and (B,C). The next and final node to be included in the

Table 4.1   Traffic Flows for
Sample Mentour Problem

Node pair	Traffic flow
AB	10
AC	10
AD	10
AW	1
AX	1
AY	1
AZ	1
BC	20
BD	30
BW	1
BX	1
BY	1
BZ	1
CD	20
CW	1
CX	1
CY	1
CZ	1
DW	1
DX	1
DY	1
DZ	1

Table 4.2   Backbone and End
Node Designations

Node	Total traffic flow	Node designation
A	34	Backbone
B	64	Backbone
C	54	Backbone
D	64	Backbone
W	4	End node
X	4	End node
Y	4	End node
Z	4	End node

loop is D. At this point, we have completed the design of the indirect routing tree. This is shown in Figure 4.6.

2. Assign direct links. This step begins by sorting the possible backbone links in decreasing order of link cost. Each node pair in the list is checked, starting at the top of the list, to see if the traffic between them is sufficient to justify a direct link. If the prespecified traffic threshold is not reached, route the traffic between the nodes on the indirect routing tree. If the traffic on the indirect routing tree reaches

Table 4.3   Potential Backbone
Link Costs

Backbone node pair	Link cost
AC	20
AB	10
AD	5
CD	4
BC	3
BD	2

Table 4.4   Link Costs to
Connect End Nodes to
Backbone Nodes

End nodes	Backbone nodes			
	A	B	C	D
W	5	2	9	1
X	3	4	10	2
Y	3	3	7	9
Z	6	4	6	8

or exceeds the traffic threshold, install a new link on the indirect routing tree to handle the overflow. *The results of this step are summarized in Table 4.5 and Figure 4.6.*

3. Size links. *The line capacities as given are sufficient for traffic flows. The final solution, including end nodes, is shown in Figure 4.7.*

4.5.3   Case study

As shown in the diagram for our case study, ATM networks can support various types of simultaneous applications by segregating traffic according to AAL category. A few of the many types of applications that ATM can support are listed below:

- Compressed H.261 video operating at 384 Kbps (AAL-2)
- Supercomputer 8 Mbps SMDS data with high-speed serial interface (HSSI) and data exchange interface (DXI) frame format (AAL-3/4)
- TDM voice/fax and serial data (AAL-1)
- Frame Relay "routed" data (AAL-5)

In our example, customer A communicates with site 1 and site 2 through an ATM UNI logical interface that runs over a SONET OC-n physical fiber optic local access loop. For illustrative purposes, customers A and B can communicate with other sites, but only through the appropriate backbone technology. For instance, customer A/site 2 can communicate with customer

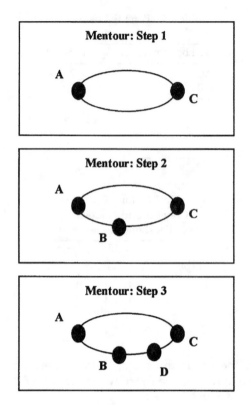

Figure 4.6   Indirect routing backbone created by Mentour.

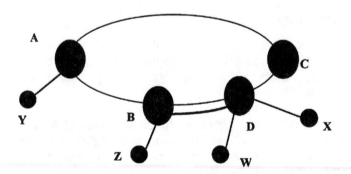

Figure 4.7   Mentour, Final Solution.

A/site 3 to exchange data relating to AAL-1 voice and fax TDM applications. Customer A/site 2 can also communicate with customer A/site 4 to exchange AAL-3/4 SMDS traffic, and with customer A/site 5 to exchange data handled by a frame relay router. The same connectivity principles hold for customer

Table 4.5 Link Sequencing and Assignment for Sample Mentour Problem

Node pair	Traffic flow	Routing
AC	10	Indirect (via AC)
AB	10	Indirect (via AB)
AD	10	Indirect (via AB to BD)
CD	20	Indirect (via CD)
BC	20	Indirect (via BD to DC)
BD	30	Direct

B applications. Note that customer C is connected through a switched multimegabit data service (SMDS) backbone. Using SMDS's E.164 public addressing scheme, customers A and B can communicate with customer C with the appropriate provisioning (see Section 3.2.7.5 for more information on SMDS and its universal addressing scheme). SMDS is a connectionless, oriented technology capable of transporting very high bandwidth data applications, such as medical imaging files, at formidable speeds. With this type of connectivity, hospitals, medical complexes, or insurance companies can share access to a variety of data files (such as magnetic resonance imaging (MRI) data files, etc.)

Since ATM can interface with and encapsulate other types of data protocols, as shown in Figure 4.8, it is theoretically possible to interconnect several backbone networks. To accomplish this degree of interconnectivity in practice is somewhat challenging. It requires, at a minimum, separate UNI interfaces for each corresponding network technology. In addition, addressing and congestion control mechanisms need to be established that can handle both ATM and non-ATM interconnections. Despite these obstacles, several major service providers and customers plan to migrate their separate, distinct network backbones onto a common SONET core backbone using ATM technology.

4.5.4 Advantages, disadvantages, and recommended uses of ATM networks

In this section we summarize major advantages, disadvantages, and recommended uses of ATM networks.

The advantages of ATM networks include:

- Good for integrating different types of user traffic
- Allows resource and application sharing
- Improved service availability and network bandwidth (particularly when using a SONET infrastructure)
- Supports high-volume and high-speed internetworking

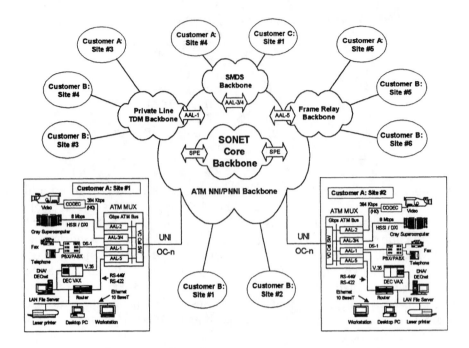

Figure 4.8   ATM network design (Source: B. Piliouras).

Some of the disadvantages associated with ATM networks include:

- ATM standards are still emerging and evolving
- Congestion control mechanisms not fully developed
- Excessive overhead (ranging from 9% to 20%) required by the network
- Requires new hardware platforms
- High cost
- Does not support wireless or broadcast communications

ATM networks are recommended for:

- Integrated voice/fax, video, and data applications
- Large-scale internetworking backbones
- Single point for network management access
- High-speed bandwidth requirements, both bursty and continuous
- B-ISDN transport mechanism for SONET-based core backbones
- Medium- to high-cost networks

4.6   SONET networks

4.6.1   Overview

Synchronous optical network (SONET) is an international standard developed by the regional Bell holding companies (RBHCs), Bell communications research (Bellcore), MCI, and various standards bodies. SONET formally became an ANSI (American National Standards Institute) standard — as ANSI T1.105 — in June 1988. Prior to the development of SONET, there had been a proliferation of proprietary fiber optic equipment that was largely incompatible across vendors. SONET was designed to establish a common standard for building fiber optical networks that would interface with existing electrical protocols and asynchronous communication devices.

Today, many service providers are installing fiber optic backbone networks based on SONET transport technology. New bandwidth intensive services — like high definition TV, high-speed computer links, local area network (LAN) links, and broadband ISDN — are being developed and are fueling the need for large-capacity SONET backbone networks. As broadband services have become more readily available and telecommunications requirements have become more time sensitive, the SONET backbone providers have continued to develop and implement "survivability" techniques to minimize the possibility of network downtime. The result is that SONET networks are extremely fault tolerant and reliable, and can recover from even catastrophic fiber cuts in a matter of seconds or less (the average recovery time in a SONET network for this type of fault is 50 milliseconds.)

SONET is based on a worldwide standard for fiber optic transmission, modified for North American asynchronous rates. In the rest of the world, this technology is known as synchronous digital hierarchy (SDH). Standards that have had a major impact on SONET development are summarized in Table 4.5.

The United States SONET standards were modified to permit internetworking with the networks of other countries. These modifications were incorporated into the CCITT standards covering SDH (synchronous digital hierarchy), which is the international equivalent of SONET. These standards, designated CCITT 707-709, define the operation of network elements down to the line signal formats, interfaces, security features, and signal structures for transmission speeds from 51.84Mbps (OC-1) to 9.95 Gbps (OC-192) and higher. SONET is the first international transmission standard that eliminates the differences between American, European, and Japanese multiplexing schemes. Table 4.6 provides a summary of the SONET signaling categories.

The lowest-level SONET signal is called the synchronous transport signal level 1 (STS-1). This is the base signal used for SONET multiplexing. The STS-1 has a signal rate of 51.84Mbps. The optical equivalent of the STS-1 is the optical carrier level 1 signal (OC-1), which is obtained by a direct electrical to

Table 4.6 SONET Standards

Organization	Standard	Description
Bellcore	TR-253	Generic requirements
Bellcore	TR-233	Wideband and broadband digital cross-connect systems (DCS)*
Bellcore	TR-303	Loop carriers
Bellcore	TR-496	Add/drop multiplexing (ADM)
Bellcore	TR-499	Transport system generic requirements
Bellcore	TR-782	SONET digital trunk interface
Bellcore	TR-917	SONET regenerator equipment
ANSI	T1.195xx	Rates, formats, jitter, etc.
ANSI	T1.106	Optical interface
ANSI	T1.119	OAM&P communications
ANSI	T1.204	Operations, administration, provision
ANSI	T1.231	In-Service performance monitoring

* DCS — Digital cross-connect systems are commonly referred to as DCSs, DXCs, and DACs, depending on the origin of the switch and the technology employed. For our purposes, we assume that all these acronyms relate to the same switching function. A digital cross-connect system electronically maps a DS0 (64 Kbps) channel into a DS1 (1.544 Mbps) line, allowing each DS0 channel to be routed and configured into different DS1 lines.

Table 4.7 SONET Signal Categories

Synchronous transport signal	Optical carrier (OC-n)	Line rate (Mbps)
STS-1	OC-1	51.84
STS-3	OC-3	155.52
STS-12	OC-12	622.08
STS-24	OC-24	1,244.16
STS-48	OC-48	2,488.32
STS-192	OC-192	9,953.28

optical conversion of the STS-1 signal. The STS-1 frame consists of 90 rows of bytes. The first three columns of data are dedicated to network management functions, while the remaining 87 columns and 9 rows are used to carry the data, i.e., the synchronous payload envelope (SPE). The STS-1 frame consists of 810 bytes, and is transmitted at the rate of one frame per 125 microseconds. The section overhead is made up of a block of 9 bytes (3 rows by 3 columns) that is transmitted 8000 times a second for a total transfer rate of 576 Kbps. The section overhead is used to perform various functions between section terminating equipment (STEs) — e.g., a SONET regenerator or other network element — including framing, error monitoring, and STS identification. Similarly, the line overhead is a block of 6 rows by 3 columns (18 bytes) transmitted 8000 times a second for a 1.152 Mbps transfer rate. The line overhead is used to indicate when a line between two LTEs (line terminating equipment) has gone bad. Line overhead data is also used to perform such functions as synchronization, multiplexing, and automatic protection switching. The path overhead is computed in a similar manner, and is used to support and maintain transport of the STS-1 synchronous payload envelope (SPE) between the terminating network elements, and to perform end-to-end error and connectivity checking. The section, path, and line overhead perform functions analogous to the functions of the section, path, and line overhead segments of the public telephone network. Layering the network functions allows different types of equipment to be built to perform specifically tailored functions.

The SONET STS-1 synchronous payload envelope (SPE) has a channel capacity of 50.122 Mbps and has been designed specifically to provide transport for a lower-speed DS3 tributary signal. Transport for a tributary signal with a signal rate lower than a DS3, such as a DS1, is provided by a virtual tributary (VT) frame structure. VTs support the transport and switching of payload capacities that are less than those provided by STS-1 SPE. By design, the VT frame structure fits neatly into the STS-1 SPE to simplify the VT multiplexing requirements. A range of different VT sizes is provided by SONET, including:

VT1.5. Each VT1.5 frame consists of 27 bytes, structured as 3 columns of 9 bytes. At a rate of 8000 frames per second, these bytes provide a transport capacity of 1.728 Mbps and will accommodate the mapping of a 1.544 Mbps DS1 signal. A total of 28 VT1.5 signals may be multiplexed into the STS-1 SPE.

VT2. Each VT2 frame consists of 36 bytes, structured as 4 columns of 9 bytes. At a rate of 8000 frames per second, these bytes provide a transport capacity of 2.304 Mbps and will accommodate the mapping of a 2.048 Mbps CEPT E1 signal. A total of 21 VT2 signals may be multiplexed into the STS-1 SPE.

VT3. Each V3 frame consists of 54 bytes, structured as 6 columns of 9 bytes. At a rate of 8000 frames per second, these bytes provide a transport capacity of 3.456 Mbps and will accommodate the mapping of a DS1C signal. A total of 14 VT3 signals may be multiplexed into the STS-1 SPE.

VT6. Each VT6 frame consists of 108 bytes, structured as 12 columns of 9 bytes. At a rate of 8000 frames per second, these bytes provide a transport capacity of 6.912 Mbps and will accommodate the mapping of a DS2 signal. A total of 7 VT3 signals may be multiplexed into the STS-1 SPE.

The concept of transporting tributary signals intact across a synchronous network has resulted in the term synchronous transport frame (STF) being applied to such synchronous signal structures. A synchronous transport frame (51.84 Mbps) is comprised of two distinct parts: the synchronous payload envelope and the transport overhead.

Synchronous payload envelope (SPE) — 50.122 Mbps (includes 576 Kbps PATH overhead). This consists of all the user data, including the PATH overhead bytes. Individual tributary signals (such as a DS3 signal) are arranged within the SPE, which is designed to transverse the network end-to-end. This signal is assembled and disassembled only once even though it may be transferred from one transport system to another many times on its route through the network.

PATH overhead — This is the overhead contained within the SPE. It allows network performance to be maintained from a customer service end-to-end perspective (576 Kbps channel). Path overhead, which contains error detection, tracing, path status, and connection establishment, is terminated by STS/OC-N end equipment.

LINE overhead — This allows the network performance to be maintained between transport nodes and is used for most of the network management reporting (1.152 Mbps channel). Line overhead, which contains error detection, pointer justification, automatic protection information (rings), alarms, and provisioning commands, is terminated by all SONET equipment capable of OC-N/STS level switching and termination.

SECTION overhead — This is used to maintain the network performance between the line generators or between a line regenerator and a SONET network element (NE) and provide fault localization (576 Kbps channel). Section overhead, which contains error detection, framing, STS identification, and other information, is usually terminated by SONET equipment that converts the optical signal into an electrical signal.

SONET lightwave terminating equipment (LTE) accepts standard optical signals and converts them into lower-level electrical signals. The electrical signal is then demultiplexed to the component STS-1s, where signal processing occurs. Signal processing may consist of checking the overhead payload for the purpose of signal performance testing, distance alarm monitoring, signal switching, and other related tasks. Once processed, STS-1 signals may be routed to various electrical or optical tributaries depending on the configuration of the LTE.

The SONET LTE is used to provide elementary connectivity between synchronous and asynchronous transmission equipment. In the most basic form, the SONET LTE will mimic many currently deployed asynchronous LTEs. However, when SONET support systems are integrated within the network, SONET LTEs offer added features and improved performance in signal performance, advanced monitoring, broadband transport capabilities (OC-N), improved cable management (optical vs. coaxial), and a more compact cable and equipment footprint.

The basic building block of a SONET network is the add/drop multiplexer (ADM). The "synchronous" functionality provided by ADM is one of the driving forces behind the development of SONET technology. The ADM offers many features, including, but not limited to, reduced cost, increased network reliability, automated ring protection, full use of SONET overhead, performance monitoring, and automated provisioning. The ADM allows selected payload channels (i.e., DS3s, STS-1s, DS1s, and other rates as allowed by vendor) to be "added" or "dropped" at a given site while allowing the rest of the signal to pass on to the next site. Due to the synchronous nature of SONET (e.g., it does not introduce timing-induced "stuff bits" as does asynchronous transport), these payload channels are mapped to higher rate signals at specific locations in the payload. In this way, the channel's exact location is always known and can be read out, and/or written over without demultiplexing the entire higher rate signal. The ability to locate specific channels within the data stream is the basis for synchronous transmission. With the deployment of ADMs, the back-to-back LTEs and manual DSX-3 patching procedure in the asynchronous environment is greatly reduced or eliminated.

The add, drop, and throughput capabilities of the ADM are controlled by a time slot assignment (TSA) feature within the ADM that allows local and remote traffic provisioning. For example, a DS3 signal coming into the ADM from the west on time slot 4 may be dropped to time slot 2 at the ADM drop site. Similarly, a DS3 signal may be added at the ADM drop site to time slot 4 and be routed to the east on time slot 7. Thus, the ADM offers flexibility in provisioning the traffic. In addition, the ADM's TSA feature increases bandwidth utilization and signal performance as compared to conventional asynchronous technology.

ADMs provide the IECs (interexchange carriers) and LECs (local exchange carriers) a great deal of flexibility in completing SONET midspan meets. Although midspan meets are not new to these backbone providers, SONET offers capabilities that have never before been possible. Some of these capabilities include:

- Differentiation of vendors
- Elimination of electrical demarcation/cross-connect points
- Improved signal performance

Ideally, ADMs should be used at drop and reinsert sites (DREI), and at other facilities that require small amounts of traffic grooming and add/drop functionality. ADM or DREI facilities provide interconnects with LECs, CAPs, and other IECs while creating a seamless network. Additionally, ADMs can be used in metropolitan and possibly regional rings where SONET capabilities and self-healing systems are desired.

The broadband DXC (BBDXC) is used to terminate synchronous optical, synchronous electrical, and asynchronous electrical signals. With this capability, the BBDXC can serve as a gateway device between synchronous and asynchronous transmission systems. A BBDXC performs six primary functions:

- Signal termination
- Grooming
- STS-1 cross-connection
- Access to asynchronous network
- Performance monitoring
- Restoration

The BBDXC is used primarily at major switch and relatively large fiber junction sites, where, in addition to restoration functions, it performs OC-12 grooming functions. A BBDXC can consolidate STS-1s at OC-12 end points, creating express trunks with high utilization rates.

Wideband DXCs (WBDXCs) have the capability of interfacing with BBDXCs at synchronous optical levels OC-3 or OC-12. The WBDXC performs the following four primary functions:

- Signal termination
- Grooming
- VT cross-connection
- Performance monitoring

The WBDXC is used primarily at switch sites, relatively large fiber junction sites with large amounts of terminating traffic, and other sites with moderate to large amounts of access/egress traffic. The WBDXC performs asynchronous to synchronous signal conversion, grooming for local express trunks, and performance monitoring of DS1 and VT level signals.

The narrowband DXC (NBDXC) — which is essentially the same as the DXC 1/0 in service today — allows SONET equipment to be interfaced at synchronous optical levels OC-3 or OC-12. The NBDXC performs four primary functions:

- Signal termination
- Grooming
- DS0 cross-connection
- Performance monitoring

The NBDXC is used primarily at switch sites, relatively large fiber junction sites with large amounts of terminating traffic, and other sites with moderate to large amounts of access and egress traffic.

OC-192 is a relatively new transmission rate equating to 9.95328 Gbps, supporting 192 STS-1s. This is four times the capacity of an OC-48. This very high-speed lightwave system provides a means to alleviate fiber constraints and congestion on network backbone routes. The use of OC-192 will better position SONET backbone providers to provide OC-N trunks for ATM-based services and to accommodate unforeseen high bandwidth traffic. It will also significantly reduce the transmission costs on a DS3 mile basis, freeing capital dollars for other projects. To accommodate OC-192 LTE equipment and increasing traffic loads, there will be a need for OC-48 express pipes. Most of these would be local in scope and would be implemented using the same guidelines used when creating OC-12 trunk express pipes for OC-48 links.

SONET technology is being developed primarily for use on fiber transmission routes, but digital radios benefit from SONET as well. Currently, many manufacturers are developing SONET digital radio products. Some manufacturers have proposed upgrades to their existing asynchronous radio systems to ease the transition into the synchronous platform. A high percentage of synchronous radio manufacturers have focused their efforts on capturing the European market, which is based on synchronous digital hierarchy (SDH) standards, which are the European version of SONET.

There is a need to accommodate SONET radios in the IEC network for a number of reasons, which include:

- Extending fiber routes into smaller markets often cannot be economically justified.
- Customers may require OC-n services at locations served only by digital radios.
- Some digital radio routes serve as backups for fibers and cannot perform this function without SONET capability.

The main options for extending SONET transmission to access areas of the network currently served via digital radio are to:

- Overbuild digital radios with SONET radios to extend the requirements for SONET bandwidth transmission
- Build a new or diverse fiber route to these sites
- Lease SONET transmission from an alternate carrier

Although the second and third options listed above may prove technically feasible in some areas, they frequently are not economically feasible. The reader should note that SONET radios may be more susceptible to failure than fiber routes. A "burst error" hit to the payload pointer bits can cause immediate payload loss. Since radios can be prone to multipath fading and dispersing conditions, which manifest themselves as burst errors, SONET radio manufacturers have been working on different schemes to alleviate this problem. One solution to the pointer vulnerability is to use a forward error correction (FEC) technique. The higher speed rates necessary for SONET transport and the FEC feature creates a need for modulation formats above 64 QAM. Some vendors are offering 128, 256, and 512 QAM schemes. It is expected that any tributary interface to a synchronous radio must be done at the OC-3 optical level.

Within the last few years, there has been an increase in the use of dispersion compensator technology in the SONET industry. Dispersion compensators are devices used on the line side of SONET LTEs to compensate for the inherent dispersion characteristics of fiber optic media. Wave division multiplexing (WDM) offers an alternative method to increase the transmission capacity of an installed fiber path by utilizing the other regions of the optical spectrum resident on each fiber. This process multiplexes optical signals from two or more transmission systems operating at different wavelengths so that they may be transmitted simultaneously over a single fiber pair. When used on single mode fiber, WDM allows multiple optical signals to be transmitted in the same direction (unidirectional) or in opposite directions (bidirectional).

Optical amplifiers boost the optical signal emerging from the line side of the LTE and allow it to travel farther before requiring regeneration. There are currently several types of optical amplifiers. Transmit and receive amplifiers have increased the spans between repeaters from about 75Km to more than 140Km (i.e., from about 45 miles to over 85 miles). Line amplifiers allow spans in excess of 500Km (300 miles) or more. Optical amplifiers operate at the optical level and are completely independent of bit rates, wavelengths, or transmission protocols, and are able to handle rates up to OC-192 and more. They eliminate the need for entire regeneration sites. This significantly reduces operations, maintenance, administration, and provisioning costs.

An optical cross-connect, which is also known as a *remotely controlled optical patch panel*, performs the same functions of a BBDXC. Optical cross-connect devices have been used to provide broadcasting services over multimode fiber transmission and to restore the network in the event of a catastrophic fiber failure. Optical cross-connect technology is still in the early stages of development; however, single-mode OPT-X devices are being deployed with single-mode fiber transmission systems in some SONET networks.

Many of the technical hurdles in fiber optics (e.g., cross talk, alignment of cross-connections, and "directivity") have been overcome, but not by all manufacturers. Many SONET backbone providers today are considering the use of an OPT-X as a vehicle for network disaster recovery, network reconfiguration, facility testing, network management, and new service offerings. The unit is completely photonic. This means that no electrical to optical signal conversion is required. It contains an MxN matrix, and it cross-connects incoming light at any one of the M ports on side A, to any of N ports on side B. This type of device does not affect the direction of the photon beam.

4.6.2   Design considerations

The survivability of telecommunications networks is becoming an increasingly important issue. In response, telecommunication service providers are building more "robustness" into their networks. Survivability is defined as the ability of a network to maintain or restore an acceptable level of performance during failure conditions using various means of network restoration. Currently, network survivability is provided by circuit diversity, equipment protection switching, and automated restoration techniques using DXC and control systems. The SONET architecture is designed specifically for survivability using self-healing ring techniques. These self-healing techniques are based on either DXC-based restoration or SHR (self-healing ring) restoration. The SHR techniques provide an alternative traffic path in the event of a fiber optic link failure, restoring service on average within 50 msecs (and 2.5 seconds in the worst case). The restoration happens so quickly that most communications functions will not be affected by the line failure and end users may not even know that there was a problem that has been corrected.

A spur site is defined as any site where there is only one network transmission route to the fiber or radio backbone. When a fiber outage occurs on a single threaded spur route, several sites can be affected, impacting a large number of users. Repair times in hours are not acceptable to many companies today because their networks play an increasingly important role in their day-to-day operations. "Spur enhancement" is the terminology used to describe an alternate transmission facility at a single threaded site. Fiber is the most reliable alternate transmission facility for spur enhancement.

Having multiple backbone sites for spur enhancement is the most reliable means to reduce the risk associated with node outages. However, it is often prohibitively expensive to use multiple backbone sites. Therefore, SONET providers usually use a single backbone site. When a single backbone site is used, the necessary precautions should be taken at the site to ensure that the alternate route equipment is located separately from the existing equipment (i.e., by locating equipment at opposite ends of the building using separate power distribution frames [PDFs], installing uninterruptable power supplies [UPS], etc.).

Since ADMs can transmit and receive traffic from two different directions simultaneously, they are ideally suited to supporting a self-healing ring topology. Two methods have evolved for implementing self-healing rings (SHRs) within the ADM: unidirectional (USHR) and bidirectional (BSHR) rings. The type of SHR used in any given type of network depends on the normal traffic routing pattern. Unidirectional SHRs are usually configured with one fiber pair. One fiber is the working channel and the other is the protection channel. Under normal conditions, all working traffic on the ring is transferred to one fiber in the same direction around the ring. Although the second fiber is reserved for protection, this does not mean it is idle until a failure occurs. The USHR uses a 1+1, or "dual-fed," configuration. This means that the working traffic is also bridged to the protection fiber and is routed in the opposite direction. So each fiber is carrying a copy of the working traffic, but in different fibers and opposite directions of transport. BSHRs can be implemented using two or four fibers. In a two-fiber BSHR, traffic originating at A and destined for B is transmitted to B on one fiber, and the responding traffic from B destined for A is returned via the other fiber in the opposite direction. In each ring type, half the channels are defined as "working," and the other half as "protection." Therefore, the working throughput of the ring is actually OC-N/2. For instance, a BSHR/2 ADM operating at OC-48 has 24 working and 24 protection STS-1 time slots on both the transmitting and receiving fibers. However, the protection time slots can be reassigned when not needed.

Protection switching uses one of two SONET overhead levels: the line or the path. A typical example of line protection occurs when normal traffic is disrupted on a working fiber and all the traffic is switched to the protection fiber. The communication involved in making this switch is coordinated using the K1 and K2 bytes of the SONET line overhead. All or none of the working traffic is switched to the protection channel (as defined in ANSI

T1.105). Path protection switching uses the path layer of the SONET over-head and avoids the complexity of the line communication process. Using path protection switching, a node can switch individual STS-1s or VTs in and out of the payload based on the path layer indications. In general, ANSI does not deal explicitly with path switching requirements or standards. SHRs generally implement protection switching in different ways. USHRs favor path switching, whereas BSHRs favor line switching. A USHR can, in fact, use line switching, but this configuration may create problems due to SS7 transport constraints. The USHR is usually implemented using path switch-ing as described in Bellcore document TR-496. USHR with path protection switching may be designated in several ways. Unidirectional path switched ring (UPSR) and unidirectional self-healing ring-path switching (USHR/P) are two examples and are essentially equivalent designations.

BSHRs can support path or line switching. While BSHR path switching is relatively new, several vendors are currently demonstrating product lines that offer it. On the other hand, line switching, which uses the line layer of the SONET overhead to initiate switching action, has been modified in ANSI T1.105 to support BSHRs. This document assumes that all BSHRs, both two and four fiber, use line switching. Line switching is further divided into ring and span switching. BSHR/2s supports only ring switching. When a ring switch occurs, the time slots that carry the working channels are switched to the corresponding empty time slots in the protection channels traveling in the opposite direction.

As illustrated in the SONET core network design (Figure 4.9), BSHR/4s support ring switching. This is similar to the ring switching already dis-cussed, except that all the time slots are placed on the protection fiber and transmitted in the opposite direction with the span switching. Span switch-ing occurs almost exactly as it does in point-to-point 1:1 systems. In the event of a working fiber failure, the traffic is switched to the protection fiber. BSHR/4 systems support several span switches in the same ring, providing full protection from multiple failures.

The unidirectional path switched and bidirectional line switched rings have subtle differences. These differences are primarily due to the adaptation of the bidirectional rings to a point-to-point, line layer-based protection, as specified in ANSI T1.105. Freedom from the line layer affords USHR/P rings several advantages:

Vendor independence: Each node of a USHR/P can be implemented with equipment from different vendors. Since vendors implementing BSHR/L use proprietary solutions to overcome protection-switching shortcomings at the line layer, each node must be from the same vendor.

Technology integration: USHR/Ps can be designed to include asynchro-nous and synchronous components.

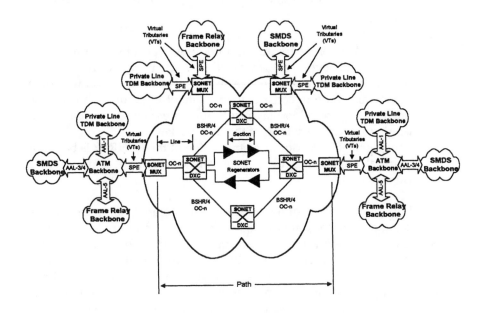

Figure 4.9 SONET network design (Source: B. Piliouras).

Line speed independence: The link speed between nodes of a BSHR/L must be at the same speed. USHR/P rings, however, can employ different speeds link-by-link, node-by-node.

Number of nodes: BSHR/L rings are typically limited to 16 to 20 nodes whereas the number of nodes in a USHR/P ring are theoretically unlimited. Some vendors claim USHR/P rings in excess of 100 nodes.

Better services: Since USHR/P rings switch at the path layer endpoints, switching of all the tributaries on the line due to a single failed tributary is avoided. For example, if the ring is an OC-12 and one STS-1 has failed, the entire OC-12 might appear bad, resulting in a line switch. Therefore, a "hit" on adjacent clean tributaries also occurs. Path switching substitutes the good STS-1 on the protection channel corresponding to the failed STS-1 on the working channel at the endpoint of the circuit. Therefore, end-to-end service performance can be provided to the customer.

However, USHR/P rings also have disadvantages, including:

Bandwidth utilization: Once bandwidth on the USHR/P ring is assigned, it cannot be reused. For example, if using an OC-12 USHR/P with

four nodes (i.e., A through D) and nine STS-1s are allocated from A
to B, there are only three STS-1 time slots available for all additional
traffic on the ring.

Difference in path lengths: Different physical lengths between the working and protection fibers may result in time-outs of some data applications in the event of a protection switch. Not only do some
applications need to see a response within a certain amount of time,
but they also require consistent amounts of time for each response.
Switching to protection may result in a longer or shorter path, thus
changing the delivery time of the data packet.

Short range design: Vendor implementation of the USHR/P is based on
a "virtual ring" concept and is usually installed on a customer-by-
customer basis.

In addition to or in lieu of the transmission level restoration techniques,
there is also a need to look at how to restore application level functionality,
in the event of a network failure. The application level, or service layer, is
defined as the individual network that provides the specific voice and data
services, such as asynchronous transfer mode (ATM). By using a grid network and employing various application diversity and restoration techniques, a single fiber cut should not adversely impact the network. Application restoration and survivability should also incorporate equipment
diversity. Specific restoration strategies for each service type/application
should be developed, reflecting priorities that have been established for
restoring each level of service in the event of a network failure.

Self-healing metro rings are used to provide survivable trunks. The
optimal connection to the ring should occur at two separate locations to
protect against local outages. ADMs using TSA functionality to switch traffic
in the event of an outage provide an effective means of restoration. The
decision to implement any or all of these restoration and survivability methods must be based on financial and business criteria.

Perhaps the most important criterion for determining which type of ring
to use is the distribution of the traffic between nodes on the ring. If the traffic
on all the nodes is destined for a single or centralized node on the ring and
the traffic from each node is nearly equal, the path USHR/P is the most
suitable method. If there is a high concentration of traffic between nodes and
the traffic is fairly heavy, the BSHR/L is the preferred choice.

Photonic-based restoration is available in two basic methods. Both methods require photonic elements at the node or junction sites. One method,
which is based on the use of optical switches, is called the optical domain
restoration (ODR) method. The second method uses optical cross-connects
(OPX-T). Both methods provide vehicles to maintain and restore an acceptable level of performance during network failures, particularly in the case
of catastrophic fiber cable cuts. These methods define a simplified network
architecture called photonic domains. This architecture breaks the network

into small subnetworks to help simplify the network management and restoration mechanism. Both methods require interoperability between the system and physical restoration layers. In either case, these methods need to use BBDXCs to efficiently perform the circuit level grooming needed for optimal interconnection of OC-12 logical facilities. The optical domain architecture applies to small independent subnetworks or optical domains (photonic domains). Therefore, the complexity of network management and restoration is reduced. ODR provides protected optical access into a specific domain and makes use of optical switches.

The optical cross-connection restoration architecture functions like end point restoration and parallels the methods used for DXC-based real-time restoration (RTR). It can be used in small independent or dependent subnetworks or photonic domains. Optical cross-connects use the restoration capacity available in the network to provide alternate routes for failed circuits. These restoration capacities are determined on either a dynamic or preplanned basis. The optical cross-connect is used as a passive cross-connect to perform network restoration at the optical level or photonic domain without converting signals to an electrical rate. The OPT-X is used in meshed networks with evenly distributed traffic. If a fiber outage takes place, the disabled section is optically switched within the domain to the planned restoration capacity. Restoration over multiple photonic domains requires more complex restoration control systems and increases the time to restore a failed link. Network restoration using optical cross-connects is applicable to single grid or multiple grid domains.

The most challenging aspects relating to the use of OPT-Xs are the network management control functions that are needed to restore failed sections of the network. Photonic-based restoration capacity is preplanned and is determined during the provisioning process. Fault correlation is another major network management process and can be specified independently of the network topology based on a business, a service, or a network element orientation. Restoration methods can be centralized or distributed. Both types of methods require a mediation device to generate alarms indicating catastrophic fiber cut failures via the detection of AIS, LOS, or BER threshold crossings. Another possibility is to use the K1/K2 bytes defined by ANSI to identify catastrophic failures.

Centralized restoration can be used in individual photonic domains, while distributed restoration can be used for restoration across domain boundaries. In a centralized restoration, an operations system (OS) controls the rerouting of traffic around the failure. Control of the restoration process is centered at the OS workstation(s). In distributed restoration systems, the photonic device controls rerouting of traffic. An algorithm controlling the restoration is programmed into each optical cross-connect. At the time of a failure, the optical cross-connects distribute control exchange messages (via signaling control channels) and coordinate activities of each device to reroute traffic. Control of the process is shared by the optical cross-connects in the network. This is essentially a sequential control process. Therefore, a sufficient amount of restoration capacity must be designed into the network for the distributed algorithms to work effectively in restoring traffic flows. This method increases the complexity of the optical cross-connect requirements and operations.

Almost all distributed algorithms use the restoration capacity available in the network to provide alternate routes for the failed circuits. When a dynamic algorithm is used, it resides in the cross-connect's system controller. In contrast, a preplanned algorithm "knows" about internal routes via "hooks" (indirect and indirect offset addressing) to preestablished connection lookup tables. A preplanned approach usually reduces the algorithm's execution time, but it increases the complexity of the network provisioning and requirements planning process.

Analysts within the industry are discussing design techniques for transmission systems operating at 40 Gbps and higher, along with plans for OC-768 solutions. With the advent of new SONET technologies such as wave division multiplexing (WDM), dispersion-shifted fiber (DSF), optical amplifiers, and optical cross-connects (OPT-X), these very high network speeds are becoming less of a future vision and more of an everyday reality.

4.6.3   Advantages, disadvantages, and recommended uses of SONET

In this section we summarize major advantages, disadvantages, and recommended uses of SONET technology.

Some advantages of SONET networks include:

- Fiber optic transmission systems (FOTS) equipment interoperability
- Operation support systems (OSS) network management capabilities
- Simpler mid-span meet handover
- Support for future high bandwidth services
- High network survivability and available bandwidth
- Excellent core backbone technology for integrating multiple network infrastructures
- Rapid worldwide growth for both SONET and SDH
- Support for very-high-volume and very-high-speed internetworking

The disadvantages of SONET include:

- Requires expensive initial outlay of capital
- Relatively high overhead (more than 4%) compared to T-1 1.544 Mbps (about 1%)
- Difference in path lengths between working and protection fibers may result in time-outs of some data applications in the event of a protection switch
- Standards are still emerging and evolving
- Requires new hardware platform

SONET networks are recommended for:

- Backbone infrastructure for LECs, IECs, PTTs, and large corporate users
- Bandwidth-intensive services like broadband ISDN, high-speed computer links, local area network (LAN) links, full-motion video, and high-definition TV (HDTV)
- Telecommunications requirements that necessitate "survivability" techniques to minimize network downtime
- Multiple, integrated, large-volume voice/fax, video, and data applications
- Very large scale internetworking backbones
- Support for SDN and B-ISDN protocols
- Medium-high to very-high cost networks

Bibliography

[Data97] DataPro, Switched Multimegabit Data Service (SMDS), *DataPro*, McGraw-Hill, Inc. March 1997.

[Dint94] Dintzis, M., X.25 packet and frame relay switches, *Datapro*, McGraw-Hill, March 1994.

[Dobr96] Dobrowski, G., and Humphrey, M., ATM and its critics: separating fact from fiction, *Telecommunications*, November 1996, pp. 31-37.

[Tayl96] Taylor, M., ATM: building an architecture for multiservice switched networks, *Telecommunications*, November 1996, pp. 43-56.

[McDy95] McDysan, D., and Spohn, D., Future directions of ATM, *ATM theory and application*, McGraw-Hill, 1994, pp. 579-595.

[Skva96] Skvarla, C., ISDN services in the U.S.: overview, *Datapro*, McGraw-Hill, April 1996.

[Summ95] Summers, C., N-ISDN, B-ISDN, and auxiliary data protocols, *ISDN Implementer's Guide: Standards, Protocols, and Services*, McGraw-Hill, 1995.

chapter five

Local area network design and planning

Contents

5.1   Management overview of LAN design and planning

Until recently, designing, planning, and implementing LANs have been, to a large extent, an intuitive process. The purchase price of LAN equipment and bandwidth has been low in comparison with modeling and design packages and the time analysts spent using the tools. At most, money has been spent determining baseline performance measures. Connectivity and compatibility issues were much more highly prioritized than properly sizing the servers and properly determining the bandwidth required for satisfactory performance. However, local area networks and the loads they carry are increasing, topologies are getting more complex, and performance is being evaluated more critically.

LAN visualization, asset management, planning, and design tools are a fairly new product area. There are many wide area network design tools on the market, but few address the LAN design marketplace. Outside of the design guidelines provided by vendors — that are usually not available to users — there are only a handful of tools that can be purchased for in-house planning and ongoing network management. The present obstacles of using such tools are:

- There is no budget for such tools.
- They require a computer that may not be part of the current installation.
- Performance information is required that may not be available.
- They require monitoring instruments that may not be part of the installation.
- Know-how to implement and use such tools may not be available.
- Very rapid changes in the technology make frequent updates necessary.
- There are separate products for the logical and physical LAN design.
- There is a lack of integration between monitoring and modeling tools.

When evaluating LAN products, the following criteria should be considered [TERP96]:

- System requirements: The processing power should be enough to meet the needs of simulation techniques, which can be very extensive for large segments or interconnected LANs. At the very least, an EGA display is generally required for PC-based systems. However, some of the design systems are based on more powerful platforms, such as SUN Microsystems workstations.
- Input data requirements: What data the user should provide prior to starting "what-if" evaluation. Input parameters may be grouped by LAN segment and interconnecting parameters. Modeling parameters may be classified into two groups:

  1. LAN segment parameters
  2. Internetwork parameters

  LAN segments parameters are:

    - Sizes — average packet and measure sizes
    - Protocols — lower, middle, and high level
    - Application and network operating systems
    - Measure of LAN power and its background load index of average workstations on a local LAN
    - Number of workstations on a remote LAN
    - Speed of the segment

  Internetwork parameters are:

    - Network architecture
    - Bridge, router, switch, and gateway transfer rates
    - Lower-level protocol on the interlink
    - Number of hops between two LANs

- ■ Background load on links between LANs
- ■ Throughput rates

- Control parameter extensions: Users may be interested in changing or extending the modeling to new operating systems, unsupported protocols, and new transmission media. This criterion checks on the openness of the modeling process. Also the ease and availability of programmability may be examined here.
- Technology used: The answers here impact the accuracy of modeling results. Queuing equations allow quick evaluations of expected performance ranges. Complex simulation allows modeling in greater detail and guarantees much higher accuracy. Some products combine both techniques.
- Applicability of live data: Once the LAN is running, LAN analyzers may be used to collect actual traffic data. Some performance models can read this data and use it to augment the modeling capabilities. It helps model calibration and validation and allows the effects of growth to be observed more accurately.
- Post-processing: The right presentation form helps to interpret the modeling results. It is extremely important to reexamine the modeling results without completely rerunning the model. Graphics and colors help to better understand the results.
- Animated displays: This capability allows designers and planners to get a feel for the impact of certain modeling parameters, such as queuing delays at congestion points or collisions in certain LAN segments. Some products provide both a step mode and an automatic mode to support of this type of visual display. In many circumstances, this graphical support accelerates the evaluation process by highlighting potential performance bottlenecks.
- Maturity of product: It is extremely important to collect implementation experiences from other users. Most products are recent developments, and just a few products are based on mature products that have been around in the WAN area for many years. Also the integration of existing solutions for LAN segments and for interconnected LANs would be a positive sign of maturity.

5.2 Historical significance

Analysis and testing using a model instead of actual networks helps to save time and money. The complexity of present LANs and interconnected LANs does not permit installations without modeling and rapid prototyping. The integration of modeling tools with management platforms and monitoring instruments enables rapid prototyping. The management platform plays the role of a broker. The broker is a standard feature of the management platform. Besides standard features of the platform, in particular, performance management applications are of prime interest (Figure 5.1).

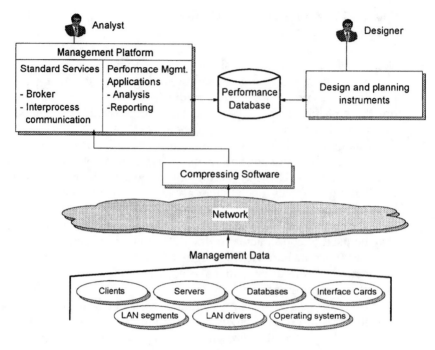

Figure 5.1

Table 5.1 Overview of
Modeling Parameters

Fixed metrics
  Transmission capacity
  Signal propagation delay
  Topology
  Frame/packet size
Variable metrics
  Access protocol
  User traffic profile
  Buffer size
  Data collision and retransmission
Performance metrics
  Resource usage
  Processing delays
  Throughput
  Availability
  Fairness of measured data
  Communigram

Performance indicators of LANs are the same for actual and modeled LANs. Table 5.1 summarizes the key performance metrics.

5.2.1 LAN modeling parameters

LAN performance indicators may be grouped into fixed, variable, and performance measurement metrics [TERP96].

Fixed metrics

- Transmission capacity — The transmission capacity is normally expressed in terms of bits/second. Although the bite rate is fixed, the total capacity can be divided into multiple smaller capacities to support different types of signals. One of the common myths regarding LAN transmission capacity is that Ethernet is saturated at an offered load (the actual data carried on the channel, excluding overhead and retransmitted bits) of 37%. Many studies have shown that Ethernet can support a 10 Mbps data rate under a distance of one kilometer with the CSMA/CD protocol.
- Signal propagation delay — Signals are limited by the speed of light, and the longer they propagate, the longer the delay. Signal propagation time is the time required to transmit a signal to its destination and generally is 5 microseconds per kilometer. Therefore, cabling distance is a factor that affects signal propagation delay. In the case of satellite communication, signal propagation delay plays an influential role, as the distance between an earth station and the satellite is about 22,500 miles. Within LANs, the internodal signal propagation delay is negligible. However, the signaling technique used (e.g., baseband or broadband) can produce different levels of delays.
- Topology — A LAN can be a star, tree, ring, bus, or combination of star and ring. The type of LAN topology will affect performance. For example, a bus LAN (e.g., Ethernet) and a token ring LAN (e.g., IBM's token ring) have a different built-in slot time — the time of acquiring network access. The topology also limits the number of workstations or hosts that can be attached to it. Ethernet limits the number of nodes per cable segment to 100, and the total number of nodes in a multiple-segment Ethernet is limited to 1024. A single IBM token ring supports 260 nodes. The higher the number, the greater the performance impact since all network traffic is generated from these nodes.
- Frame/packet size — Most LANs are designed to support only a specific, fixed size of frame or packet. If the message is larger than the frame size, it must be broken into smaller sizes occupying multiple frames. The greater the number of frames per message, the longer the delay a message can experience. Like every other LAN, Ethernet, for example, has a minimum packet size requirement: it must not be shorter than the slot time (51.2 microseconds) in order to be able to detect a collision. This limit is equivalent to a minimum length of 64 bytes, including headers and other control bytes. Similarly, Ethernet has a maximum of 1518 bytes as the upper boundary, in order to minimize access time.

Variable metrics

- Access protocol — The type of access protocol used by a LAN is probably the most influential metric that affects performance. IBM's token ring uses a proprietary token access control scheme, in which a circulating token is passed sequentially from node to node to grant transmissions. A node must release a token after each transmission and is not allowed to transmit continuously on single ring architecture. Ethernet, on the other hand, employs the I-persistent CSMA/CD access control in which a node that waits for a free channel can transmit as soon as the channel is free with a probability of 1 (i.e., 100% chance to transmit).
- User traffic profile — A computer system and network is lifeless without users. Many factors constitute a user's traffic profile: message/data arrival rate (how many key entries a user makes per minute), message size distribution (how many small, medium, and large messages are generated by a user), type of messages (to a single user, multiple users, or all receivers), and the number of simultaneous users (all active, 50% active, or 10% active.)
- Buffer size — A buffer is a piece of reserved memory used to receive, store, process, and forward messages. If the number of buffers is too small, data may suffer delays or be discarded. Some LANs have a fixed number of buffers, and some use a dynamic expansion scheme based on the volume of the messages and the rate of processing. In particular, LAN internetworking devices are likely sources of buffer problems.
- Data collision and retransmission — Data collision is inevitable, especially in a bus LAN, unless the transmission is controlled in an orderly manner. Two factors need to be considered: how long it takes nodes to detect a data collision and how long it takes to actually transmit the collided messages. Various detection schemes are used by different topologies. For example, Ethernet employs a "jam" time, which is the time to transmit 32 to 48 more bits after a collision is detected by a transmitting station so that other stations can reliably detect the collision. The more influential factor is the time to actually transmit the data after collision. Many LANs use a binary exponential backoff scheme to avoid a situation in which the same two colliding nodes collide again at the next interval. Both collision detection and retransmission contribute delays to the overall processing delay. Generally, waiting time is dependent on network load and may become unacceptably long in some extreme cases.

Performance measurement metrics

The performance of a LAN cannot be quantified with a single dimension. It is very hard to interpret measured metrics without knowing what applications (users) are involved. The following measurement metrics are generally obtainable:

- Resource usage — Processor, memory, transmission medium, and in some cases, peripheral devices all contribute to the processing of a user request (e.g., open a file, send a message, or compile a program). How much of their respective capacities are used and how much reserved capacities are left need to be evaluated in conjunction with processing delay information (in some cases, user's service level goals).
- Processing delays — A user's request is likely to suffer delays at each processing point. Both host and network can cause processing delays. Host delays can be divided into system processing and application processing delays. Network delays can be viewed as a combination of delays due to hardware and software. However, at the end user level, a total processing delay (or response time) is the only meaningful performance metric.
- Throughput — Transmission capacity can be measured in terms of throughput — the number of messages or bytes transmitted per unit of time. In LAN measurement, throughput is an indication of the fraction of the nominal network capacity that is actually used for carrying data. In general, packet headers are considered useful in estimating throughput, if no other measurement facilities are available, since the header contains the number of bytes in a frame. A metric related to throughput is channel capacity. Each transmission medium has a given maximum capacity (e.g., bits/second), which is a function of message volume and message size.
- Availability — From an end user's point of view, service availability is determined by its availability and consistency. A network can be in operation, but if a user suffers long delays, as far as the user is concerned, the network is virtually unavailable since it is seen as unreliable. Therefore, reliability measurement is a permanent measurement metric. However, most LAN measurement tools are only able to measure availability (up and down time), since timing measurement may add several orders of magnitude of complexity to measurement tools
- Fairness of measured data — Since network traffic tends to be sporadic, the measured period and the internal data-recording rate are quite important. An hourly averaged measured data rate may not be able to reveal any performance bottlenecks; a 1-second recording rate can generate an enormous amount of data that requires both processor time and storage. As a general practice, a peak-to-average ratio is used in which data in short intervals with known high activity are collected. The ratio between the high activity periods and the average periods can be established for studying network capacity requirements.

- Communigram — In order to quantify the traffic between communication partners, the volume is quite important. The measured and reported intervals are very important. An hourly averaged rate may not be able to reveal any performance bottleneck; a 1-second recording rate can generate an enormous amount of data that requires both processor time and storage. As a general practice, a peak-to-average ratio is used in which data in short intervals with known high activity are collected. The ratio between the high activity periods and the average periods can be used for sizing resources supporting the communication between partners.

5.2.2 LAN hubs as source of modeling parameters

Hubs play a very important role in managing the local area network. Using hubs to collect and preprocess performance-related information that may be used to build baseline models can even emphasize the role. This segment shows a few examples for Ethernet and token ring segments.

Ethernet indicators

Hub level indicators provide information about the data transmission results on a logical hub. These statistics actually consist of multiple physical level port statistics. Typical statistics required for diagnostics and performance analyses are:

- Peak traffic in the segment within a specified time window
- Average traffic in the segment for a specified period of time
- Current traffic at the time of the last sample
- Total packets received
- Total bytes received
- Missed packets
- Number of cyclic redundancy check errors on the segment
- Frame alignment errors on the segment
- Collision rate in the segment for a specific period of time

Token ring indicators

Token ring indicators consist of hard and soft error and of general performance statistics. The hub vendors usually support most of the indicators referenced as:

Hard error indicators

- Ring purges by the active monitor
- Number of times input signal is lost
- Beacons in the ring for a specified period of time

Soft error indicators

- Number of line errors
- Number of burst errors
- Number of AC errors
- Number of abort sequences
- Number of lost frames
- Number of receive data congestion errors
- Number of frame-copied errors
- Number of token errors due to token loss or frame circulation
- Number of frequency errors

General indicators:

- Cumulative number of bytes on the ring for a specified period of time
- Frame count
- Average and peak utilization level
- Average and peak frame rate
- Bytes per second on average or peak
- Average frame size for a specified period of time
- Current and peak number of stations on the ring
- Current operating speed of the ring

The modeling, design and planning of LANs is based on monitoring all or at least some of these indicators. Principal data collection sources include:

- LAN analyzers and monitors
- SNMP MIBs
- RMON MIBs and
- Device-dependent applications from equipment vendors.

These tools will be described in more depth in segment 5.3.

5.2.3 *Modeling objects in LANs*

The targets, in other words, the managed objects (MOs) are the same for performance analysis, tuning, modeling, and design. These targets in local area networks are the following:

- LAN servers, in particular their CPUs and I/O-devices
- LAN drivers
- LAN interface cards
- LAN operating systems
- Workstations
- Peripherals

LAN servers

LAN servers have a major impact on the performance of the network. Almost independently, whether the servers are high-end or low-end, they are equipped with systems management software, automatic server recovery, remote maintenance, and predictive diagnostics. They use complex instruction set computing (CISC) or reduced instruction set computing (RISC) chips and support industry-standard PC-I/O buses, such as extended industry standard architecture (EISA), micro channel architecture (MCA) and peripheral component interconnect (PCI). Despite technological advances, bottlenecks may occur at the CPU and at I/O-devices.

Increasing the speed or number of CPUs improves performance, but provides only a partial solution. Without sufficient work, the CPU is in waiting state. This is why high-performance servers use special cache designs, bus designs, memory management, and other architectural features to keep the CPU busy. These enhancements apply to servers that run in asymmetric multiprocessing (ASMP) mode, which dedicates individual CPUs to independent tasks SMP-mode, which enables multiple CPUs to share processing tasks and memory; and clustering mode, which enables CPUs on multiple servers to work in ASMP mode. To prevent the CPU from accessing main memory too often, high-speed cache memory is placed in between. Cache memory is actually placed within and also outside the CPU. Measurement results confirm efficiency improvements of the CPU in the range of 10% to 20%.

There is another technique, called pipelining, that prevents the CPU from waiting unnecessarily while data is being transported from memory. The transport process requires a cycle of time on the CPU-to-memory bus. In a nonpipelined architecture, a second cycle is not started until the first one is completed, and there is a time delay before the second cycle starts. In pipelined bus architecture, the second cycle begins before the first cycle completes. This way, the data from the second cycle is available immediately after the completion of the first cycle.

The vendor usually determines the size of the cache; the LAN analyst is expected to set the systems parameters accordingly.

Cache systems are designed to keep the CPU supplied with instructions and data by managing access to main memory. However, bus controlled I/O-devices also contend for access to main memory, and they run at a much slower speed than the CPU. Therefore, CPU-to-memory operations should take priority over I/O-to-memory operations, but not at the expense of interrupting these I/O-operations. This is why most high-performance servers are engineered to let the CPU and I/O devices simultaneously access main memory by maximizing concurrency and minimizing contention. Maximizing concurrency is supported by placing buffers between high-speed system buses and the I/O-to-memory bus. These buffers capture data reads and writes between buses to prevent one device, such as a CPU or I/O-card, from waiting for another to finish. Vendors use these buffers in segmented bus architectures that can segregate different devices on various buses. This technique is expected to be enhanced continuously.

Disk I/O bottlenecks are identified by such applications as high-volume transaction processing that moves many small transactions between the CPU and disks and decision support systems with many record moves. In these and similar cases, performance is affected by disk speed, the number of disk drives, and the intelligence and speed of drive array controllers. The number of disk drives in the system has a greater effect on server performance than the speed of individual drives. This is because of reduced latency in the positioning of drive heads and because more than one set of read/write-heads may be active at a given time. The greater the number of drives, the greater the performance of drive array. There are still other ways to improve disk I/O performance. Intelligent array controllers support more than one disk channel per bus interface and implement support for multiple redundant array of inexpensive disks (RAID) levels of hardware.

LAN drivers

Another important set of elements affecting local area network performance are the software routines called drivers. Drivers accept requests from the network and provide the interface between the physical devices (disk drives, printers, network interface cards) and the operating system. The drivers also control the movement of data throughout the network and verify that the data have been received at the appropriate address.

The critical role that drivers play, however, means that driver problems can have a large impact on the performance of the overall network. Drivers have traditionally been supplied by the LAN vendors and have been tailored to the operating systems and varied according to size. Today, it is more likely third-party software developers will provide customized drivers for networks.

These customized drivers, however, can be rather detailed and lengthy. If a driver takes up too much RAM, other applications will have insufficient room in which to operate, causing them to alter their normal operating procedures in order to reduce memory requirements. Also, the larger a driver is, the more code it has to execute, causing a network to delay when responding to additional requests, such as requests for printer services or requests from other users for processing jobs.

Interface cards can also affect the performance. Memory management is crucial to speed and performance. Factors such as DMA vs. shared memory and onboard processors and buffers can mean large differences in the actual throughput between two cards on the network. The performance difference, for example, between Ethernet cards can be as high as 50%.

LAN interface cards

When data from the CPU is being sent to a network port of a disk, it can cause a different bottleneck and limit the number of users that can simultaneously make server requests. The network is expected to become the bottleneck with file and print services, video servers, and imaging systems. In other cases, high-performance, low-utilization network interface cards are used. The performance of a server-to-LAN channel is affected by NIC driver optimization, the bus mastering capabilities of the controller, concurrent access to server memory, and the number of LAN channels per bus interface.

But, in cases where CPU use is an issue, there is a critical limit where placing additional NICs in the server will not improve the performance due to the overhead associated with routing and servicing the NIC. The practical threshold is around three NICs per server. Vendors will address how to connect servers to high-speed technologies such as ATM, FDDI, frame relay, and 100 Mbps Ethernet. They will also address the concept of placing servers on dedicated high-speed LANs and of using switching to keep the server and end users from waiting for the network.

LAN operating systems

Usually, the LAN operating system is the most prominent factor affecting the performance of a local area network. The operating system performs many functions, including communication with operating systems of servers and clients, the support of interprocess communication, the maintenance of network-wide addressing, and the moving of data within the network to manage the files and control the input and output requirements from interconnected LANs. The more efficiently the LAN operating system is able to perform these tasks, the more efficiently the LAN will operate.

High performance combined with low purchase costs and operating expenses cannot be met by many vendors. The LAN operating systems market concentrates gradually around a very few powerful products, such as NetWare, Windows NT, LAN Manager, Vines, and LANtastic.

Workstations

Another element affecting performance is the network workstation, which is often called a client. The performance of a workstation has more impact on both the perceived system performance and the actual system performance than any other component. The operating systems are frequently the same as with servers. In this case, coordination between the two is easier over the LAN. In other cases, with a powerful file server on a 10 Mbps LAN, an older PC with DOS and with limited RAM may become the bottleneck since it cannot accept or display data as fast as the file server, and the network hardware can supply it. At times, it is cheaper and more practical to upgrade the workstation, rather than the LAN itself. Adding more RAM or a coprocessor could improve the overall performance substantially. The protocol software can also affect workstation performance. A full seven-layer OSI stack requires considerable resources to run. Even with user-friendlier protocols it may have major effects on performance, depending on how they select packet sizes, transfer buffers, and translate addresses.

The network workstation is just as important to the overall performance of the LAN as the server and operating systems. The workstation executes the network's protocols through its driver software; a faster workstation will improve the performance of the LAN. One factor to consider is whether the workstation should contain a disk drive of its own or not. Obviously, a diskless workstation will ease the budget and improve security. But diskless workstations have their own set of costs. For one, these workstations are dependent on shared resources. If the work being performed at the station does not involve sharing resources, a workstation with its own disk may be more appropriate. Moreover, diskless workstations add to the traffic load on the LAN. This could be significant, especially if the workstations are for programmers who typically do not need to share files but who often work on files that are extremely large. The bandwidth on the LAN may become a performance bottleneck in the future.

Peripherals

Printing requirements also affect LAN performance in a variety of ways. Modern printers provide much more advanced printing capabilities than were available just a few years ago. Complete pages are transmitted all at once with improved fonts and high-end graphics. These printing capabilities, however, if not handled properly, can degrade network performance. If the user runs into such a performance problem, and if enough printers are not available, redirecting the printing job to a local printer may help. Also another server dedicated to handling printer functions can be added.

Another way to avoid bottlenecks caused by heavy printing requirements is to use a network operating system that incorporates a spooler to control these requirements. Spoolers are designed to accept a printing request from the network logically and in order, and they complete print requests without additional help from workstations.

Effective filtering across bridges and routers ensures that traffic volume is not too high throughout the network and keeps performance constantly high. If high volumes indicate the need for partitioning, it is recommended to use bridges, routers, or virtual LANs.

These examples represent a certain type of client/server-oriented network. Other rules apply in a peer-to-peer LAN and in interconnected LANs. In the latter, routers take over a lot of responsibilities by directing traffic to the targeted destinations. But, routers have a limited performance in terms of receiving, interpreting, and forwarding packets, frames, and messages.

Figure 5.2 shows the typical three operating areas of routers:

- Below A: normal performance
- B: stress point
- Beyond B in the area of C: degraded performance

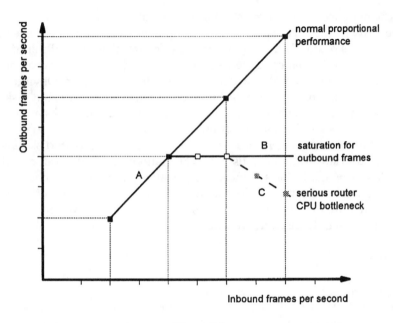

Figure 5.2

Other types of performance bottlenecks may include:

- Too many collisions due to too many stations and or messages.
- Bandwidth of the LAN is too narrow.
- Retransmissions required by high error rates.
- Message storms.
- An uncontrolled chain of confirmations and reconfirmations in the LAN

5.3 Information sources for baseline LAN models

The design and planning of LANs should not start from scratch. In most cases, networks are running and therefore measurement data about actual performance is available. This could be the beginning of the network optimization or a new design or a combination of both. There are many information sources available to the LAN designer and planner. This chapter outlines monitors, SNMP-based managers, RMON-based information collection, and device-specific management applications. Each of these can provide a baseline model for the designer and capacity planner.

5.3.1 LAN analyzers and monitors

These instruments are very widely used in the industry. In this section two examples are shown that represent this industry: the Sniffer family from Network General and NetMetrix from Hewlett-Packard. These two are not the only LAN analyzers; others are available from Wandel & Golterman, Shomiti Systems, Technically Elite, Kaspia Systems, Net2Net, ADC Kentrox, and Netscout Systems.

Distributed sniffer system from Network General

Today's client/server networks demand a cost-effective network analysis solution. Users need to solve network problems and monitor multiple network segments across the country and around the globe from a central office. The distributed sniffer system (DSS) delivers remote monitoring (RMON) and expert analysis solutions for proactively managing network performance and solving network problems. With DSS, the user can view network activity from a central location using one of the popular network management platforms. As a result, users can leverage their expertise across every segment, save valuable time and money, and manage the entire network from end to end without leaving the central office.

The distributed sniffer system offers the following benefits:

Proactive management.   DSS gives LAN managers proactive control over their entire network. They can baseline the network's normal behavior to help identify incremental changes and potential problems. Because DSS notifies them of potential problems before an end user calls to complain to LAN managers, DSS can solve underlying issues before they become headaches.

Fast problem resolution.   DSS targets the underlying cause of problems on Ethernet and token ring LANs, as well as bridged and routed internetworks. The user can look at detailed interpretations of over 140 protocols. DSS identifies network problems and provides LAN managers with suggestions for corrective action to help expedite problem solving.

Maximizes performance.   DSS reports a number of statistics from all seven OSI layers in real time. This information helps keep the network running, enabling operators to solve problems before they cause downtime. Using historical statistics, DSS helps planning server placement and segmentation to maximize network performance.

Reduces network management costs.   Users can manage remote locations without the time and expense of traveling. DSS automates software distribution and provides out-of-band serial support. This helps centrally monitor and analyze network performance and resolve problems without delay.

The product family consists of the following elements:

- Analysis solutions
    SniffMaster Console
    Sniff Servers
    Sniffer Internetwork Servers

- Monitoring solutions
    Foundation Manager Console
    Cornerstone Agent
    Cornerstone Probe

Expert Analysis provides three types of information: symptoms, diagnoses, and explanations to help quickly resolve performance, protocol, internetwork, and physical layer problems.

- Symptoms are clues to potential problems that might be lurking on the networks. Expert Analysis symptoms provide a means to proactively manage your network.
- For serious problems, Expert Analysis delivers automatic diagnoses to help LAN managers resolve bottlenecks and performance problems that may bring your network down.
- For every potential and existing problem, Expert Analysis provides supporting data and automatically recommends solutions.

With a single keystroke, Expert Analysis displays packets associated with the discovered problem so LAN analysts can investigate it in even greater detail. A complete record of Expert Analysis information can be written in CSV format and uploaded to a console for use in spreadsheet, database, reporting, and plotting applications. This information can be used to analyze statistics and generate management reports.

DSS with Expert Analysis alerts LAN managers to many common network problems including:

- Performance bottlenecks: The Expert Analysis application helps detect distributed segment inefficiencies by analyzing and interpreting network data such as multiple retransmissions that consume bandwidth.
- Internetworking problems: Expert Analysis identifies problems that impact the performance of internetwork links. Duplicate network addresses are readily detected and the nodes involved are identified automatically. This helps improve application throughput, avoid the purchase of unnecessary equipment, and boost end-user productivity on multisegment enterprise networks.
- Protocol violations: As part of DSS, the Expert Analysis application recognizes nonstandard protocol activity and suggests solutions for protocol problems.
- Physical layer problems: DSS with Expert Analysis speeds the process of identifying and solving Ethernet and token ring lower-layer problems, including congestion and ring beaconing.

The Protocol Interpretation application enables LAN operators to troubleshoot complex network problems by providing full decodes for over 140

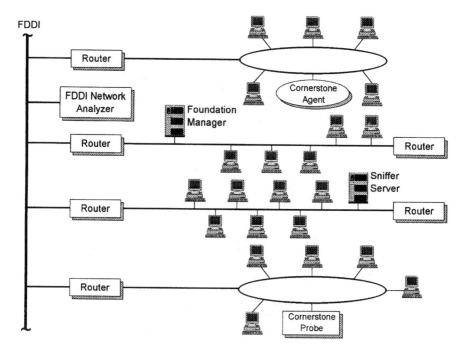

Figure 5.3

protocols. With a detailed view of the composition of data packets, the user can see the actual communication occurring on the network and get to the cause of network communication problems. Protocol Interpretation displays the contents of each frame in English-language text for each of the seven OSI layers.

The DSS RMON application offers standards-based monitoring and enhanced performance analysis capabilities for distributed network segments. To protect investments in other RMON technologies, Network General's standards-based monitoring solution fully supports all the RMON MIB groups for Ethernet and token ring LANs.

LAN managers can proactively monitor network activity on every segment with this DSS monitoring solution. DSS also integrates trouble prevention, early detection, and problem resolution for distributed network segments. Users can quickly transition from DSS monitoring to analysis on the same console by saving RMON data in Sniffer trace file format. Afterwards, LAN performance analysts can analyze RMON data using DSS Expert Analysis and Protocol interpretation applications.

Figures 5.3 and 5.4 illustrate DSS components in local and remote networks.

DSS analysis solutions consist of two components: Sniffer Servers and SniffMaster Consoles. DSS standards-based monitoring solutions consist of three components: Foundation Manager Consoles, Cornerstone Agents, and

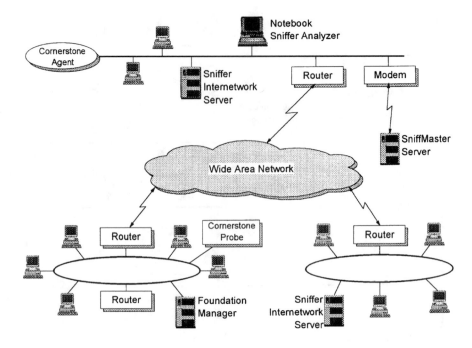

Figure 5.4

Cornerstone Probes. Scalable DSS components can be combined to proactively monitor and quickly troubleshoot your entire network from a central location.

Sniffer Servers support the Expert Analysis and Protocol Interpretation applications. When attached to mission-critical Ethernet, token ring, or internetwork segments, Sniffer Servers communicate information back to centrally located SniffMaster Consoles.

Sniffer Servers can communicate with more than one SniffMaster Console, allowing multiple technicians to view network activity simultaneously. Advanced features such as automatic uploads from remote Sniffer Servers ensure that data are collected from key network segments, regardless of the server location or the time of day.

SniffMaster Consoles display the information gathered by Sniffer Servers. Expert Analysis and Protocol Interpretation applications are downloaded, launched, and controlled from SniffMaster Consoles. For maximum flexibility, Network General offers both Unix and Microsoft Windows versions of the SniffMaster Console.

Foundation Manager Consoles consolidate and synthesize RMON information collected from Cornerstone Agents and Cornerstone Probes on a distributed network. LAN analysts can automatically save RMON data in Sniffer trace file format to quickly troubleshoot alarms with Network General's Expert Analysis and Protocol Interpretation applications. This provides a seamless, icon-based transition from monitoring to analysis from a centralized console.

Cornerstone Agents and Cornerstone Probes gather statistical information specific to the standard RMON Management Information Base (MIB) groups for both Ethernet and token ring LANs. When attached to remote segments, Cornerstone Agents and Probes gather RMON statistics, generate alarms, and relay information back to centrally located Foundation Manager Consoles. Agents and Probes can also send SNMP alarms to SNMP management systems.

The Cornerstone Agent has a local user interface for on-site monitoring. It also communicates RMON data to centralized Foundation Manager Consoles. Cornerstone Agent software can be installed in any 486+ PC running Microsoft Windows or OS/2 to leverage the user's existing hardware investment.

Cornerstone Probe is comprised of an RMON agent preinstalled in a server. Cornerstone Probes collect and send RMON data to Foundation Manager Consoles to enable centralized monitoring and performance analysis of remote LAN segments.

The Network General Reporter application helps to collect and present DSS analysis and monitoring data. The Reporter offers over 20 standard report formats to document and graphically depict network usage over time, error summaries, baseline comparisons, and enterprise-level trends. LAN managers can use the Reporter in conjunction with DSS analysis and monitoring solutions to effectively solve network problems, justify support costs, and plan for the future.

Network General developed the first analysis solution designed to solve problems on router-based internetworks and to manage bandwidth costs. By providing a comprehensive view of communications traveling over internetwork links, Expert Sniffer Internetwork Analyzers pinpoint internetwork bottlenecks and offer recommendations to improve bandwidth efficiency, application throughput, and response times.

Network General's portable analysis tools offer a wide range of features for analyzing data traffic across internetworks including:

- Expert analysis of internetworks for automatic problem identification
- Seven-layer analysis of over 140 LAN protocols encapsulated within leased-line (HDLC), frame relay, and X.25 protocols
- Utilization statistics of internetwork bandwidth for performance analysis
- Capture of network traffic at data rates of up to 2.048 Mbps

Expert Sniffer Internetwork Analyzers also can be used to assess traditional terminal-to-host WAN configurations. This portable tool addresses a range of traditional wide area communications modes and protocols from async, bisync, and frame relay to X.25 and SNA. Supporting speeds from 50 bps up to 2.048 Mbps, Expert Sniffer Internetwork Analyzers meet existing and future internetwork speed requirements.

Data collected by the Sniffer family of products can also support performance modeling and capacity planning. Optimal Performance uses Expert Analysis information to automatically generate a baseline model of an organization's network, including topology, LAN and WAN traffic, protocols and applications. Predefined elements in the model include bridges, routers, hubs, switches, and protocols.

NetMetrix from Hewlett Packard

The product offers workstation-based traffic monitoring and analysis on multiple platforms, including HP OpenView, SunNet Manager, and NetView for AIX. The integration methods used are: SNMP traps, command line interface, and enhanced command line interface.

HP NetMetrix incorporates features of the traditional traffic monitors and analyzers — such as the ability to simulate network loads, gather statistics, perform traces, and provide seven-layer packet decode — and removes the encumbrances of proprietary hardware probes and clumsy user interfaces. The result is a workstation-based (UNIX/SPARC) segment monitor and an X-windows graphic user interface (GUI) supporting a suite of five very useful NetMetrix applications:

1. Traffic generator simulates network load, generates user-defined packet streams, and can respond to decoded packets in real-time.
2. Protocol analyzer is providing seven-layer packet decode on most major protocols.
3. Load monitor correlates traffic statistics to help users optimize bridge/router placement and answer other critical questions for fine-tuning.
4. NFS monitor measures NFS load and response time (by server, client, NFS procedure, or time interval), client/server distribution analysis, and server performance comparisons.
5. Internetwork monitor coordinates multiple agents across the network to provide a cohesive picture, such as displaying which groups and hosts are talking to each other.

NetMetrix provides a distributed architecture, supporting continuous monitoring and analysis of all segments in real time. While several other products provide similar capabilities, NetMetrix is less expensive and more comprehensive, particularly in analyzing NFS traffic. Hewlett Packard's aggressive pursuit of simple, effective methods for platform-application and application-application integration has put the company ahead of its competitors in providing automated network management for its customers.

The standard features include a protocol analyzer for various protocols, such as AppleTalk, DECnet, IPX, SNMP, SNA, TCP/IP, and Vines. Figure 5.5 shows a typical NetMetrix configuration.

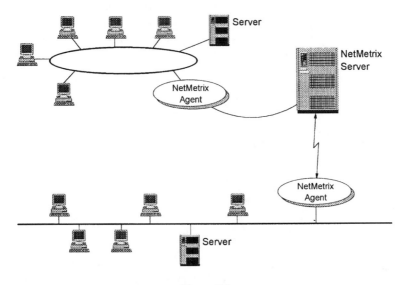

Figure 5.5

NetMetrix is now an important member of MeasureWare, which is a suite of performance-related tools from Hewlett-Packard.

5.3.2   SNMP MIBs as information source

The Simple Network Management Protocol (SNMP) originated in the Internet community as a means for managing TCP/IP and Ethernet networks. SNMP's appeal broadened rapidly beyond the Internet, attracting waves of users searching for a proven, available method of monitoring multivendor networks. SNMP's monitoring and control transactions are actually completely independent of TCP/IP. SNMP only requires the datagram transport mechanism to operate. It can, therefore, be implemented over any network media or protocol suite, including OSI.

SNMP is based on three basic concepts: manager, agent, and management information base (MIB) (see Figure 5.6).

An agent is a software program housed within a managed network device (such as a host, gateway, bridge, router, brouter, hub, or server). An agent stores management data and responds to the manager's request for this data.

A manager is a software program housed within a network management platform. The manager has the ability to query agents using various SNMP commands.

The management information base (MIB) is a virtual database of managed objects, accessible to an agent and manipulated via SNMP to achieve network management.

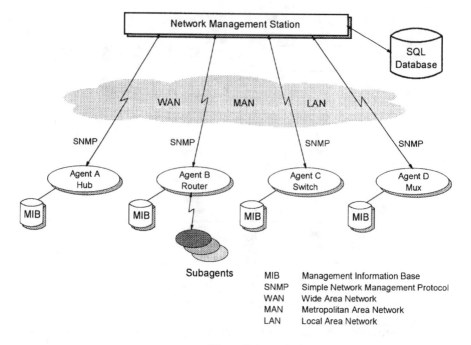

Figure 5.6

Agent responsibilities

Each agent possesses its own MIB view, which includes the Internet standard MIB and, typically, other extensions. However, the agent's MIB does not have to implement every group of defined variables in MIB specification. This means, for example, that gateways need not support objects applicable only to hosts, and vice versa. This eliminates unnecessary overhead, facilitating SNMP implementation in smaller LAN components that have little excess capacity for bearing overhead. An agent performs two basic functions:

- Inspecting variables in its MIB
- Altering variables in its MIB

Inspecting variables usually means examining the values of counters, thresholds, states, and other parameters.

Altering variables may mean resetting these counters, thresholds, and so on. It is possible to actually reboot a node, for example, by setting a variable.

A MIB implementation can be hosted on several types of platforms [TERP96]:

- Object-oriented databases
- Relational databases

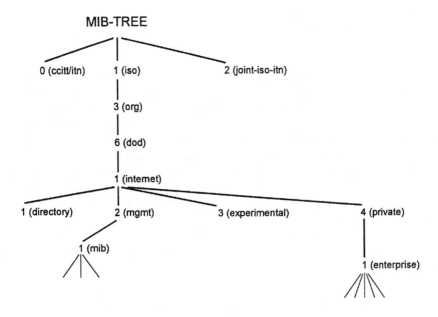

Figure 5.7

- Flat file databases
- Proprietary format databases
- Firmware

Basically, MIB information is distributed in agents. A typical configuration may include at the agent level a disk-based relational database, or a combination of PROM with static object attributes, and a RAM with dynamically changing information.

The MIB (management information base)

The MIB conforms to the structure of management information (SMI) for TCP/IP-based Internets. The SMI, in turn, is modeled after OSI's SMI. While the SMI is equivalent for both SNMP and OSI environments, the actual objects defined in the MIB are different. SMI conformance is important, since it means that the MIB is capable of functioning in both current and future SNMP environments. In fact, the Internet SMI and the MIB are completely independent of any specific network management protocol, including SNMP.

MIB is documented as a chain of numbers in hierarchical order. Internet is in the middle of this hierarchy, as shown in Figure 5.7.

The SNMP-MIB repository is divided into four areas:

- Management attributes
- Private attributes

- Experimental attributes
- Directory attributes

MIB I includes a limited list of objects only. These objects deal with IP internetworking routing variables. MIB II extends the capabilities to a variety of media types, network devices, and SNMP statistics, not limited to the territory of TCP/IP. There have been many attempts to improve the performance of MIB accesses. A query language interface seems to offer a number of new capabilities such as a relational mask, and fast access.

The management attributes are expected to be the same for all agents supporting SNMP. Thus, SNMP managers may work with agents from various manufacturers. In order to offer more functionality, vendors populate the private attributes as well.

Real MIB implementations may require extensions, incorporating other types of information to be stored for supporting operations. Such extensions may include [TERP96]:

- Improving security management by including security information about access to various managed objects, to the network management system itself, and to manipulating the MIB.
- Extended user data which may help to administer user and groups of users. Users and user groups can themselves be managed objects that can be created, maintained, and eventually deleted in the MIB.
- Configuration histories and profiles support to retrieve past information for reporting or for supporting backup, and alternate routing as part of LAN fault and performance management.
- Trouble tracking helps to resolve networking problems more rapidly. The augmented MIB would provide the ability to translate event reports into trouble tickets, assign work codes to staff, recognize and categorize types of problems, relate problems, escalate problems by severity, and close trouble tickets after eliminating the problem and its causes.
- Extended set of performance indicators supports more advanced performance management by reporting on resource utilization, on threshold violations, on trends, and on the quality of managing bandwidth between interconnected LANs.

The performance of MIBs and their counterparts must be observed very carefully at the network management station. Parameters to be observed include the registration of actual command frequencies, number of traps, frequency of information retrievals by GetNextRequest, and the relation of positive to negative poll responses. MIBs perform optimally only in special environments.

Manager responsibilities

Managers execute network manager station (NMS) applications and often provide a graphical user interface that depicts a network map of agents. Typically, the manager also archives MIB data for trend analysis. It may be implemented in two different ways.

- Each agent's MIB entries are copied into the dedicated MIB segment.
- MIB entries are copied into a common area for immediate correlation and analysis.

At the manager level, presentation services and database services are offered. It is very important to decide about the right database at the manager level. Some issues involved are discussed here:

- Use of object-oriented databases

  The advantages are as follows [TERP96]:

  - They are naturally suited, because the MIB itself is formally described as a set of abstract data types in an object-oriented hierarchy.
  - They are able to model interface behavior through the ability to store methods.

  The disadvantages are:

  - This technique is not yet mature for product implementations.
  - There are no standards yet for query and manipulation languages.
  - The MIB object class hierarchy is broad and shallow in an inheritance tree.
  - Performance characteristics are not yet well documented.

- Use of relational databases

  The advantages are:

  - The relational techniques are mature, stable, and have many supporters.
  - There is a standard access language (SQL).
  - There are good-quality translators for translating ER models into relational schema.
  - There are many choices of applications.

  The disadvantages are:

  - They are not well suited to storing Object Oriented (OO) models.
  - Performance depends on the tuning the database and is highly application dependent.

- Use of other databases

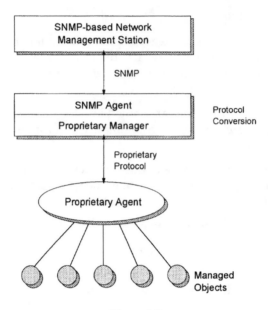

Figure 5.8

Flat-file databases and other proprietary formats can be tailored very specifically to MIBs and can be optimized for performance. But the design and implementation of network management applications may then become more complex and time-consuming to develop and to maintain.

SNMP proxy agents

Proxy agent software permits an SNMP manager to monitor and control network elements that are otherwise not addressable using SNMP. For example, a vendor wishes to migrate its network management scheme to SNMP, but has devices on the network that use a proprietary network management scheme. An SNMP proxy can manage those devices in their native mode. The SNMP proxy acts as a protocol converter to translate the SNMP manager's commands into the proprietary scheme. This strategy facilitates migration from the current proprietary environment, which is prevalent today, to the open SNMP equipment (Figure 5.8).

Proxy agents are well suited for vendors with an existing base of non-SNMP devices communicating efficiently under a proprietary scheme. By using a proxy agent, the vendor can reduce the investment risk of putting SNMP equipment in the field.

In summary, SNMP's major advantages are:

- Its simplicity eases vendor implementation effort.
- It requires less memory and fewer CPU cycles than CMIP.
- It has been used and tested on the Internet.

- SNMP products are available now and are affordable.
- Development kits are available free of charge.
- It offers the best direct manager-agent interface.
- It is robust and extensible.
- Polling approach is good for LAN-based managed objects.

SNMP has several disadvantages, including:

- Weak security features
- Lack of global vision
- Problems with the Trap command
- No object-oriented data view
- Unique semantics, making integration with other approaches difficult
- High polling overhead
- Too many private MIB extensions

5.3.3 RMON-probes and RMON-MIBs as information source

The remote MONitoring (RMON) MIB will help to bridge the gap between the limited services provided by management platforms and the rich sets of data and statistics provided by traffic monitors and analyzers. RMON defines the next generation of network monitoring with more comprehensive network fault diagnosis, planning, and performance tuning features than any current monitoring solution. The design goals for RMON are (STAL93):

- Off-line operation: In order to reduce overhead over communication links, it may be necessary to limit or halt polling of a monitor by the manager. In general, the monitor should collect fault, performance, and configuration information continuously, even if it is not being polled by a manager. The monitor simply continues to accumulate statistics that may be retrieved by the manager at a later time. The monitor may also attempt to notify the manager if an exceptional event occurs.
- Preemptive monitoring: If the monitor has sufficient resources, and the process is not disruptive, the monitor can continuously run diagnostics and log performance. In the event of a failure somewhere in the network, the monitor may be able to notify the manager and provide useful information for diagnosing the failure.
- Problem detection and reporting: Preemptive monitoring involves an active probing of the network and the consumption of network resources to check for error and exception conditions. Alternatively, the monitor can passively — without polling — recognize certain error conditions and other conditions, such as congestion and collisions, on the basis of the traffic that it observes. The monitor can be configured to check continuously for such conditions. When one of these conditions occurs, the monitor can log the condition and notify the manager.

- Value-added data: The network monitor can perform analyses specific to the data collected on its subnetworks, thus offloading the manager of this responsibility. The monitor can, for instance, observe which station generates the most traffic or errors in network segments. This type of information is not otherwise accessible to the manager who is not directly attached to the segment.
- Multiple managers: An internetworking configuration may have more than one manager in order to achieve reliability, perform different functions, and provide management capability to different units within an organization. The monitor can be configured to deal with more than one manager concurrently.

Table 5.2 summarizes the RMON MIB groups for Ethernet segments. Table 5.3 defines the RMON MIB groups for token ring segments. At the present time, there are just a few monitors that can measure both types of segments using the same probe.

Table 5.2   RMON MIB Groups for Ethernet

Statistics group	Features a table that tracks about 20 different characteristics of traffic on the Ethernet LAN segment, including total octets and packets, oversized packets and errors.
History group	Allows a manager to establish the frequency and duration of traffic-observation intervals, called "buckets". The agent can then record the characteristics of traffic according to these bucket intervals.
Alarm group	Permits the user to establish the criteria and thresholds that will prompt the agent to issue alarms.
Host group	Organizes traffic statistics by each LAN node, based on time intervals set by the manager.
HostTopN group	Allows the user to set up ordered lists and reports based on the highest statistics generated via the host group.
Matrix group	Maintains two tables of traffic statistics based on pairs of communicating nodes; one is organized by sending node addresses, the other by receiving node addresses.
Filter group	Allows a manager to define, by channel, particular characteristics of packets. A filter might instruct the agent, for example, to record packets with a value that indicates they contain DECnet messages.
Packet capture group	This group works with the filter group and lets the manager specify the memory resources to be used for recording packets that meet the filter criteria.
Event group	Allows the manager to specify a set of parameters or conditions to be observed by the agent. Whenever these parameters or conditions occur, the agent will record an event into a log.

Table 5.3   RMON MIB Groups for Token Ring

Statistics group	This group includes packets, octets, broadcasts, dropped packets, soft errors, and packet distribution statistics. Statistics are at two levels: MAC for the protocol level and LLC statistics to measure traffic flow.
History group	Long-term historical data for segment trend analysis. Histories include both MAC and LLC statistics.
Host group	Collects information on each host discovered on the segment.
HostTopN Group	Provides sorted statistics that allow reduction of network overhead by looking only at the most active hosts on each segment.
Matrix group	Reports on traffic errors between any host pair for correlating conversations on the most active nodes.
Ring station group	Collects general ring information and specific information for each station. General information includes: ring state (normal, beacon, claim token, purge); active monitor; number of active stations. Ring station information includes a variety of error counters, station status, insertion time, and last enter/exit time.
Ring station order	Maps station MAC addresses to their order in the ring.
Source routing statistics	In source-routing bridges, information is provided on the number frames and octets transmitted to and from the local ring. Other data includes broadcasts per route and frame counter per hop.
Alarm group	Reports changes in network characteristics based on thresholds for any or all MIBs. This allows RMON to be used as a proactive tool.
Event group	Logging of events on the basis of thresholds. Events may be used to initiate functions such as data capture or instance counts to isolate specific segments of the network.
Filter group	Definitions of packet matches for selective information capture. These include logical operations (AND, OR, NOT) so network events can be specified for data capture, alarms, and statistics.
Packet capture group	Stores packets that match filtering specifications.

RMON is very rich in features and there is the very real risk of over-loading the monitor, the communication links, and the manager when all the details are recorded, processed, and reported. The preferred solution is to do as much of the analysis as possible locally, at the monitor, and send just the aggregated data to the manager. This assumes powerful monitors. In other applications, the managers may reprogram monitors during oper-ations. This is very useful when diagnosing problems. Even if the manager can define specific RMON requests, it is still necessary to be aware of the trade-offs involved. A complex filter will allow the monitor to capture and report a limited amount of data, thus avoiding overhead on the network. However, complex filters consume processing power at the monitor; if too many filters are implemented, the monitor will become overloaded. This is particularly true if the network segments are busy, which is probably the time when measurements are most valuable.

The existing and widely used RMON version is basically a MAC-standard. It does not give LAN managers insight into conversations across the network or connectivity between various network segments. The extended standard is targeting the network layer and higher. This will give visibility across the enterprise. With remote access and distributed workgroups, there is substantial intersegment traffic. The following functions are supported (Table 5.4):

- Protocol distribution
- Address mapping
- Network layer host table
- Network layer metrics table
- Application layer host table
- Application layer matrix table
- User history
- Protocol configuration

After implementation, more and more complete information is available for performance analysis and capacity planning.

Remote monitoring is a new technology for continuously monitoring LAN segments. Thus, problems and performance bottlenecks can be highlighted in real-time or in near-real-time. There are basically two components:

- Client that consists of hardware and software components and is called monitor or probe
- Server that is responsible for centrally collecting and processing monitored data

The server is usually implemented on a management platform and communicates with the clients using SNMP. RMON offers multiple benefits:

- Reduction of problem resolution time

  Monitors can be programmed for various problem conditions. If conditions are met and thresholds are violated, the management station is immediately notified. Trouble tickets are opened and dispatched to the workforce.

- Reduction of outage time for the network and its components

  Actions to problem resolution are initiated earlier because information is available in real-time. Triggering is automatic; support personnel do not need to be on-site.

Table 5.4 RMON2 Extensions

Protocol distribution and protocol directory table	The issue here is to support multiple protocols running on any one network. Current implementations of RMON employ a protocol filter which analyzes only the essential protocols. RMON2 employs a protocol directory system which allows an RMON2 application to define which protocols an agent employs. The protocol directory table will specify the various protocols an RMON2 probe can interpret.
Address mapping	This feature matches each network address with a specific port to which the hosts are attached. Also identifies traffic-generating nodes/hosts by MAC, token ring or Ethernet address. It helps identify specific patterns of network traffic. Useful in node discovery and network toplogy configurations. In addition, the address translation feature adds duplicate IP address detection resolving a common troublespot with network routers and virtual LANs.
Network-layer host table	Tracks packets, errors, and bytes for each host according to a network-layer protocol. It permits decoding of packets based on their network layer address. In essence it permits network managers to look beyond the router at each of the hosts configured on the network.
Network-layer matrix table	Tracks the number of packets sent between a pair of hosts by the network layer protocol. The network manager can identify network problems quicker using this matrix table, which shows the protocol-specific traffic between communicating pairs of systems.
Application-layer host table	Tracks packets, errors, and bytes by host on an application-specific basis. Both the application-layer host table and matrix table trace packet activity of a particular application. This feature can be used by network managers to charge back users on the basis of how much network bandwidth was used by their applications.
Application-layer matrix table	Tracks packet activity between pairs of hosts by application.
Probe configuration	Based on the Aspen MIB, it defines standard parameters for remotely configuring probes — parameters such as network address, SNMP error trap destinations, modem communications with probes, serial line information and downloading of data to probes. It provides enhanced interoperability between probes by specifying standard parameters for operations, permitting one vendor's RMON application the ability to remotely configure another vendor's RMON probe.
User history collection group	The RMON2 history group polls, filters, and stores statistics based on user-defined variables creating a log of the data for use as a historical tracking tool. This is in contrast to RMON1 where historical data is gathered on a predefined set of statistics.

- Reduction of traveling expenses

  Portable monitors require that technicians and engineers travel to the LAN site. Using RMON-probes it is not necessary. This is very beneficial in case of many distributed LAN segments.

- Better scheduling for expensive engineers and technicians

  Due to centralized data processing for multiple LAN-segments, technical personnel can be more economically utilized. In addition, the quality of problem resolution is improving.

- Availability of history data

  RMON probes provide detailed information on multiple indicators. At the central site, data can be compressed and maintained for future use in performance databases. Future use is meant for performance analysis and network design.
  But, there are certain areas that must not be ignored:

- RMON requires the TCP/IP protocol.

  Most of the RMON products are based on TCP/IP. In case of using legacy protocols such as SNA, DSA, DNA or Novell, probes cannot be connected without protocol converters. But protocol converters increase the complexity of the productive networks.

- RMON probes must be installed in each LAN-segment.

  In order to guarantee the best results, each segment is expected to be continuously measured. Rotating measurements offer cost benefits, but do not guarantee complete data sets. RMON1 is segment-oriented; RMON2 offers cross-segment statistics as well. In case of continuously measuring LAN-switches, even more probes are required.

- Cost of probes

  The price/performance ratio is continuously improving. But the probes cost money. In case of many single segments, dedicated probes are required. In general, probes could be continuously measured up to four segments in any combination of Ethernet and token ring.
  There are three basic ways of implementing RMON probes in LAN-segments (Figure 5.9):

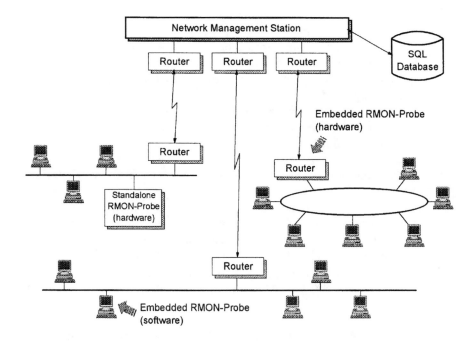

Figure 5.9

Probe as standalone monitor

Advantages:
- Excellent performance
- Full functionality
- Availability in stackable or rack-mountable forms

Disadvantages:
- High expenses for an average segment
- Many probes are required for LAN switches with probe-ports
- Standalone hardware to be maintained by the vendor
- Advanced LAN technologies are supported after delays

Embedded probe into hubs and routers

Advantages:
- Not much individual investments are required
- Costs are lower than with standalone probes
- Integration into switches is less expensive than providing probes for each switched segment

Disadvantages:
- Hubs and routers should use the latest version to accommodate probes

- Upgrades are not always economic
- The performance is not always good
- Conformance to standards is not always guaranteed
- Probes may cause breakdowns and performance bottlenecks
- In order not to impact performance, not all RMON indicator groups are supported
- Probes may come from different vendors
- Integration with management platforms is not obvious
- Functionality is usually limited

Probe is implemented as software module into UNIX- or PC-workstations
Advantages:
- Much lower costs in comparison to other alternatives
- Performance is comparable with standalone probes
- Scalability and expandability are well supported
- Advanced LAN technology is supported more rapidly
- Combination of Ethernet and token ring is possible
- Outband-support is possible by special configuration of the probes

Disadvantages:
- Purchase of adapters and additional workstations is required
- The user is in charge of maintenance
- Functionality may be limited

It is assumed that a meaningful combination of all three alternatives will be implemented. From the design and planning perspective, the emphasis is on the data gathered, but not on the implementation alternative chosen.

5.3.4   Input from device-dependent applications

Not only raw SNMP (MIB-II and RMON), but also device-specific data may help to build the basis for network design and planning. Practically, all vendors offer management applications for processing SNMP data. The emphasis is, however, real-time fault management, But, processed data can be saved, further processed and maintained (Figure 5.10).

This segment introduces a few examples, represented by Bay Networks, Optivity, Cisco, and 3Com.

The following section briefly describes key characteristics of a couple of leading applications of Optivity from Bay Networks.

The main function is the management for Bay Networks's smart hubs and routing/bridging modules. Optivity runs on SunNet Manager, HP OpenView, and NetView for AIX by using icon launching, "SuperAgent," Global Enterprise Management (GEM) applications as integration techniques.

Optivity collects hub, board, and port-level data such as MAC-layer diagnostics, errors, and utilization information. Optivity also provides

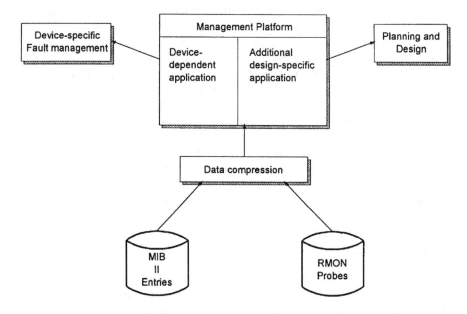

Figure 5.10

detailed ring and segment views, as well as real-time views of SynOptics's LattisHub, LattisRing, or System 3000 concentrators, including active LEDs and configuration status. Optivity also includes an Autotopology Plus feature, which employs a heuristic, recurring algorithm for dynamically creating and updating a hierarchical network map.

In addition to these features, there are several other aspects that make Optivity a unique product offering. On the UNIX platforms (SunNet Manager, IBM NetView for AIX), and HP OpenView UNIX), it supports distributed domain management by employing what Bay Networks calls "Super-Agent" software residing in critical hubs throughout the network. Each SuperAgent uses polling to collect a wealth of data in a given domain (such as a single Ethernet segment) and processes that data locally, forwarding on concise packages of information to the central Optivity application. By distributing intelligence to the SuperAgents scattered throughout the network helps to reduce traffic overhead and speed up the management process.

Network "intelligence" requires processing power, and Bay Networks supply this in the form of "Network Control Engines" or NCEs. An NCE is essentially a Sun SPARCstation residing on a Bay Networks concentrator module. Since the NCE comes preloaded with UNIX, many customers use NCEs to distribute SunNet Manager functions as well, such as collecting and processing data locally to avoid traffic overhead. NCEs also support HP NetMetrix traffic monitoring and analysis applications. Customers may choose to run Optivity in a centralized mode without NCEs. However, NCEs are required to support Optivity's distributed domain management capabilities.

Optivity and the SuperAgents form a foundation for Bay Network's value-added "LattisWare" applications such as RouterMan, PathMan, Meter-Man, and BridgeMan. For example, RouterMan displays all protocols and interfaces supported by routers on the network, including performance statistics. PathMan can determine and display the path between any two stations in the network, to assist in troubleshooting.

Optivity also supports another class of applications, called global enterprise management (GEM) developed by third parties. The first GEM applications available are NetLabs's Vision Desktop and Asset Manager. Both Vision Desktop and Asset Manager support the IETF Host Resources MIB and emerging desktop management task force (DMTF) specifications. Asset Manager supports automatic collection of PC component data; Vision Desktop provides a graphical interface and can assist in monitoring software-licensing violations.

3Com Transcend

The main function of this product is integrated management for 3Com adapters, hubs, and routers. The application is running on SunNet Manager, HP OpenView, and IBM NetView for AIX by using menu bar integration.

In an effort to simplify the router configuration process even further, 3Com has introduced an architecture called "boundary routing," which takes standard routing software for n-way local routing and extends the LAN interface portion over the wide area. The goal of this is to simplify routing software functions for remote routers, which are often located at branch offices where there are no technical support staff. 3Com Transcend takes advantage of boundary routing to simplify router management.

Transcend software obtains information from SmartAgent intelligent device agents embedded in 3Com adapters, hubs, and routers. 3Com SmartAgents localize polling and organize collected data to reduce bandwidth overhead of management data. SmartAgents are capable of correlating information from multiple 3Com devices to provide a more integrated view of network status, and assist in the creation of baselines for performance management.

Cisco CiscoWorks

The main function of this product line is configuration, performance, fault, and security management for Cisco's routers. This application is running on SunNet Manager, HP OpenView, and NetView for AIX. The integration methods include icon launching from SunNet Manager map; command line integration possible with HP NetMetrix (traffic monitoring), and Remedy ARS (trouble ticketing).

CiscoWorks is a suite of SNMP-based operations and management applications for users of Cisco's routers. CiscoWorks includes significant enhancements for easing remote installation and router software management. In particular, it provides a group-editing feature. The CiscoWorks global command capability allows managers to specify a group of routers and apply common configuration changes or software updates to the entire group. In addition, the CiscoWorks menu specifically calls out frequently used commands, such as enable passwords, SNMP community strings, and access lists. CiscoWorks includes a feature for checking a router's configuration against information stored in an SQL database.

CiscoWorks provides a series of applications for day-to-day router monitoring and troubleshooting. In addition, Cisco offers a "management series" for off-line analysis of network traffic and trends. Together, these utilities help satisfy both real-time immediate concerns of network managers, as well as requirements for long-term planning and trend analysis.

CiscoWorks' operations series includes six major aspects:

Configuration File Management
Path Tool
Health Monitor
Environmental Monitor
Device Management Database
Security Management

In particular, CiscoWork's Path Tool provides visualization of the actual path taken by the data; it detects interface changes much more efficiently than SNMP monitoring alone. The Management Series portion of Cisco-Works assists managers in achieving long-term goals of network management, such as historical trend analysis for determining the cost-effectiveness of transmission options, determining usage, identifying potential problem areas, and isolating chronic problems. A Data WorkBench feature is actually a Sybase report writer tool that allows administrators to create reports. These reports can display traffic through every router interface, including throughput and error rates, traffic peaks, and percentage of broadcast traffic vs. total traffic.

CiscoWorks Blue with Native Service Point helps to improve the management of routed networks in SNA environments. Cisco routers are defined as physical units (PU) from a VTAM environment, using NetView for MVS. Developed with support from IBM, Native Service Point preserves the security features of the mainframe environment. It will provide physical and logical maps to identify SNA and IP resources and determine which SNA PUS and LUS are located on which router ports. The result is less administration effort and more visibility for status, performance and capacity management.

5.4 LAN visualization instruments

Graphical LAN documentation tools

Changes, ongoing maintenance, troubleshooting, design, planning as well as technical and customer support, all need accurate network documentation.

Listed below are a few easy-to-use instruments that incorporate graphical capabilities. They are used to visualize networks and design alternatives. However, they are not considered design and planning tools.

ClickNet from PinPoint Software Corporation

ClickNet visually represents any size network with a few mouse clicks. The interface allows diagrams to be created simply by dragging and dropping an icon from the object window. With its comprehensive, professionally drawn library of over 1300 images, networks can be constructed within a short period of time.

ClickNet includes the following modules:

- A powerful drawing utility: This drawing program is the backbone of ClickNet Professional. It was optimized specifically for creating network diagrams.
- An indispensable database: Each network symbol comes with its own data set of information about the equipment symbolized. This gives the user an automatic way to keep track of all the elements on the network.
- A complete set of predesigned reports: As the user diagrams its network, the information in the database becomes automatically available in any of 25 predefined network management reports — from a simple segment/mode listing to full network load analysis.

In particular, the following features are important to the network professional:

- Smart lines move with symbols
- Store multiple levels of information
- Dynamic database functionality
- 25 predefined, ready-to-run management reports
- Extensive library of over 1300 network and computer images
- Intuitive icon-based interface that runs under Windows
- Easy to use toolbar for quick access to common functions
- Handy palettes for instant changes in colors and styles
- Self-correcting lines and grid background assure total precision in length, angle, and alignment
- On-line help, plus context-sensitive "pop-up" menus
- Complete undo functionality; up to 20 steps can be undone if necessary

In addition to all these features, ClickNet allows "what-if" scenarios to evaluate various configuration alternatives.

NetViz from Quyen Systems, Inc.

Accurately and throughly documenting a network is a multidimensional task that involves various technologies, geographical access, networking services, business applications, and users. It consists of graphics (across two dimensions) and data (a variety of details about nodes and links) that may be captured at different levels within the network. Effective network documentation must establish and maintain the interrelationships between these two kinds of information. Thus, traditional drawing programs and database management programs, even when used together, are inadequate for documenting networks. This product takes an innovative approach to the network documentation challange. This Windows program integrates data with graphics, broadening the user's ability to organize, manage, access, and use information about the network. It provides a broad array of capabilities:

- Drag and drop graphics
- Graphics completely under user's control
- Multilevel documentation capability
- Integrated data manager
- On-screen data display
- Graphics export capability

NetViz is ideal for diagramming and documenting network, systems, and processes. The networks can be divided into manageable pieces and linked together to increase visibility. A diagram with multiple dimensions is shown in Figure 5.11. The product allows the user to drill down to any desired depth within the network documentation.

This product is distinguished by:

- Object-oriented business graphics plus integrated dynamic data management.
- Drag and drop simplicity in the creation or change phase
- Support of multilevel network topologies; by double-clicking on any node, subnetworks can be viewed in unlimited depth.
- Imports text files so the user can use data stored in existing databases and spreadsheets.
- Imports map and floor plan graphics for backgrounds, imports custom node symbols.
- Exports diagrams and data for use in presentations and word processing, exports data for use in other programs.

Considering all these features, NetViz offers an easy-to-use alternative to document and visualize relatively simple networks.

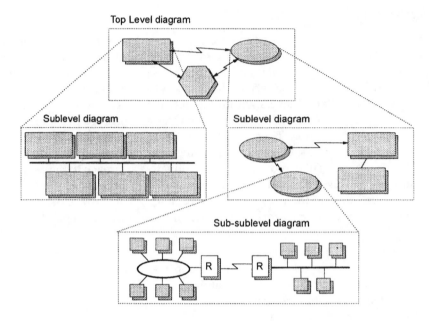

Figure 5.11

GrafBASE from Network Dimensions, Inc.

GrafBASE is a graphical database for network information and configuration management. Through multiple nested views of network and equipment layouts, the user can access configuration information for interconnected WANs, MANs, campus networks, or for LANs. GrafBASE maintains the network information in a single application, for instant visual representation, reporting, data access, presentations, planning, and tracking.

The product's features concentrate on network mapping, network documentation, exporting/importing graphics and data, and real-time alarm management.

GrafBASE's object-oriented graphics allows the creation of specific network elements of different classes and types that uniquely define the network. These elements define the different devices, links, and LANs that comprise the network. Network elements are represented pictorially in the GrafBASE tools box. The users define these elements with icon representations of their choice and provide each element its own data attributes.

To create networks, the users pick a network device from the tools box and place it in a panel with the network view. To connect devices, a network link is selected from a tool box. With a click, the devices can be connected. GrafBASE can create the links, even if the two end nodes are within different subnetwork panels.

Users can create multiple panels with different subnetwork views and then relate these hierarchically for easy navigation. Each network element that makes up the networks can be documented with detailed information, which can be accessed from GrafBASE's formatted reports.

GrafBASE can simplify the daily tasks for network engineering, planning, tracking, and support with the following tools:

Network mapping

- Map the wide-area network (WAN) hierarchically from a worldview to a specific country. Then zoom for further expansion to a county or metropolitan area.
- Map the local area network (LAN) from a campus view, to a building, floor, or a single room.
- Interrelate all views, so the user can access any view simply by traversing the network configuration hierarchy.
- Include GrafBASE's high-quality maps for geographical backgrounds or bring in externally created bit-maps or CAD (DXF) backgrounds.
- Specify a latitude/longitude coordinate and GrafBASE will correctly locate nodes on a geographical map. Alternatively, include Graf-BASE's area code/prefix data (in the United States and Canada), for automatic placement of nodes on geographical maps.

Network documentation

- Creates the user's own classes and types of network objects for the GrafBASE tools box.
- Defines the user's own attributes for each network object appearing in the GrafBASE tools box.
- Specifies detailed descriptions for each network node and link.
- Accesses detailed formatted reports for managing network information.

Export/Import graphics and data

- Enters network data from a text file easily created through current applications, a database, or word processor.
- Exports graphics to word processors, or other graphics drawing tools with Windows clipboard.
- Imports graphics in bit-mapped (BMP) format, or from CAD programs that creates DXF files.
- Imports user's owned icons for representing network objects.

With GrafBASE the user can

- Define networks visually in detail.
- Track network equipment and facilities.
- Plan and present different network layouts to management and customers.

- Quickly create and access network information for troubleshooting engineering, reporting, service, and support.

GrafBASE saves time and money by letting the users consolidate network information and access it both visually and in report format from a single, MS-Windows, PC 386/486-based application.

GrafBASE runs on low-cost IBM 386- or 486-compatible computers with up to 6 MB available on hard disk. It uses the familiar graphical interface of Microsoft Windows™, version 3.x (running standard or enhanced mode). A mouse and an EGA or higher resolution color monitor are required. The hard disk requirements vary from 2 to 6 Mb, depending on the data options installed. 2 MB is the minimum hard disk space required for program files. Print and plot to any device supported by Microsoft Windows.

Besides supporting visualization and documentation, GrafBASE can also be used as a planning instrument. Users can combine GrafBASE with analytical and simulation models to quantify the results of network planning.

5.5   LAN planning and design tools

Modeling, design, and capacity planning always starts with the intelligent interpretation of the measurement results. Performance visualization and topology layout help the analyst to understand how the current network is functioning. Related to topology is the maintenance of assets, including all kinds of managed objects. Due to the very dynamic nature of LANs and interconnected LANs, the rate of change is very high. Tools of this category are tightly connected to asset management, change management, and network monitoring. Figure 5.12 shows the interfaces of the network design and planning process to these processes.

This section introduces a number of modeling alternatives. The sequence of the product overviews is not related to the priority, applicability, or quality of the products.

Bestnet from BGS

For companies with a strong IBM presence — particularly where SNA is the internetworking solution — this tool can help to model and optimize performance. The modeling part consists of BESTnet Boundary and BETnet MSNF. The modeling part uses accurate and up-to-date configuration and workload data collected by Capture. The actual data are collected in VTAM and in the network and are merged together to guarantee unique visualization of the configuration, its load, and utilization.

Modeling is supported for SNA-sessions with token rings, SNA-sessions over X.25 and SNA sessions across SNI-gateways.

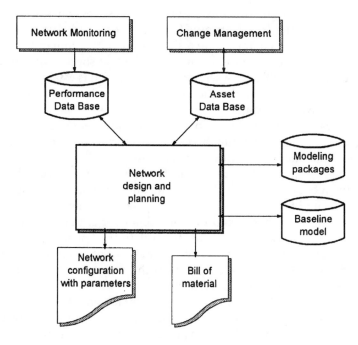

Figure 5.12

COMNET III from C.A.C.I.

COMNET III is an off-the-shelf modeling tool that predicts the performance of LANs, MANs, and WANs. This product is capable of modeling the following technologies: X.25, ISDN, SS7, frame relay, cell relay, FDDI, SNA, TCP/IP, DECnet, CSMA/CD, token passing, polling, radio and satellite networks.

Nodes and links define the network topology. They can be organized hierarchically into subnetworks. Nodes perform processing functions and contain parts for connecting to links. Ports provide input or output buffering and processing. A link is a physical transmission facility connecting parts on two or more nodes.

The library of node objects includes a generic node model as well as specific configurations for modeling computer groups, routers, and ATM switches. The library of link objects includes point-to-point, CSMA/CD, CSMA, token ring, FDDI, token bus, polling, and ALOHA models.

A node has a list of commands that it can execute. The command library includes process data, read file, write file, transport message, answer message, and setup session objects.

Sequences of these commands are associated with application sources attached to a node. Applications can be scheduled to occur according to an interarrival time distribution or contingent on satisfaction of an incoming message requirement.

Attaching message sources, response sources, session sources, or call sources to a node can also contribute to the network load. A message source produces applications that execute a single transport command; a response source executes a single answer command. A session source executes a setup command, which establishes a session. Once a session is established, a series of messages is sent for the duration of the session. A call source is used to model circuit-switched traffic.

Messages are transported from source to destination using transport, routing, data link, and medium access control protocol objects. The source and destination nodes provide transport services. The transport service breaks a message into packets at the source node and reassembles the packets into a delivered message at the destination node. The packets are transported by an end-to-end transport connection. A node selects an outgoing port for a packet using a routing protocol. For subnetworks that use connection-oriented routing, packets follow the path established at session setup time. The data link service at the port creates frames out of incoming packets and then uses the medium access controller at the port to contend for the link. Physical layer services are provided by links.

Users can add their own variations to any of the object libraries of nodes, links, traffic sources, protocols, and distributions by copying an object and changing its parameters or by developing a new object. Adding objects by changing parameters requires no programming; developing a new object requires a minimal amount of programming using MODSIM II.

In particular, for LANs, the following reports are provided:

- LAN statistics, including packet/frame transfer time, transit time, system waiting time, overhead percentages, throughput rates, message losses, interarrival distribution diagrams.
- Client statistics showing queue lengths, waiting time, and eventual blocking.
- Server statistics, concentrating on queue lengths, waiting time, blocking, and storage utilization.
- Link statistics, including expected utilization rates and throughput.

NetSolve from Quantessential Solutions

NetSolve is designed for interconnected LANs. It is a wide-area network modeling software that contains several WAN pricing and design applications. The product supports a maximum of 250 switching locations with a maximum of 25,000 switch traffic routes.

NetSolve has a modular structure. It starts with simple point-to-point pricing and performance calculations, and ends with Mesh Designer. Mesh Designer contains a set of modules that automate mesh network design and optimize the key criteria of price, performance, and reliability. Standard features include billing ratios and the ability to create custom tariff and point-of-presence databases. NetSolve can interact with COMNET III from C.A.C.I.

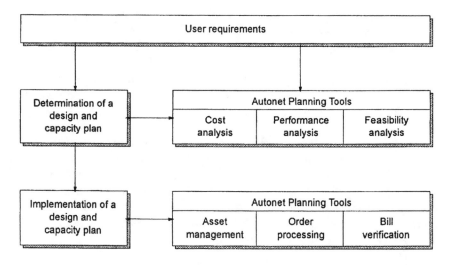

Figure 5.13

After Mesh Designer is used to produce a least-cost LAN interconnection model, the model file can be exported to COMNET III, allowing analysts to evaluate the performance of the model. The result is an optimal WAN performance at reasonable costs.

Autonet from Network Design and Analysis Corporation

Autonet consists of planning and operational tools (Figure 5.13).

These planning tools help with optimizing the interconnection of LANs. In particular, the following functions are supported:

- Design and maintenance of cost effective network
- Effective use of networking resources
- Identifying billing errors
- Building a contingency plan
- Optimizing network performance
- Connecting the models to change management
- Connecting the models to asset management

Performance-1 provides performance analysis for WANs and LANs using analytical modeling techniques with the convenience of a Windows graphical user interface. The WAN component includes a queuing model for most known WAN protocols. The LAN component is based on a discrete time analysis model. Both Ethernet and token ring are supported. The main goals with this module are:

- Evaluating the impact of increased traffic by performing a sensitivity analysis

- Analyzing the performance of the existing network using indicators like response time and resource utilization
- Determining the maximal allowable traffic for facilities and equipment
- Planning future growth to optimize network design and
- Evaluating LAN performance of Ethernet and token ring networks to ensure ideal operating conditions

Performance-3 is a comprehensive performance analysis tool for multiprotocol networks comprised of LAN-to-LAN connections and WAN transmission facilities. By entering the network topology, equipment specifications, and workload profiles, the model will generate application response times and network utilization. Link and node failures can also be simulated to create robust networks using the Windows graphical user interface. The main goals with this module are:

- Determining network bottlenecks in facilities and in equipment
- Achieving high accuracy and quick results to review
- Predicting future performance problems to avoid stressed utilization of resources
- Performing and modeling failure analysis to evaluate whether the network has adequate spare capacity for emergencies
- Importing traffic data from LAN probes, which gather information on network traffic
- Evaluating local LAN performance in Ethernets, token rings, and FDDIs to ensure ideal operating conditions under various load mixes
- Performing what-if analyses to determine the behavior under changing configurations and workload conditions

Autonet, in combination with special LAN modeling tools, can support an enterprise modeling solution.

OPNET from Mil3

OPNET is a comprehensive software environment for modeling, simulating, and analyzing the performance of communication networks, processors, and applications, and also distributed systems. OpNet presents an advanced graphical user interface that supports multiwindowing, makes use of menus and icons, and runs under XWindows. Supported platforms include widely used engineering workstations from Sun, HP, DEC, IBM, and Silicon Graphics.

Graphical object-oriented editors for defining topologies and architectures directly parallel actual systems, allowing an intuitive mapping between a system and its model. Its hierarchical approach simplifies the specification and presentation of large and complex interconnected LANs.

The process editor provides a powerful and flexible language to design models of protocols, resources, applications, algorithms, queuing policies, and other processes. Design specification is performed in the Proto-C language, which combines a graphical state-transition-diagram approach with a library of approximately 300 communication- and simulation-specific functions. The full generality and power of the C language is also available.

OPNET simulations generate user-selected performance and behavioral data. Simulation results can be plotted as time series graphs, scatter plots, histograms, and probability functions. Standard statistics and confidence intervals can be generated and additional insight can be obtained by applying mathematical operators to the collected data. The product provides an advanced animation capability for visualizing simulation events. Both automatic and user-customized animations can be displayed interactively during or after a simulation. Animations can depict messages flowing between objects, control flow in a process, paths of mobile nodes, and dynamic values such as queue size or resource status.

OPNET provides open system features including interfaces to standard languages, the ability to take advantage of third-party libraries, an application program interface, access to databases and data files such as those generated by network analyzers, and PostScript and TIFF export for desktop publishing.

Their modeling technology is based on a series of hierarchically related editors that parallel the structure of actual networks. The usual approach starts at the wide area and ends in process models of selected nodes.

Node objects are created to represent the various communicating sites of a network. Each node belongs to a class that defines its internal structure and attributes. Either built-in or user-defined properties of node models can be used as attributes. Attribute examples include the processing speed of a switch, the buffer capacity of a gateway, and the traffic generation rate of a workstation.

Nodes communicate with each other via various communication links. All versions of OPNET support fixed-position nodes with point-to-point and bus links, e.g., Ethernet LANs. In addition to these, OPNET Modeler/Radio supports mobile and satellite node communications over radio links. Link models are used to account for delays and transmission error rates. Each type of link model can be customized, if necessary, to fit particular applications. Models of local or metropolitan area networks can be created within the context of a subnetwork. They can be designed to contain any combination of node and link objects deployed within their local coordinate systems. Communication links can be defined between subnetworks to form internetworks on a larger scale. Subnetworks can contain other subnetworks to form unlimited network hierarchies.

The Process Editor combines a state-transition-diagram representation with an extensive procedure library to support rapid development of process models. Since protocol and algorithm specifications are often defined using these diagrams, this development environment gives the modeler a critical head start in transforming specifications into operational models.

The Analysis Tool provides a graphical environment to view and manipulate data and statistics collected during the simulation. It supports the evaluation of complex network performance and behavior. Standard and user-specified probes can be inserted at any point in a model to collect data and statistics. Simulation output collected by probes can be displayed graphically, viewed numerically, or exported to other software packages. In addition, results from a series of simulation runs can be automatically collated into a single OPNET output file, facilitating sensitivity analysis. OPNET graphs can be printed in color or black and white on PostScript output devices. Graphs can also be exported using standard TIFF or Encapsulated PostScript formats for placement in desktop publishing packages. Using menus, network variables can be inspected and compared in a multiwindow environment that includes time series plots, histograms, probability and cumulative distribution functions, scattergrams, and animated displays of process model execution.

Typical performance indicators include:

- Response time and its distribution
- Transfer delays in nodes
- Communigrams between selected stations and applications
- Bandwidth utilization
- Packet length distributions
- Throughput of interconnecting devices
- End-to-end delay analysis

Data exports facilitate the use of other tools (e.g., SAS to validate OPNET models with measurement data collected by RMON-probes or other monitors).

BONeS network modules from the AltaGroup

The BONeS network modeling modules comprise a comprehensive set of libraries of network devices and applications. The BONeS network modules support the following activities:

- Determining the effect of additional traffic on existing configurations
- Evaluating the impact of new applications before the software is purchased or upgraded and installed
- Determining when to upgrade LAN segments to different protocols or higher speeds
- Configuring interconnections with various types of routers
- Sizing WAN links and plan for link failure
- Optimizing WAN capacities and determine current and projected bandwidth requirements
- Performing client/server studies to determine the hardware capacity required at the server for the requisite number of users, as well as the impact of the additional traffic on the network

- Analyzing router performance and experimenting by simulating various hardware interconnecting devices and trying out various routing protocols

The BONeS network modules can be used with the BONeS PlanNet simulation engine for plug-and-play simulation using standard components. When used with the BONeS Designer engine, the user can develop new modules or make detailed adjustments to existing network modules. Simulation of the network results in plots of various performance metrics such as delay, utilization, and throughput. These plots can be used to make decisions on the WAN, MAN, and LAN design.

LAN segment modules are used to model network segments using specific network protocols. When a traffic generator is attached to the network segment, packets are generated in accordance with the specified LAN protocol, and packet transmission is simulated with one of these modules. Each module acts as a bus or traffic concentrator with traffic modules connected to represent user nodes. LAN segment models include:

- Token ring with 4 Mbps, 16 Mbps, and 16 Mbps with early token release.
- Ethernet represents the standard Ethernet — 10Base-5 with the CSMA/CD family of protocols.
- 10Base-T module models the behavior of 10Base-T with the CSMA/CD family of protocols.
- 100Base-T module models the behavior of 100Base-T with CSMA/CD family of protocols.
- FDDI module models the high-speed token passing protocol that operates over fiber media. FDDI includes a complex capacity allocation scheme to accommodate both bursty and stream traffic consisting of synchronous (voice and real time video) and asynchronous (typically data packets) portions.

Table 5.5 summarizes the input parameters and output results for each of the local area networks modeled.

BONeS interconnect modules assist in identifying bottlenecks, evaluating throughput of interconnected components, discovering impact of various routing protocols, and developing spanning tree bridge configurations. The router library consists of two components:

1. Multiprotocol router with vendor specific parameters: it represents a bridge/router that can be used to connect LAN segments to each other or to a WAN. This module supports Apple ARP, DECnet, Novell IPX, OSPF, RIP, and XNS protocols. To allow users to support other routing protocols, there is a generic routing protocol available. Using these features, users can add other routing protocols to the module by

Table 5.5 Input and Output Parameters for Modeling LANs by BONeS

LANs	Ethernet	10-BaseT	100-BaseT	Token ring	FDDI
Input parameters					
Segment/ring length	x			x	x
Mean cable length	x	x	x		
Repeated signal distance		x	x		
Propagation delay	x				
Target token rotation time					x
Priority thresholds					x
Synch allocation/node					x
Buffer size/node (frames)				x	x
Hardware latency				x	x
Label	x	x	x	x	x
Collect statistics	x	x	x	x	x
Output parameters					
Delay	x	x	x	x	x
Packets/sec	x	x	x	x	x
Collisions/sec	x	x	x		
Transmission attempts per packet	x	x	x		
Bits/sec	x	x	x	x	x

specifying the protocol characteristics. The multiprotocol router also supports spanning tree bridging. It models the performance characteristics of approximately 20 router devices, including those from 3Com, Cisco, Chipcom, DEC, IBM, Proteon, Vitalink, Wellfleet, and Xyplex. The performance characteristics used for modeling the devices are obtained from published benchmarks of the vendors. The multiprotocol router consists of two types of components: processors and ports. Data is received or transmitted via the ports, and the processors make routing decisions. Statistics can be collected for the ports and/or the processors.

2. Single protocol router: It is a streamlined version of the multiprotocol router and is optimized for ease of use. This module supports only one routing protocol. The vendor-specific performance parameters are not available with this module.

The bridge module provides the spanning tree algorithms. It consists of two types of components: processors and ports. Data are received and transmitted via the ports, and the processors make bridging decisions. Statistics can be collected for the ports and/or for the processors.

Other modules support WAN-modeling and traffic generation for applications running under TCP/IP, FTP, NFS, and NetBIOS.

NetReview design service from IBM

IBM has packaged design and modeling tools into a consulting package. This package contains the following modules:

- Traffic survey: This module analyzes data extracted directly from the network components and reports what is happening in the network, revealing bottlenecks or potential problems.
- Cost optimization: Even in a network with a technically sound design, there are still many choices to make that have significant cost impacts; for example, decisions on which service to use from which suppliers, what devices to use where in the network, use of more centralized or distributed structures, etc. This module produces a cost-optimal data, voice, or integrated traffic network solution that delivers the correct level of service to support business requirements while minimizing the investment in communication facilities and equipment.
- Backbone design: This module considers traffic volumes, performance and availability requirements, and produces an optimal backbone design and routing strategy for the networks.
- Performance simulation: This module allows one to project network design into the future and analyze its probable performance. Simulation provides maximum reassurance that the design choice will keep the network's future proof.
- Design consultancy: In addition to the above modules, network design specialists are available on a daily consulting basis to help users with network design problems using various types of instruments from IBM and from third parties.

This professional service package is very IBM-oriented. The input parameters are populated by IBM network monitors or application packages. IBM does not offer specific modules for LAN design and monitoring. The package is limited to SNA backbone design and performance prediction and is useful in cases where SNA is used as the LAN interconnection architecture.

NetMaker from Make Systems

NetMaker is an object-oriented suite of tools for designing, simulating, and analyzing internetworks. It allows users to visualize network data incorporating simulation of specific vendor's devices and tariffs from specific carriers. The product is a UNIX-based client-server software package that helps managers design their networks, judge how specific equipment or services will affect their installations, predict the need for new equipment, services, or bandwidth and foresee potential performance bottlenecks.

Users can populate the database by adding predefined templates, loading log files from network devices, and importing files that define characteristics of various pieces of equipment [JAND94]. All applications, menus, and graphics are melded together seamlessly and transparently for the end user.

In addition to parameters about network devices, NetMaker can channel other types of data into its reports and displays, including an optional database of tariff information for both dial-up and leased lines from carriers such as AT&T, MCI, and Sprint. Users also can plug in any other tariff information they have on hand.

NetMaker supports various multiplexers and routers that interconnect LANs. Make Systems connects the modeling device to monitors from Frontier Software Development, Inc. and Concord Communications. The monitors will load the database with accurate performance data on service and utilization indicators.

NetMaker includes six core applications. Visualizer is a prerequisite for all other applications. It furnishes the graphical user interface, a network map, assorted report templates, and network device templates. The Planner works with plug-in libraries to perform network simulation, predicting the behavior of networks using parameters for specific equipment. Analyzer gives managers a clue as to what will happen if specific connections or devices fail. Accountant supplies the tariff data that is integral to many NetMaker reports. Designer exports information from the Accountant to explore alternative network topologies. Interpreter analyzes traffic in router-based internetworks.

Visualizer is optimized for managing large and complex networks. It solves the problems of acquiring, organizing, viewing, and reporting on large, complex network data. The users can request many sublevel views, to display accurate, detailed information about the status and configuration of the network. Visualizer provides management reports and a report writer so the user can create custom reports. The network topology can be viewed using geographical or logical layouts. Management reporting is enhanced with printed layouts showing the network's actual connectivity. Visualizer provides ASCII import facilities to acquire and baseline data about the network, and an Object Editor to facilitate making changes to network objects and to understand the impact of changes on the network. Visualizer plug-ins are used to gather network topology and view data.

Interpreter extracts and organizes information about network traffic patterns and loads this information onto the physical topology of the network. Reports organize traffic by protocol type, protocol mix, and location. Traffic profiles can be built to model effects of adding or moving users or of merging or adding LANs. Data is then used in application planning and capacity planning to see effects of traffic pattern changes on your LAN/WAN utilization. Interpreter's traffic modeling capabilities are facilitated by plug-ins to acquire traffic data.

Using the simulation engine and libraries, which include device-specific routing behaviors, Planner lets users model "what-if" scenarios, such as adding new users, adding new applications, changing the topology by adding new network elements or changing bandwidth, and reconfiguring attributes that affect traffic routing, routing metric conversions or quality of service. Planner helps to improve LAN/WAN utilization, reduce the time and uncertainty involved in planning and implementing changes, and identifies the most optimal and cost effective deployment of network resources. Planner also provides data input to report on LAN/WAN use, to graphically display traffic paths, and to report on demand-specific failure messages. Combined with Planner simulation capabilities, Analyzer's survivability and sensitivity analysis is useful in developing disaster recovery plans. Analyzer helps measure network sensitivity to changes or potential failures. It provides an early warning system to avoid loss of mission-critical applications caused by transmission and equipment failures, and to identify problems that may result from changes in network loading. Analyzer identifies problems related to date and time of day use of network resources.

Accountant helps determine the least-cost providers of bandwidth, and the best allocation of bandwidth and equipment costs among network users. Accountant helps to identify cost-effective alternatives to WAN backbone designs and to determine choices among service providers to ensure cost savings are realized. The user can query its plug-in databases for service provider alternatives and edit custom databases so exact cost data can be used to analyze network utilization. Accountant uses plug-ins to calculate carrier costs and to assign network costs to users.

All the applications work together behind the scenes, contributing functions to one another. For instance, if a department is planning to move to a new location, a manager might query the package about new bandwidth requirements and the associated costs; to display the answer entails the teamwork of the Planner and Accountant applications. To present a report about reducing costs while still fulfilling user's bandwidth needs, the Designer and Accountant applications work in concert. Administrators can call up a network map and click on an icon that represents a network device or connection. They can drag that icon to an open management report to bring up detailed information about the item. Users can edit this data and run new analyses to predict what might happen if, for example, more bandwidth were added to a network connection. Managers can obtain information about the most cost-efficient type of carrier services to use with particular vendors' devices.

NetMaker offers the combination of simulating various equipment and a decision making tool at a reasonable price.

Coronet Management System from Compuware

Coronet Management System is a software suite that automatically identifies the specific applications in a network and denotes their bandwidth consumption and response time. The three Windows-based packages of the system not only monitor LAN traffic without requiring specialized probes or hardware but also automatically identify the applications generating that traffic.

With Coronet's SingleView, Super Monitor, and Quick Model software, network troubleshooters can diagnose and pinpoint application-layer problems on Ethernet and token ring LANs for a range of network protocols, including TCP/IP, IPX, LAN Manager, Appletalk, DECnet, and Vines (JAND95). In addition to spotting performance problems, the Coronet packages can help network managers determine the throughput needs of specific applications. This, in turn, makes network planning easier. The package runs either as a standalone monitor or as part of ManageWise from Novell. SingleView and Super Monitor are the core components of the management suite. SingleView functions as the management console. It uses SNMP to discover network devices — including workstations, servers, bridges, and routers — in a multisegment internetwork and generates a network map of those devices. Super Monitor serves as the monitoring device at strategic points throughout the network. It turns the host PC into a management agent that listens in on LAN traffic and sends data back to a SingleView console. The software monitors the so-called conversation path between specific clients and servers, showing exactly which applications are running on a particular segment, where they are coming from, and how much bandwidth they are consuming. Super Monitor also gauges related response times for each application. Super Monitor does not have to be placed on every server or workstation on the LAN to monitor application traffic. Instead, network managers can deploy the software on workstations situated near key network gateways on LAN segments, such as next to routers, servers, or bridges.

Super Monitor uses the PC adapter card as its point of entry into the network. The software resets the adapter to what Coronet calls promiscuous mode, which enables Super Monitor to capture all the packets traversing the LAN for analysis. Packets captured by the monitor go through several different analyses. First, Super Monitor looks at the port configurations, or port bindings, associated with specific applications. Network applications — like Notes from IBM-Lotus or the Oracle database management system from Oracle — incorporate unique port bindings into the network protocol. These port bindings identify the application to the network operating system. By reading the port bindings, the monitor can identify which applications are responsible for captured packets. Users can configure port bindings for custom applications.

Super Monitor also identifies LAN traffic associated with specific applications by tracking "open file" requests on NetWare IPX LANs. These requests are included within IPX packets sent from clients to servers. When Super Monitor decodes the packet, it reads the application name from the open file request. In addition to decoding the packets, the monitor can analyze files of packet data captured by protocol analyzers made by Novell and Network General. Data gathered by distributed Super Monitors is forwarded to the network manager's SingleView console. The information is stored in a database at the console for use by Quick Model, the third component of the package. Quick Model contains a variety of rules that are applied to captured data to perform "what-if" analyses.

The combination of an application monitor and modeling package is very useful. With this product, the criticality of component outages can be quantified. In addition to physical and logical views, application views can be displayed. Based on accurate data, the modeling part helps to make short-term decisions about configuration changes in the LANs.

Optimal Performance from Optimal Networks Corporation

Optimal Performance imports real network topology and traffic collected by network monitors such as Expert Sniffer Network Analyzer and Distributed Sniffer System from Network General to automatically create an accurate model of the networks. Using a high-performance engine designed specifically for network analysis and simulation, this product then analyzes the dynamics of the networks.

With an analysis of performance, the product delivers specific problem-solving recommendations that address common operational decisions. Based on the optimization criteria that the user specifies, recommendations are made on:

- LAN segmentation
- Hub partitioning
- LAN switch deployment
- Optimal server and workstation positioning
- Optimal application positioning
- Capacity requirements for WAN links
- Survivability of bridges, routers, switches, and WAN links.

The predictive "what-if" capability lets the user evaluate proposed network designs and changes before financial commitments are made. The product simulates all seven layers of the OSI reference model to ensure a realistic and accurate reflection of a network's design and performance. The software's event simulator gives the user an animated picture of the proposed network, making it easier to understand its operation. Optimizing the existing network and providing previews of future designs and changes helps to save time and money and an optimal return on investment.

Optimal Performance analyzes networks in order to make specific optimization recommendations. The user sets the goals that govern the analysis of the network. Optimization goals include LAN segmentation constraints, utilization thresholds for LAN segments and WAN links, and whether the positioning of servers and applications should be based on achieving the least average delay for all clients, or minimizing overall network traffic. The user defines specific optimization settings for how he or she wants to tune the networks and to focus on areas of greatest interest.

Given a model of the network and the optimization goals, Quick Analysis reviews the network traffic using computational methods to produce a set of recommendations and reports. An Event Simulation can then be used to

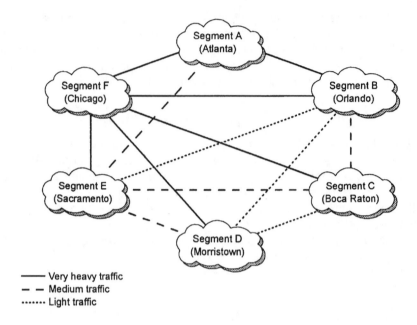

Figure 5.14

characterize timing-dependent qualities of traffic flow and network capacity. The user can view the traffic dynamics of the model, monitor activity on segments, bridges, routers, Ethernet switches, and WAN links. For in-depth analysis, the user can capture operational details of any model element and analyze its performance following a simulation run. Event simulation allows one to receive both optimization recommendations and detailed, time-dependent performance analysis.

Optimal Performance offers comprehensive reports, traffic matrixes and graphs. Included are reports on optimization, survivability, traffic, and network performance. Following an event simulation, the user can also generate various statistical graphs. The reports enable the network managers to present a business case for network changes and major investments to non-technical managers with data, graphs, and easy-to-understand reports.

Figure 5.14 displays a "traffic matrix of segments" traffic. It can also be broken into a traffic matrix between workstations in the LANs.

Other available reports include:

- Each recommendation for a LAN segment includes a summary of its configuration, its performance, and the utilization objectives for optimization.
- If the recommendation is to split a LAN into two or more segments, the workstations and servers to be attached to each segment are specifically listed. The user can use the report as an action list for changes that should be made to the network configuration.

- The optimal segment for a server or an application is specified and the relative performance improvements on every other segment are listed. Recommendations for the optimal positioning of a server provide its traffic load and optimization setting. Similarly, recommendations for the optimal positioning of an application identify the number of clients, the total number of messages transmitted and received, and its optimization setting.
- For WAN links, recommendations define specific bandwidth requirements to achieve a desired utilization. A summary identifies the connections of a WAN link, as well as its use in each direction of a connection.
- An optimization report provides a summary of recommendations for LAN segmentation, the optimal positioning of servers, workstations and applications, and the capacity requirements for WAN links. Optimization recommendations for LAN segments include the splitting of some segments and the replacement of others with higher-speed alternatives in order to achieve specified goals for utilization.

With this product, it is easy to change a network design. Starting with a model of the existing network, the user can make changes to evaluate their impact and can automatically implement recommendations. LAN segments, workstations and servers, bridges, routers, Ethernet switches, WAN links and applications can be added, moved, deleted, and reconfigured. One can point-and-click on the icon for a LAN segment to change it from an Ethernet to a FDDI segment. Similarly, a server can be moved easily from one segment to another. In addition, network protocols or routing protocols can be examined and their impacts determined, or an additional T1 link can be added. The traffic generated by a client/server application can also be modeled and modified to see the change in end-to-end response time and the demand on network capacity. Through these types of simulations, changes to the network can be modeled before the changes are actually implemented.

When evaluating design and planning products, LAN managers should consider many issues, such as how the network model is created, how the network traffic is characterized, and what platforms the applications run on. Managers should be sure to find out whether the products they are considering perform event simulation or not. With event simulation, an application will simulate each step of a network process dynamically, such as setting up a call over a WAN, allocating bandwidth, and taking down the link. In contrast, other products perform analytical modeling and rely on static routing tables, so they do not give a complete picture of actual network utilization or reflect changes that occur dynamically on the network.

Optivity from Bay Networks

A major problem in network planning is the lack of adequate data on the network behavior. Until now, long-term planning issues critical to managing network growth and expense have been largely based on "gut feeling" and wishful thinking. Optivity Planning from Bay Networks gives network managers the empirical information and simulation tools they need to help reduce cost of ownership through long-term trending analysis and rational network design. The new package includes NetReporter, an SNMP/RMON-based data collection and reporting tool, and DesignMan.

NetReporter is a powerful data collection and reporting tool. Data collection is based on SNMP/RMON, enabling users to take advantage of the widest possible range of Bay Networks–embedded RMON agents, OnSite standalone probes, StackProbes, and MIB-II devices for long-term trending based on actual traffic.

Using NetReporter, the network manager can define and run a number of data collectors. Each data collector may be defined for a domain of interest — individual hosts, a group of segments, or the entire enterprise network. The user can also specify the types of data collected from a selection of predefined report types. Sampling and storage intervals are also configurable for each data collector to maximize report resolution and minimize disk space use.

Once defined, the data collector may be activated to gather data over any time span — hours, days, weeks, months, or years. NetReporter operates independently of real-time Optivity applications, and multiple data collectors may be active simultaneously. All data is stored in a bundled SQL database for use with NetReporter report generation. The database is transparent to the user and requires no special expertise for installation or ongoing maintenance. NetReporter also provides an easy migration path to a full client/server RDBMS environment for storage of all network related data.

The functions supported include:

- Unattended, long-term data collection for Ethernet, token ring, and FDDI networks using Bay Networks–embedded RMON agents, OnSite standalone RMON probes, StackProbes, NMMs, and MIB-II devices.
- The ability to run multiple data collectors (probes) simultaneously.
- Wide range of built-in reports for measuring network performance characteristics over time.
- Automatic, periodic report generation based on user-specified frequency and display criteria. An autosave to GIF feature allows automatic posting of reports to the Web.

Presentation-quality reports can be produced at any time while a data collector is running or following its termination. The user can select from a wide range of two-dimensional and three-dimensional report formats, including pie charts, bar charts, line graphs, and tabular reports. Data may be cross-sectioned to view either all variables for a single segment or to view a single segment for all variables in the data set. Data is automatically extracted from the database, and the user can adjust the report interactively before sending it to a printer. NetReporter includes an automatic report feature allowing the user to specify in advance the frequency for generating reports from a specific data collector. Generation, either direct-to-printer or to a file, is fully automatic.

Another key output option is the ability to save reports to GIF (graphical interchange format), allowing immediate World Wide Web access of any report. Reports are automatically stored in a directory hierarchy, utilizing a naming convention that uniquely identifies each report, and allows easy posting from any Web server.

In addition to producing formatted reports, the user may also export collected data for use with other applications. This effectively extends the use of data gathered by NetReporter for use as input to other products, including popular spreadsheet programs.

Managing the growth of network data after collection is another task at which the NetReporter excels. The database size is fully user configurable, and the application keeps track of the size of every report and how much total disk space is being used by reports. Data aggregation is critical for managing the growth of collected data used for long-term reporting. The NetReporter can perform user-configurable data aggregation on the way into the database and can be "rolled-up" after collection.

DesignMan adds Layer 3 support, using pre-RMON2 extensions to display data flows at the logical subnet level. This is in addition to the physical segment level information available in previous versions. The application uses historical data from the network to assist network managers in identifying the optimal configuration for segmenting LANs, or for placing nodes in a client/server environment. Real-time data is first collected over a period of time, and then traffic analysis tools help identify either the segments that would benefit from further segmentation, or the nodes that are best candidates for moving. Finally, a switch simulator lets the network manager experiment with various configurations to determine the best possible distribution of switches, hubs, and end nodes. It is a graphics-based network design tool for planning network segmentation. It offers quick analysis of end-station and segment traffic patterns.

DesignMan uses data sets called collections. Data can be collected over a user-defined period from a single Ethernet or token ring segment, or from multiple segments connected to a switch, bridge, or router. This data, including byte and frame counts by MAC or IP-layer source and destination address pairs, completely characterizes traffic on the segment. The data is saved to a file on the local workstation, enabling the network manager to "play" this actual LAN traffic back through the DesignMan interface.

DesignMan allows the network manager to hypothetically "assign" discovered individual end stations to new ports by simply clicking on a given address and typing in the port number. Data can then be replayed to see the simulated effects of the move before any changes are actually made. With DesignMan, network managers can test new designs before moving a single wire, virtually eliminating the need to make physical network changes to find optimal LAN configurations through trial and error.

DesignMan simulates 10 megabit-per-second (Mbps) Ethernet, 100 Mbps Fast Ethernet or FDDI, 4 or 16 Mbps token ring, and 155 Mbps ATM, as well as other user-configurable speeds. The DesignMan main window includes VCR-style controls for playing back data, including buttons for play, pause, rewind, and stop. Playback can be interrupted at any time, or when performance thresholds are reached.

With the addition of Layer 3 support, the user now has the choice of using DesignMan to collect and manipulate data at the logical subnet level, in addition to the physical segment information available in the previous version. This means that DesignMan now provides meaningful information in routed as well as switched or bridged environments, since Network Layer addresses are used in the data collection. Thus, subnet utilization graphs are possible, in addition to segment utilization graphs.

Summary views are available for segments and subnets. These views graphically display station-to-station traffic. Filters are supplied to display a user-defined number of top talkers, or stations with greater than a configurable percentage of total traffic.

Two distinct algorithms are provided for distributing stations around the ports of a simulated bridge or switch in DesignMan's Layer 2 mode.

The bridge algorithm tries to group stations that spend the most time talking to each other on a common LAN and also tries to prevent top-talker pairs from using the same cross-LAN link.

The switch algorithm attempts to maximize the throughput of the switch fabric. As a result, top talkers are placed on separate dedicated LANs, and devices that generate little traffic are distributed to other LANs. An attempt is made to identify servers and to dedicate LANs to them.

DesignMan also allows the user to manually drag and drop nodes from one segment or subnet to another and then replay the collection to explore the effects of the move. For example, when the main screen shows that traffic across a router is high due to users on subnet (or segment) A accessing a server on subnet B, users can drag the server from the Subnet Summary screen onto the subnet A icon. When Rewind is selected, the matrix data will be recalculated taking into account the effects of the move. The utilization graphs for both subnets will change based on the move, as will the main screen's matrix display and table.

Optivity Planning eliminates the guesswork in network configuration and planning by producing reports on network patterns over time and providing simulation modeling based on playback of actual network traffic. Optivity Planning allows network managers to identify and respond to changes over time and to maximize the distribution of switches, hubs, and end users to maintain and improve network performance.

Optivity Planning runs on management platforms, such as HP Open-View Network Node Manager, SunNet Manager, and IBM NetView for AIX. In addition to the above platform support, a standalone UNIX version is also available. This platform-independent version can run under SunOS,

Solaris, HP-UX, and AIX operating systems without having to install a network management platform.

Performance Tools from Netsys

The NETSYS Enterprise/Solver™ family of network intelligent tools aids problem solving, management, and planning of an overall network. The Connectivity Tools™ manage the network as a system rather than as independent devices. The Connectivity Tools use the VISTA (View, Isolate, Solve, Test, and Apply) network change methodology. VISTA enables the user to manage quickly and effectively change on the network, helping control the network-wide effects of any moves, adds, or changes. The integrated simulation capabilities instantly identify end-to-end configuration errors. Because users can test the changes on off-line, the risks to the live network are minimized.

The NETSYS Connectivity Tools are used to view protocol-sensitive topologies, letting one visually navigate networks to quickly gain a complete understanding of how they work. The topologies are derived from actual configuration files and are automatically drawn — users do not have to manually link interfaces and group objects. All physical and logical relationships between routers on the network are presented so the user can focus on solving problems, not drawing pictures. Topology views such as Campus, Virtual Ring Group, and OSPF areas are only a few of the options that give this feature. All protocol-sensitive topologies support:

- Visual queries highlighting devices running combinations of network or routing protocols
- Pop-up menus accessing configuration information on each LAN, link, router, and circuit
- Graphical views of routing table coverage
- Graphical displays of round-trip paths between end systems

The NETSYS Connectivity Tools speed problem resolution by validating network configurations and round-trip connectivity between devices. The Integrity Checks Report highlights occurrences of more than 75 types of common configuration problems within and between routers, including:

- Access list errors
- Overlapping subnet masks, redundant addresses, unmapped WAN circuits, bad subnet masks
- Unnecessary routing updates
- Incomplete peer definitions for RSRB, DLSw+, BGP, etc.
- Mismatched encapsulations

Other reports that are automatically generated include: Blocked Routing Updates, Partitioned Major IP Networks, IP and IPX Routing Exceptions, and Device/Address Inventory. Combined, these reports provide a quick overview of the network, pinpointing error sources.

The NETSYS Connectivity Tools calculate routing tables for all routers in network models. The user can create connectivity tests to quickly identify problems between any IP, IPX, RSRB, or DLSw+ devices. Automatic tests also check for problems with indirectly connected static routes and at routing protocol boundaries. Users can view complete or partial round-trip paths in the topology and obtain detailed step-by-step path information from an accompanying report.

The NETSYS Connectivity Tools' comprehensive topologies and reports are tools to solve or prevent problems on a network. Having isolated problems, the user can easily correct router configurations by, for instance, tuning an access list to seal a security loophole or modifying a routing protocol to advertise routes correctly. Since the users make changes to copies of configurations off-line, the users minimize risk to the live network.

The NETSYS Connectivity Tools offer integrated simulation capabilities. This enables the users to test planned changes to networks. Users can identify and fix hard-to-find downstream connectivity and security problems before rolling them out on live networks. This capability significantly reduces rollbacks and downtime, resulting in substantial savings in time and cost.

Using the "what-if" capabilities also allows you to quickly determine the effects of planned changes such as:

- Access list modifications
- Route filtering
- Protocol migrations
- Address changes

"What-if" capabilities also enable the user to verify that key connectivity and security requirements are met, even in the face of device or network failures. It is virtually impossible to verify this manually in even moderately sized networks. The final step of the VISTA network change methodology is applying configuration changes to live networks. The NETSYS Connectivity Tools simplify this process by exporting "router-ready" command delta files. The delta files contain all changed commands plus appropriate "no" commands to remove previous settings, thereby eliminating a time-consuming and error-prone manual task. The delta files can be easily applied to networks using telnet, ftp, or CiscoWorks. The NETSYS Enterprise/Solver product line is an integrated network management toolkit designed to troubleshoot, manage, and plan end-to-end networks.

The NETSYS Connectivity Tools parse actual router configuration files and generate routing tables to diagnose and solve connectivity, routing, and security problems. By regarding the set of configurations as a system, the Connectivity Tools bring a unique, end-to-end network perspective to trouble-shooting, management, and planning, thereby improving network uptime while reducing the time and cost required to manage the network.

The NETSYS Performance Tools can be added to the Connectivity Tools to diagnose and solve network performance problems, tune existing networks, and plan network changes. The Performance Tools add vendor-certified device performance models and actual traffic data collected from network probes to analyze interactions between traffic flows, topology, routing parameters, router configurations, and Cisco IOS features.

The NETSYS Advisor can be added to the Connectivity and Performance Tools to give network operations staff and managers greater control of their router configurations and network performance. NETSYS Advisor solves a wide variety of routing and access-list errors, providing specific commands and advice on the errors it detects in networks. Automated collection and reporting functions execute connectivity and performance tests on up-to-date network information. A new Web interface also allows wider distribution of NETSYS reports and protocol-sensitive topologies.

The Multi-Vendor Router module extends Connectivity Tools to mixed networks of Bay Networks and Cisco routers. The Multi-Vendor Router module parses Bay Networks router configurations and generates protocol-sensitive topologies; it also diagnoses and solves connectivity and routing problems.

Protocols and topologies supported are:

- LAN Topologies: Ethernet, Fast Ethernet, token ring, FDDI, ATM Interface
- WAN Connectivity: Serial, HSSI, BRI, frame relay, IP Unnumbered, SMDS, X.25, Async, Dialer, Channelized T1
- Protocols: IP, IPX, SNA, AppleTalk, VINES, DECnet
- Routing Protocols: RIP, IGRP, EIGRP, OSPF, IPX-RIP, SRB, RSRB, DLSw+, BGP

The requirements for this product are such that most users have the hardware and software required to deploy NETSYS advisor.

Advanced Professional Design from NetSuite

NetSuite Advanced Professional Design offers features that enhance and simplify the network design process for network equipment manufacturers, value-added resellers, system integrators, and end users. In addition, NetSuite also supports the *NetSuite Toolkit*, which provides users with a collection of problem-solving functions (such as a Web interface and Intelligent Floorplan Designer).

NetSuite approaches network design from a process-oriented rather than task-oriented perspective. It is important to note that NetSuite has built a solid foundation from the wiring closet up to ensure that network designs meet the complex needs of today's businesses. With NetSuite Advanced Professional Design and NetSuite Toolkit, the efficiency of the entire design process can be substantially improved.

New features of NetSuite Advanced Professional Design include:

Whiteboard Mode   The "whiteboard environment" temporarily suspends validation, allowing users to quickly capture a design concept. With whiteboard mode, the systems engineer, network manager, or designer can quickly generate a network design "on-the-fly" and validate the design at a later date.

Automatic Device Replacement   This new drag-and-drop feature allows the user to provide intelligent global replacement of network devices in one easy step. By dragging the device and dropping it onto the design sheet, devices can be "transformed" into new devices within the network design, while retaining all of the appropriate network connections.

With the NetSuite Toolkit, NetSuite offers users the ability to customize their network design process by providing a bundle of applications in a separate product offering. These key applications include:

DesignView — Provides users with the capability to convert network diagrams, designs, and device details developed in NetSuite Advanced Professional Design 2.0 into a series of hyperlinked HTML documents on the customer's Intranet or Internet server for sharing with authorized staff, contractors, and customers.

Foundry — Allows users to model their own device chassis and adapter cards and merge them within the over 3000 devices that can be found within the NetSuite Library.

Floorplan Designer — Enables users to easily overlay the background maps of network designs with detailed information on wiring structures and physical location labels, including distribution frames and cable layout channels.

IP Calculator — Aids users in the tedious tasks of planning IP subnets and generating node addresses and masks for networks. The IP Calculator computes allowable node address ranges, and subnet and broadcast masks based on user-specified criteria.

Virtual Agents Professional Simulation Tool from Network Tools

Virtual Agent Professional from Network Tools is a dynamic network modeling and management product designed to demonstrate and validate real-time network management capabilities using captured MIB and RMON data. Virtual Agent Pro builds a replicated LAN/LAN internetwork that can be monitored and controlled by any SNMP management application. With Virtual Agent, users can compare the management and agent capabilities of alternative vendor products or assess the potential management impacts of network changes. Network sales and marketing staff can use this capability to add new devices to a virtual network to demonstrate solutions in advance of product availability, speeding end-user acceptance of new technologies. Engineering staff can use it to develop, test, and debug management functionality before product shipment.

Key Virtual Agent benefits can be grouped in accordance with multiple functional areas, such as

Sales and marketing:

- Presents an entire product line on a portable PC
- Demonstrates new products and features before release
- Enables more effective customer network proposals
- Decreases the sales cycle for new technologies

Engineering and quality assurance:

- Allows new agent proof-of-concept testing from a functional spec
- Validates management and agent capabilities across diverse networks
- Enables parallel development and testing
- Speeds the availability of robust device management

End users:

- Allows evaluation of new products in the context of existing networks
- Supports "what-if" testing of network changes
- Enables management comparisons of vendor solutions before purchase
- Ensures manageability of networks before deployment

Training and support:

- Enables low-cost, mobile, in-time training
- Simulates network manageability for quicker problem resolution
- Eliminates costly training hardware, installation, and set-up
- Allows in-house and partner training before product release

Virtual Agent begins by scanning and capturing MIB and RMON data from SNMP devices on an existing network. With this information, the tool builds a software-based virtual network, comprised of hubs, switches, routers, and other devices, which serves as a starting point for subsequent SNMP modeling. Each of the simulated agents maintains all the individual characteristics and configuration attributes captured during collection, such as unique IP address, MAC address, and statistical information. Error logging during capture ensures the integrity of collected data before beginning simulations.

Virtual networks can be monitored and controlled in real time by any network management application, such as HP OpenView or SunNet Manager. Viewed from a network management screen, a replicated network appears and behaves exactly like its physical counterpart. Management queries such as PING, ARP and SNMP elicit the required responses from each.

Virtual Agent's built-in scripting scenarios allow users to modify replicated networks to test different product and management behaviors and to evaluate the impacts of changes. By creating routines that change MIB values of simulated agents, users can script and test "what-if" situations, such as increasing traffic, network errors, alerts, traps, and failed devices. Scenarios can be designed to focus on the management response of a particular device or to test system-wide configuration and manageability. Users can also add or delete simulated agents to evaluate the manageability of proposed network upgrades. This might include testing the impact of adding new devices, design and configuration changes, and alternative central/remote site schemes. By testing changes first, users can be confident that new network configurations will function as planned. Extensive error logging and tracing facilitates diagnosis and troubleshooting of unexpected interactions during simulations.

With Virtual Agent, the management functionality of new products can be developed, tested, debugged, and refined before physical hardware is available, significantly reducing product development time. Developers can define new agents with unique management behaviors and add them to existing virtual networks to test capabilities and interactions with other intelligent devices. In this manner, new management functionality can be "field-tested" in various diverse and complex environments throughout the development process, for more robust, feature-rich products.

Virtual Agent provides a mechanism for vendors to shorten sales cycles and for end users to reduce the time needed for network planning and deployment. By adding new devices into virtual networks, sales personnel can incorporate both emerging and existing products into customer demonstrations. They can then show the benefits of new features and advanced network management capabilities and thus accelerate end user acceptance of emerging technologies. Customers can evaluate the management capabilities of alternative vendor solutions within the context of their own networks before purchase. By allowing customers to "try before they buy," Virtual Agent demystifies new technologies and simplifies migration, resulting in fewer management problems over the long term.

The following networks and devices are supported:

- Any device with an SNMP MIB
- Agent MIB libraries available for products from 3Com Corporation, Ascend Communications, Bay Networks, Cisco Systems, and Cabletron Systems
- Other MIBs downloadable from the Internet

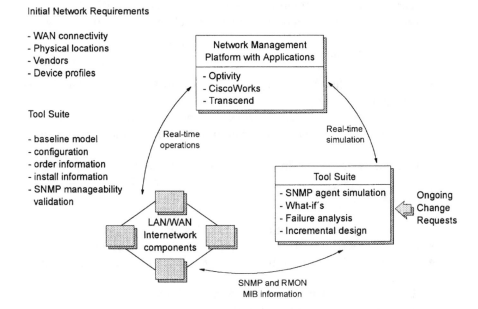

Figure 5.15

Works with any SNMP network management application, such as Bay Optivity Enterprise, Cabletron Spectrum Enterprise Manager, 3ComTranscend Enterprise Manager, Cisco CiscoWorks, HP OpenView, Sun-Net Manager, and IBM NetView.

Caliper Design and Configuration Tool from Network Tools

The Caliper Design and Configuration Tool from Network Tools is a fully automated solution for designing multivendor, multiple technology inter-networks. With minimal user effort and input, Caliper automatically designs low-cost LAN/WAN internetworks in minutes. From a single Windows screen, novice and expert users alike can explore multiple network designs and pricing alternatives and tailor design to meet their needs.

Figure 5.15 shows the position of LAN design tool suites in the dynamic environment of network operations. Ongoing change requests drive the design and planning process.

Key Caliper benefits can be identified in accordance with the following groups:

Internetworking vendors:

- Automates proposal process to drastically reduce hands-on corporate cycles
- Equips partners and channels with corporate capabilities

- Increases revenue potential by enabling hands-off RFPs
- Improves the effectiveness of new and veteran sales staff

Partners and channels:

- Increases ability to handle complex network proposals for higher margins
- Enables custom-tailored network designs in minutes
- Increases revenue potential through faster turnaround and enhanced bid capability
- Eliminates need to train staff on multiple product lines

End users:

- Takes the guesswork out of network design
- Enables practical vendor and technology design comparisons
- Gives users the appropriate information to best evaluate vendor proposals
- Ensures smooth installation by preventing network configuration errors
- Automatically documents and draws the network

Caliper's interface guides users through the definition of their infrastructure and service requirements, such as the number of buildings in the network, number of floors per building, data center locations, wiring requirements, vendor and technology preferences, and user population sizes. Before the design process begins, Caliper checks all inputs for completeness and consistency and allows the user to make corrections as needed. Users are not required to choose specific network components. Instead Caliper performs the selection automatically, enabling even novice users without extensive knowledge of networking technologies and equipment to use the tool effectively.

Drawing from a knowledge base of multiple vendors' products and technologies, Caliper automatically builds network solutions that include both topology designs and equipment hardware configurations. Caliper's goal is to minimize equipment costs, given the user-specified requirements and inherent hardware-related constraints. Its top-down approach satisfies system-level requirements first, and then works down toward device-specific solutions. Caliper verifies overall design feasibility to prevent compatibility errors, such as wiring infrastructures that cannot support desired LAN media. It also supports equipment assembly rules and dependencies and validates devices for proper configuration of chassis, slots, bus allocation, ports, and the like.

Caliper enables users to automate and significantly reduce the time needed to explore design alternatives. By simply changing requirements or vendor/technology preferences, users can examine different LAN/MAN topologies, compare equipment from multiple vendors, and tailor designs to special needs. Vendors can respond on-site to customer "what-if" questions and last-minute requirement changes. Without vendor involvement, end users can evaluate multiple technologies, products, and designs before purchase to determine the configuration that best meets their network and budget requirements.

For each network design, Caliper automatically provides all the information necessary to take the customer from purchase order through installation. A Bill of Materials report specifies exactly what components to order, where they should be installed, and associated costs. (See Fig. 3.33 in Chapter 3). Detailed rack/device drawings to the port level, rendered through industry-standard Visio-graphics, provide all the information needed to assemble and configure each component. Wiring reports are available to show how to interconnect locations. (See Fig. 3.34 in Chapter 3). By providing this level of detail, Caliper ensures that designs can be implemented as ordered, without risk of costly configuration errors.

Caliper's comprehensive database includes parts catalogs, network stencils, and price lists for the most popular internetworking vendors. New equipment can be added to the database as needed and existing network stencils and shapes can be edited. Caliper automatically verifies all database additions and modifications to ensure new parts are properly defined. In addition, users can configure Caliper to utilize only a subset of the full product inventory, enabling, for example, custom databases for products sold through specific channels.

Based on simple facility infrastructure and service requirements — such as the number of buildings in the network, number of floors per building, data center locations, wiring requirements, and size of user population — Caliper automatically creates low-cost internetwork designs (see Figure 3.31 in Chapter 3).

Chisel from Network Tools

Chisel is a multifunctional tool that tests the end-to-end behavior and performance of applications in a live network. From a central personal computer (PC) console, Chisel provides the functionality to proactively baseline application performance and monitor application quality of service; load and stress test the scalability of network architectures, devices, and applications; and verify the effectiveness of corporate firewalls across all seven OSI layers, for both Internet and Intranet configurations.

5.6   Summary

The number of alternatives for designing and sizing LANs and interconnected LANs is continuously increasing. LAN design and planning should start with connectivity and compatibility questions. Certain technologies and equipment cannot be combined with each other. Unsophisticated design instruments can filter out incompatible solutions.

After completing the design with compatible components, sizing the components helps to guarantee reasonable performance. In order to quantify performance from the very beginning, models are expected to compute multiple alternatives by using "What-If" techniques. Parameters that can be changed include the topology, load estimates, configuration settings, and computing power of networking components.

This analysis allows the impacts of changes to the LAN components to be assessed before they are actually implemented.

chapter six

Intranets

Contents

6.1 Management overview

An Intranet is a company-specific, private network based on Internet technology, and as such, it is a form of local area network. However, one of the major distinctions between traditional LANs and Intranets is the reliance of the latter on TCP/IP, packet switching, and Internet technologies. In the case of the Internet, the technology is deployed over a public network, while in the case of Intranets, the technology is deployed within a private network.

According to George Eckel, author of *Intranet Working*, one of the important benefits of Intranets is that they provide a cost-effective vehicle for communication, since the expense of reaching one person or 1 million people is essentially the same. Intranets are becoming the corporate world's equivalent of a town hall where people can meet, chat, and exchange information.

The emergence of Intranets promises to change the way companies communicate with their employees and how they conduct their business. For example, after years of using satellite feeds to disseminate information to its 208 network affiliates, CBS News now uses an Intranet to provide its affiliates with point-and-click access to information on upcoming news stories. Access to this information is provided through the CBS Newspath World Wide Web home page.

6.1.1 Benefits of Intranets

Intranets offer many potential benefits, including:

- Reduced operating costs
- Improved employee productivity
- Streamlined processing flows
- Improved internal and external communication
- New and improved customer service
- Cross-platform capability

We discuss some of the ways these benefits can be achieved in the discussion that follows.

6.1.1.1 The paperless office

Many companies find that Intranets simplify corporate-wide communications and reduce printed material costs by eliminating the need for many paper-based processes. For example, some organizations offer complete manuals on their corporate Web site in electronic form, instead of distributing the information in printed form. Companies can benefit immediately from an Intranet by replacing their printed materials, little by little, with electronic versions. Electronic media is cheaper to produce, update, and distribute than printed material. Often, printed material is out of date by the time it is distributed. Electronic documents, however, can be easily modified and updated as the need arises.

6.1.1.2 Improved Customer Service

For many organizations, having the right information at the right time can make a significant difference in their ability to close a sale or meet a deadline. In today's competitive business environment, companies are under constant pressure to improve productivity while reducing costs. To achieve these productivity gains, companies must constantly improve their relationships with employees, customers, vendors, and suppliers. Intranets provide an important avenue for making these improvements.

Using an Intranet, vendors, employees, and customers can access information as it is needed, alleviating delays associated with mailing or distributing printed materials. For example, Intranets have been used to:

- Distribute software updates to customers, reducing the need to distribute printed materials and floppy diskettes or CD-ROMs
- Process customer orders on-line

- Process and respond to customer inquiries and questions about products and services
- Collect customer and survey data

Using an Intranet, all these activities can be completed electronically in a matter of minutes.

6.1.1.3 Improved help desks

Intranets have been used to augment help desk services. For example, when someone in the organization learns about a new technology or how to perform a new task (for example, running virus software), he or she can put information and instructions for others on a personal Web page. Others within the organization, including help desk staff, can then access this information as needed. In an organization empowered by an Intranet, every employee can leave the imprint of his or her expertise.

6.1.1.4 Improved corporate culture

Intranets help to cultivate a corporate culture that encourages the free flow of information. Intranets place information directly into the hands of employees, promoting a more democratic company structure. The danger of "information democracy" is that once it is in place and taken for granted, management cannot easily revert to older, more controlled forms of communication without seriously damaging employee morale and cooperation. Every individual in an Intranet environment is empowered to access and distribute information, both good and bad, on a scale heretofore unknown in the corporate environment.

Intranets dissolve barriers to communication created by departmental walls, geographic location, and decentralized organizations. Placing information directly in the hands of those who need it allows organizations to decentralize and flatten decision making and organizational processes, while maintaining control over the information exchange. Individuals and groups can distribute ideas freely, without having to observe traditional channels of information (i.e., an individual, a printed document, etc.) that are far less effective in reaching geographically dispersed individuals.

6.1.1.5 Cross-platform compatibility

Since the early 1980s, organizations with private networks have struggled with connecting and disseminating information between different types of computers, such as PCs, Macintoshes, and UNIX-based machines. To help manage potential barriers to electronic communication posed by hardware and software incompatibilities, many companies have instituted strict standards limiting corporate users to specific hardware and software platforms. Even today, if a company uses Macs, PCs, or UNIX-based machines, sharing a simple text document can be a challenge.

Intranets provide a means to overcome many of these software and hardware incompatibilities, since Internet technologies (such as TCP/IP) are platform independent. Thus, companies using Intranets no longer need to settle on one operating system, since users working with Macintosh, PC, or UNIX-based computers can freely share and distribute information. In the sections that follow, we will explain why this is so.

6.1.2 Intranet planning and management

To implement an Intranet, a company needs a dedicated Web server, communications links to the Intranet, and browser software. *Unfortunately, Intranets do not come prepackaged and fully assembled.* They require careful planning and construction, if they are to be effective in meeting the needs of the organization. In the sections that follow, we discuss recommendations for planning and implementing an Intranet.

6.1.2.1 Gaining support

The first step toward a successful Intranet implementation is to obtain company-wide support for the project, including support from upper management. A quality presentation should be made to both management and staff to explain the benefits of the Intranet project. Some of the benefits of the Intranet are tangible and easy to measure, whereas others are intangible and difficult to measure. To gain widespread support for the Intranet project, decision makers must be shown what an Intranet is and how it will benefit the organization. There are many resources (including complete presentations) available on the World Wide Web to help promote the Internet in a corporate environment.

6.1.2.2 Planning the intranet strategy

After selling upper management on the idea of an Intranet, the next step is to define the goals, purpose, and objectives for the Intranet. This is an essential part of the Intranet project planning.

The Intranet project plan should include an overview of the organizational structure and its technical capabilities. The current communication model used to control information flow within the organization should be examined with respect to its strengths and weaknesses in supporting workflow processes, document management, training needs, and other key business requirements. It is important to understand and document existing systems within the organization before implementing the Intranet.

The Intranet plan should clearly define the business objectives to be achieved. The objectives should reflect the needs of the potential users of the Intranet. Conducting interviews with employees and managers can help identify these needs. For example, the Human Resource department may wish to use the Intranet to display job opportunities available within the organization. If this need is to be satisfied, the Intranet should be designed to display job information and job application forms on a Web server, so applicants can apply for positions electronically. The Human Resource department might also wish to offer employees the ability to change their 401K information using the Intranet. Each goal identifies shapes and defines the functionality that the Intranet must support. An employee survey is also an excellent way to collect ideas on how to employ the Intranet within the organization.

In summary, the following questions are helpful in defining the requirements of the Intranet project:

- Will Intranet users need to access existing (legacy) databases?
- What type of training and support will Intranet users require?
- Who will manage, create, and update the content made available through the Intranet?
- Will individual departments create their own Web pages autonomously?
- Will there be a central authority that manages changes to the content offered on the Intranet?
- Do users need a way to access the Intranet remotely?
- Will the Intranet need to restrict access to certain users and content?
- Will a Webmaster or a team of technicians/managers be assigned to coordinate and manage the maintenance of the Intranet?
- Will the Intranet be managed internally or will it be outsourced?

6.1.2.3   Selecting the implementation team

After the Intranet project plan has been developed and approved, the implementation team should be assembled. If the organization does not have an infrastructure in place that is capable of implementing the Intranet, additional staff and resources will need to be hired or the project will need to be outsourced to a qualified vendor.

It is important that the Intranet team assembled has the requisite skills to successfully execute the project plan. A number of skill assessment checklists are provided below to help evaluate the resources available within an organization and their ability to successfully support the Intranet implementation.

Technical Support Skills Checklist.   The Intranet project will require staff with the technical skills needed to solve network problems, understand network design, troubleshoot hardware and software compatibility problems, and implement client-server solutions (such as integrating network databases). Thus, the following skills are required to support an Intranet:

- Knowledge of network hardware and software
- Understanding of TCP/IP and related protocols
- Experience implementing network security
- Knowledge of client-server operations
- Experience with custom programming
- Knowledge of database management

Content development and design checklist. A typical organization has many sources of information: Human Resource manuals, corporate statements, telephone directories, departmental information, work instructions, procedures, employee records, and much more. To simplify the collection of information that will be made available through the Intranet, it is advisable to involve people familiar with the original documentation and also those who can author content for Intranet Web pages. If possible, the original authors of the printed material should work closely with Intranet content developers to ensure that nothing is lost in translation.

The following technical skills are needed to organize and present information (content) in browser-readable format:

- Experience in graphic design and content presentation
- Basic understanding of copyright law
- Knowledge of document conversion techniques (to convert spreadsheet data, for example, into a text document for HTML editing)
- Experience in page layout and design
- Experience with Web browsers and HTML document creation
- Knowledge of image-conversion techniques and related software
- Knowledge of programming languages and programming skills
- CGI programming and server interaction

Management support skills checklist. As previously discussed, the company's management should be involved in the planning and implementation of the Intranet. Ideally, management should have a good understanding of the Intranet benefits, and the expected costs and time frames needed for the project completion. Managers with skills relating to quality-control techniques, process-management approaches, and effective communication are highly desirable. Thus, the following management skills are recommended:

- Understanding of the organization's document and information flow
- Experience with the reengineering process
- Knowledge of quality-control techniques
- Knowledge of the company's informal flow of information
- Experience with training and project coordination

6.1.2.4 Funding growth

The initial cost of setting up a simple Intranet is often quite low and may not require top management's approval. However, when complex document management systems are needed to integrate database access, automate workflow systems, implement interactive training and other advanced features, the Intranet should be funded with the approval of top management. To gain approval for the project, upper management must be convinced that the Intranet is an integral part of the company's total information-technology deployment strategy. This involves quantifying the tangible benefits of the Intranet to the organization. Management also needs to understand how the

Intranet will change the way people work and communicate. As shown in the Appendix, an ROI analysis is an important step in justifying the economic benefits of the Intranet project.

6.1.2.5 Total quality management

Effective deployment of an Intranet often involves reengineering current process flows within the organization. Employees are usually most receptive to changes that make their jobs easier. To avoid perceptions that the Intranet is an intimidating intrusion of yet another technology, it is advisable to involve staff as early as possible in the deployment planning. This will facilitate the transition to the Intranet and encourage employee participation in the Intranet's success.

After migrating the company's work processes to the Intranet, it is up to managers and employees to adhere to the procedures that have been put in place to improve productivity and teamwork. Management should not assume that because employees have a new tool — the Intranet — this alone is sufficient to ensure that the desired attitudes and service levels will be attained. Instead, managers should view the Intranet as one aspect of their quest for total quality management (TQM).

TQM involves creating systems and workflows that promote superior products and services. TQM also involves instilling a respect for quality throughout the organization. TQM and the successful deployment of Intranets represent a large-scale organizational commitment, which upper management must support.

6.1.2.6 Training employees

If employees are expected to contribute content to the Intranet, they will need to be given tools and training so they can author HTML documents. In general, it is a good idea to encourage employees to contribute to the content on display through the Intranet. To do otherwise means that the organization may have to depend on only a few people to create HTML documents.

After being given initial training, users should be surveyed to determine if the tools they have been provided satisfy their needs. Many users find that creating HTML documents is difficult. If the users have difficulty creating HTML documents, the training efforts may need to be improved or modified. For example, this might involve training one person in each department to assume the responsibility of training the rest of the department. Patience and diligence are needed when introducing employees to new HTML authoring skills.

In summary, the following actions are recommended to help develop an effective program for training employees to author high-quality HTML documents:

- Conduct a survey to assess user training needs and wants
- Train users how to develop HTML content

- Provide users with HTML authoring tools that complement what they already know (for example, the Internet Assistant for Microsoft Word is a good choice for users already familiar with Microsoft Word)
- Review the design and flow of material that will be "published" on the Intranet
- Provide feedback to HTML authors on ways to improve the site appearance and ease of use

6.1.2.7 Organizational challenges

In addition to technological challenges, companies may also face the following organizational challenges after the initial release of an Intranet:

- Marketing the Intranet within the organization so that all employees will support its growth and continued use
- Obtaining additional funding on an ongoing basis to implement new capabilities
- Encouraging an information-sharing culture within the company so that all employees will contribute to building a learning organization
- Merging a paper-based culture with the new culture of electronic documentation
- Ensuring that the content on the Intranet is updated on a regular basis
- Preventing one person or group from controlling (monopolizing) the content on the Intranet
- Training employees to author HTML content so they can contribute material to the Intranet
- Training employees on Intranet etiquette, thereby facilitating on-line discussion forums and other forms of user interaction on the Intranet
- Using the Intranet as an integral part of working with customers and vendors
- Measuring the Intranet's overall effectiveness and contribution to the organization

As is the case when introducing any information technology to an enterprise, Intranet deployment requires careful planning, effective implementation, and employee training. In the short term, most of the organizational focus is usually on the technical aspects of the Intranet deployment. But as time goes on, organizational issues relating to how the Intranet is used within the organization must be managed. When an organization actively examines and works toward resolving these issues, they are better able to achieve a culture of teamwork and collaboration.

6.1.3 Intranets vs. groupware

There are several key differences between groupware products (e.g., Lotus Notes from IBM) and Intranets. The decision to use either or both should be predicated on the company's needs, size, and budget. A primary difference between Lotus Notes and Intranets is that Lotus Notes uses a proprietary system to distribute and track information across a network. The cost of a Lotus Notes solution is based on the total number of system users. As the number of system users increase, so do the costs. On the other hand, Intranets use open systems technology to distribute information. The only per-client cost associated with Intranets is the cost of the browsers needed to view information.[1] If a company has less than 100 employees, it may be more cost effective to use groupware products. However, as the company size grows, Intranets become more attractive from a cost perspective.

In addition to cost considerations, Intranets are attractive because employees can easily update information content. Partly in response to the competitive threat that Intranets pose to Lotus Notes, IBM is now offering new Intranet-compatible features in its product line.

6.1.4 Management summary

In this section, we introduced the notion of an Intranet and discussed some of the major issues surrounding the use of Intranets. The following list summarizes key points:

- An Intranet is a company-based version of the Internet. Intranets provide an inexpensive solution for information sharing and user communication.
- Companies that have installed Intranets have found that installation costs are generally low and that system versatility is high.
- An Intranet provides an easy way for users to communicate and share common documents, even if they are using different machines, such as IBM-compatible and Macintosh personal computers.
- Some organizations have expanded their Intranet to allow customers to access internal databases and documents.
- Many companies can establish a functional Intranet using in-house personnel with a minimal amount of new equipment.
- Intranet solutions are open and are shaped by competitive forces, whereas groupware products tend to be closed and proprietary.

Personal computers have moved computing power away from the Information Services Department and into the hands of users. Today, Intranet technology is taking this one step further, giving users even more control over the creation and distribution of information throughout the enterprise.

Internet technology adheres to open standards that are well documented. This, in turn, encourages the development of cost-effective and easy-to-implement Intranet solutions. As the popularity of Intranets has increased, so has the demand for new tools and Web-based solutions. This demand has fueled competition among software manufacturers, which, in turn, has

---

[1] The Netscape Navigator browser is available for around $50, and Microsoft Internet Explorer is currently free.

resulted in better and cheaper Intranet products. Groupware products — such as Lotus Notes — are already dropping in price because of the increased popularity of Intranets.

In summary, Intranets can be used to improve productivity, simplify workflows, and gain a competitive advantage over those who have yet to learn how to capitalize on the benefits of Intranets.

6.2   Technical overview

6.2.1   Internet basics

6.2.1.1   Packet Switching

Packet switching was introduced in the late 1960s. In a packet-switched network, programs break data into pieces, called packets, which are transmitted between computers. Each packet contains the sender's address, the destination address, and a portion of the data to be transmitted. For example, when an e-mail message is sent over a packet-switched network, the e-mail is first split into packets. Each packet intermingles with other packets sent by other computers on the network. Network switches examine the destination address contained in each packet and route the packets to the appropriate recipient. Upon reaching their destination, the packets are collected and reassembled to reconstitute the e-mail message.

6.2.1.2   TCP/IP

The U.S. Advanced Research Projects Agency (ARPA) was a major driving force in the development and adoption of packet-switched networking. The earliest packet switched network was called the ARPAnet. The ARPAnet was the progenitor to today's Internet. By the early 1980s, ARPA needed a better protocol for handling the packets produced and sent by various network types. The original ARPAnet was based on the network control protocol (NCP). In January 1983, NCP was replaced by the transport control protocol/internet protocol (TCP/IP). TCP/IP specifies the rules for the exchange of information within the Internet or an Intranet, allowing packets from many different types of networks to be sent over the same network.

6.2.1.3   Connecting to the internet

One way to connect to the Internet is to install a link from the company network to the closest computer already connected to the Internet. This approach is illustrated in Figure 6.1. When this method is chosen, the company must pay to install and maintain the communications link (which might consist of a copper wire, a satellite connection, or a fiber optic cable) to the Internet. This method was very popular with early adopters of the Internet, which included universities, large companies, and government agencies. However, the costs to install and maintain the communications link to the Internet can be prohibitive for smaller companies.

Company Office Building   **Internet**

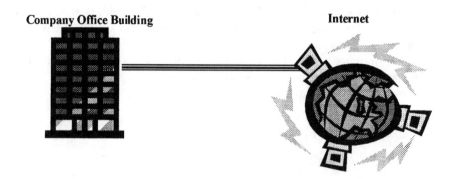

Figure 6.1   Direct connection to the internet.

Fortunately, specialized companies — called Internet service providers (ISPs) — are available to provide a low-cost solution for accessing the Internet. ISPs pay for an (expensive) connection to the Internet, which they make accessible to others through the installation of high-performance servers, data lines, and modems. Acting as middlemen, the ISPs rent time to other users who want to access the Internet. Figure 6.2 illustrates how small offices and a schoolhouse might connect to the Internet using an ISP.

Two important decisions must be made when deciding what type of Internet connection is the most appropriate. The first decision is the company budget allocated for Internet connectivity, and the second is the Internet connection speed needed to support business requirements. Both decisions are interrelated. ISPs offer a variety of options for connecting to the Internet, ranging from a simple dial-up account over phone wires to high-speed leased lines from the company to the ISP. Dial-up accounts are typically available for a low, flat monthly fee, and are generally much cheaper than a leased line connection. However, the leased line connection is usually much faster than the dial-up connection.

When a dial-up account is used, a modem and a phone line are used to call and log into the ISP server (or computer), which in turn acts as the doorway to the Internet. The transmission speed of the connection is limited by the speed of the modems used by the user and the ISP. A modem is not needed when a leased line connection is available to the ISP. Leased lines are available in many different configurations with a variety of options. The most common link types are ISDN (which support transmission speeds from 56 Kbps to 128 Kbps), T1 (transmitting at speeds up to 1.54 Mbps), and T3 (transmitting at speeds up to 45 Mbps).

If a company needs to make only an occasional connection to the Internet — for example, less than 20 to 50 hours per month for all users — a dial-up account should be sufficient. The costs for a basic dial-up account with unlimited access are in the range of approximately $250 per year. However, if a company needs faster data transfer speeds or has several users who must access the Internet for substantial periods of time over the course

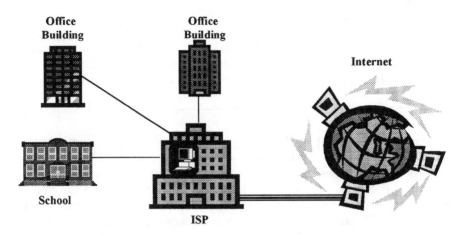

Figure 6.2 Internet access through ISP.

of a month, a leased line connection should be considered. Typical monthly charges and special equipment fees for a leased line connection might average $200 to $1,200 per month for an ISDN connection, $1,000 to $3,000 per month for a T1 connection, and $2,000 to $10,000 per month for a T3 connection. Note that the actual line costs will vary according to the service provider and specific options the company has chosen to implement.

The fastest growing segment of Internet users are those who connect to the Internet through an Internet service provider via an ordinary telephone connection. There are two major protocols for connecting to the Internet in this way: serial line internet protocol (SLIP) and point-to-point protocol (PPP). SLIP is the older protocol and is available in many communications packages. The faster PPP is newer and therefore it is not as widely supported.

Principles of queuing analysis can be applied to the problem of sizing the links needed to support the Internet access, whether or not that access is to an ISP or to a direct Internet connection. The reader is referred to Chapter 2 for specific techniques on how to estimate the throughput and performance characteristics associated with using different size link capacities. This analysis can be used to determine whether or not a dial-up or leased line connection is sufficient to support the bandwidth requirements with tolerable transmission delays.

6.2.1.4 Basic terminology
In this section, we define commonly used Internet and Intranet terminology.

The World Wide Web

The World Wide Web — or Web — is a collection of seamlessly interlinked documents that reside on Internet servers. The Web is so named because it links documents to form a web of information across computers worldwide.

The "documents" available through the World Wide Web can support text, pictures, sounds, and animation. The Web makes it very easy for users to locate and access information contained within multiple documents and computers. "Surfing" is the term used to describe accessing (through a Web browser) a chain of documents through a series of links on the Web.

Web browsers

To access and fully utilize all the features of the Web, special software — called a Web browser — is necessary. The main function of a Web browser is to allow you to traverse and view documents on the World Wide Web. Browser software is widely available for free, either through a download from the Internet or from ISPs. Commercial on-line services — such as America Online and Prodigy — also supply browsers as part of their subscription service. The two most commonly used browsers are Netscape Navigator and Microsoft Internet Explorer. Some of the common tasks that Netscape Navigator and Microsoft Internet Explorer support include:

- Viewing documents created on a variety of platforms
- Creating and revising content
- Participating in threaded discussions and news groups
- Viewing and interacting with multimedia presentations
- Interfacing with existing legacy data (non-HTML based data) and applications
- Gaining seamless access to the Internet

It should be noted that the same Web browser software used for accessing the Internet is also used for accessing documents within an Intranet.

Uniform resource locator (URL)

The Web consists of millions of documents that are distinguished by a unique name called a URL (uniform resource locator), or more simply, a Web address. The Uniform resource locator is used by Web browsers use to access Internet information. Examples of URLs include:

http://www.netscape.com
ftp://ftp.microsoft.com

A URL consists of three main parts:

1. A service identifier (such as http)
2. A domain name (such as www.ups.com)
3. A path name (such as www.ups.com/tracking)

The first part of the URL, the service identifier, tells the browser software which protocol to use to access the file requested. The service identifier can take one of the following forms:

- http: //— This service identifier indicates that the connection will use the hypertext transport protocol (HTTP). HTTP defines the rules that software programs must follow to exchange information across the Web. This is the most common type of connection. Thus, when Web addresses start with the letters "http" it indicates that the documents are retrieved according to the conventions of the HTTP protocol (hypertext transport protocol).
- ftp: //— This service identifier indicates that the connection will use the file transfer protocol (FTP). This service identifier is typically used to download and copy files from one computer to another.
- gopher: //— This service identifier indicates that the connection will utilize a gopher server to provide a graphical list of accessible files.
- telnet: //— This service identifier indicates that a telnet session will be used to run programs from a remote computer.

The second part of the URL, the domain name, specifies which computer is to be accessed when running server software. An example of a domain name is: www.tcrinc.com.

The final part of the URL, the path name, specifies the directory path to the specific file to be accessed. If the path name is missing from the URL, the server assumes that the default page (typically, the homepage) should be accessed. Large, multipage Web sites can have fairly long path names. For example, these URLs request specific pages within a given Web site.

- http://www.apple.com/documents/productsupport.html
- http://www.bmwusa.com/ultimate/5series/5series.html
- ftp://ftp.ncsa.uiuc.edu/Mac/Mosaic
- http://www.microsoft.com/Misc/WhatsNew.htm

Home pages

Companies, individuals, and governments that publish information on the Internet usually organize that information into "pages," much like the pages of a book or a sales brochure. The first page that people see in a sales brochure is the cover page, which may contain an index and summary of the brochure contents. Similarly, a home page is the first page that users see when they access a particular Web site. The home page is to the Web site what the cover page is to a sales brochure. Both must be appealing, concise, informative, and well organized to succeed in maintaining the reader's interest. The home page is usually used to convey basic information about the company and what it is offering in the way of products and/or services.

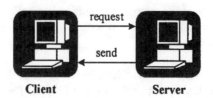

Figure 6.3 Typical client-server interaction.

Many companies publish the Internet address (known as a uniform resource locator or URL) of their home page on business cards, television, magazines, and radio. To access a Web site, a user has merely to type the URL into the appropriate area on the Web browser screen.

Client programs and browsers

Across the Internet, information (i.e., programs and data) is stored on the hard disks of thousands of computers called servers. The servers are so named because, upon request, they serve (or provide) users with information. A server is a remote computer that may be configured to run several different types of server programs (such as Web server, mail server, and ftp server programs).

A client program is used to initiate a session with a server. Client programs are so named because they ask the server for service. In the case of the Web, the client program is the Web browser. All client-server interactions take the same form. To start, the client connects to the server and asks the server for information. The server, in turn, examines the request and then provides (serves) the client with the requested information. The client and server may perform many request-response interactions in a typical session. Figure 6.3 illustrates a common type of client-server interaction.

Software programs — such as a browser — use HTTP commands to request services from an HTTP server. An HTTP transaction consists of four parts: a connection, a request, a response, and a close.

Where Web documents reside

When users publish Web pages, they actually store the pages as files that are accessible through a file server. Typically, Web pages reside on the same computer on which the server program is running, but this is not necessarily true. For security reasons, it may be necessary to limit accessibility to various files on the Web server. Obviously, it might be disastrous if internal documents and data were made available to competitors. To prevent this type of security risk, a WebMaster (or Systems Administrator) can configure the Web server so it only allows specific clients to access confidential information, based on a need to know basis. The WebMaster can control access to the server by requiring users to log-in with a username and password that has predetermined access privileges.

HTML — The language of the World Wide Web

The European Particle Physics Laboratory at CERN, in Geneva, Switzerland developed HTML in the late 1980s and early 1990s. The hypertext markup language (HTML) is the language of the World Wide Web. Every site on the Web uses HTML to display information.

Each Web document contains a set of HTML instructions that tell the browser program (e.g., Netscape Navigator or Microsoft Internet Explorer) how to display the Web page. When you connect to a Web page using a browser, the Web server sends the HTML document to your browser across the Internet. Any computer running a browser program can read and display HTML, regardless of whether that computer is a personal computer running Windows, a UNIX-based system, or a Mac.

If word processor formatted files — such as Microsoft Word — were used to create Web pages, only users with access to Microsoft Word would be able to view the Web page. HTML was designed to overcome this potential source of incompatibility. All users can access Web pages from their browser since all Web pages conform to HTML standards. An (HTML) Web page is a plain text file (i.e., an ASCII text file) that can be created and read by any text editor. There are many software programs available to convert document files to HTML equivalents. In addition, many standard presentation and word processing packages offer built-in routines to convert a standard document to a Web-ready HTML file. This type of conversion might be helpful, for example, if you wanted to convert a Microsoft PowerPoint presentation into a set of HTML files for display on the Web.

After HTML files are transferred to a Web site, anyone with a browser can view them.

HTML provides the browser with two types of information:

1. "Mark-up" information that controls the text display characteristics and specifies Web links to other documents.
2. "Content" information consisting of the text, graphics, and sounds that the browser displays.

Hypertext and hyperlinks

Documents on the Web can be interconnected by specifying links (called hyperlinks) that allow the user to jump from one document to another. The hypertext markup language (HTML) code, which drives all Web pages, supports hypertext. Hypertext, in turn, supports the creation of multimedia documents (containing pictures, text, animation, sound, and links) on the Web.

Hyperlinks (or simply, links) are visually displayed on the Web pages as pictures or underlined text. When a user clicks on a hyperlink displayed on their browser screen, the browser responds by searching for and then loading the document specified by the hyperlink. The document specified in the hyperlink may reside on the same computer as the Web page on display or it may reside on a different computer on the other side of the world. Much of the Web's success has been attributed to the simplicity of the hyperlink point-and-click user interface.

There are four basic layouts for linking Web pages with hyperlinks: linear, hierarchical, web, and combination. Which layout is the most appropriate depends on the type of information that is being presented and the intended audience. Section 6.3.2. presents examples of each of these layouts, in the context of structuring the flow in which information is presented on the Intranet Web pages.

FTP — the file transfer protocol

FTP (file transfer protocol) is a standard protocol for transferring and copying files from one computer to another. Depending on the configuration of the FTP server program, you may or may not need an account on the remote machine to access systems files. In many cases, you can access a remote computer with FTP by logging on with a username of "anonymous," and by entering your e-mail address as the password. This type of connection is referred to as "anonymous FTP session."

After logging in to the remote FTP server, it is possible to list a directory of the files that are available for viewing and/or copying. The systems administrator determines which files can be accessed on the remote server and who has access privileges. When system security is a major concern, the system administrator may require a specific username and password (as opposed to allowing an anonymous log-on procedure) to gain access to system files.

FTP is very useful in accessing the millions of files available on the World Wide Web. Most browsers have built-in FTP capabilities to facilitate downloading files stored at FTP sites. To access an FTP site using your browser, you type in the FTP site address, much like entering a Web address. For example, to access the Microsoft FTP site, the address ftp://ftp.microsoft.com would be entered into the browser address window.

Java

Java is a new programming language released by Sun Microsystems that closely resembles C++. Java is designed for creating animated Web sites. Java can be used to create small application programs, called applets, which browsers download and execute. For example, a company might develop a Java applet for their Web site to spin the company's logo, to play music or audio clips, or to provide other forms of animation to improve the appeal and effectiveness of the Web page.

Network computers

Although personal computers are becoming more and more common, the number of households with a personal computer is still only about one third the number of households with a television. The main reason for this is that personal computers are still too expensive for the masses. The network computer is a scaled down, cheaper version (under $500) of the personal computer. A network computer is designed to operate exclusively with the Internet and Java applets.

6.2.2   Intranet components

This section will provide an overview of the components necessary to create an Intranet. The final selection of the Intranet components depends on on the company's size, level of expertise, user needs, and future Intranet expansion plans. In addition, we also examine some of the costs associated with the various Intranet components. A detailed cost analysis of an Intranet deployment project is provided in the Appendix.

An Intranet requires the same basic components found on the Internet. These components include:

1. A computer network for resource sharing
2. A network operating system that supports the TCP/IP protocol
3. A server computer that can run Internet server software
4. Server software that supports HyperText Transport Protocol (HTTP) requests from browsers (clients)
5. Desktop client computers equipped with network software capable of sending and receiving TCP/IP packet data
6. Browser software installed on each client computer

It should be noted that if a company does *not* want to use an internal server, an Internet service provider (ISP) can be used to support the Intranet. It is very common for organizations to use an ISP, especially when there is little information content or interest in maintaining a corporate operated Intranet server. ISPs are also used when the organizational facilities cannot support the housing of an Intranet server.

In addition to the software and hardware components listed above, HTML documents must be prepared to provide information displays on the Intranet. The creation and conversion of documents to HTML format is very easy using commercial software packages, such as Microsoft's FrontPage. Third-party sources are also available to provide this service at a reasonable cost.

| STAR | TOKEN RING | BUS |

Figure 6.4 Common network topologies.

6.2.2.1 Network requirements

The first requirement for an Intranet is a computer network. For the purpose of this discussion, we assume that a basic computer network is in place. We focus, therefore, in the discussion that follows on the hardware and software modifications needed to support an Intranet.

Most computer networks are local area networks (LANs). LANs are based on a client-server computing model that uses a central, dedicated computer — called the server — to fulfill client requests. The client-server computing model divides the network communication into two sides: a client side and a server side. By definition, the client requests information or services from the server. The server, in turn, responds to the client's requests. In many cases, each side of a client-server connection can perform both client and server functions.

Network servers are commonly used to send and receive e-mail and to allow printers and files to be shared by multiple users. In addition, network servers normally have a storage area for server programs and to back up file copies. Server applications provide specific services. For example, a corporate-wide e-mail system typically uses a server process that is accessible from any computer within the company's network.

A server application (or server process) usually initializes itself and then goes to sleep, spending much of its time simply waiting for a request from a client application. Typically, a client process will transmit a request (across the network) for a connection to the server, and then it will request some type of service through the connection. The server can be located at either a local or remote site.

Every computer network has a physical topology by which it is connected. The most common topologies used to connect computers are the star, token ring, and bus topologies. Figure 6.4 shows three commonly used network topologies.

A network-interface card is needed to physically connect a computer to the network. The network-interface card resides in the computer and provides a connector to plug into the network. Depending on the network, twisted-pair wiring, fiber optic, or coaxial cable may be used to physically connect the network components. The network-interface card must be compatible with the underlying network technology employed in the network (e.g., Ethernet or token ring).

The principles of queuing analysis and the minimal cost design techniques presented earlier in this text are applicable to the design of Intranets. Therefore, the reader is referred to Chapter 2 and Chapter 3 for specific guidelines on how to design and size the computer network and network components that will support the Intranet.

6.2.2.2 Network operating systems

The Internet supports connectivity between various hardware platforms running various operating systems. In theory, there is no reason why an organization must stay with one type of machine or operating system when implementing an Intranet. However, in practice, many organizations use only one network operating system to simplify the task of managing the network.

The primary choices for network operating systems are UNIX, Windows NT, and Novell NetWare. We now discuss each of these operating systems and important considerations surrounding their use.

UNIX

Many larger companies use UNIX-based machines as their primary business application server platform. UNIX is a proven operating system that is well suited for the Internet's open system model. Unfortunately, learning how to use UNIX is not easy. Also, using a UNIX-based machine limits the choices available for developing interactive Intranets and other software applications. Many programmers, for example, prefer to develop applications using Windows-based machines and programming languages (such as Microsoft's Visual Basic or Borland's Delphi).

Windows NT

Many companies choose Windows NT over UNIX because NT is easy to install, maintain, and administer. Windows NT, like UNIX and OS/2, provides a high-performance, multitasking workstation operating system. It also supports advanced server functions (including HTTP, FTP, and Gopher) and communications with clients running under MS-DOS, Windows 3.1, Windows 95, Windows for Workgroups, Windows NT Workstation, UNIX, or Macintosh operating systems. The latest version of Windows NT Server includes a free Internet information server (IIS) and a free Web browser (Internet Explorer). Microsoft designed the Internet information server so that it can be installed and up and running on a Windows NT workstation in less than 10 minutes. The Windows NT Server comes with a built-in remote access services feature that supports remote access to the Intranet through a dial-up phone connection. A Windows NT server can be purchased for about $5,500 (to support approximately 150 users).

Novell NetWare (IPX/SPX)

The NetWare operating system provides network-wide file and printer sharing for Ethernet or token ring networks. It runs on all major computer platforms, including UNIX, DOS, Macintosh, and Windows. However, behind the scenes, NetWare sends and receives data packets based on the IPX/SPX protocol. IPX is an acronym for internetwork packet exchange, and SPX is an acronym for sequenced packet exchange. Like TCP/IP, the IPX/SPX protocol defines a set of rules for coordinating network communication between network components.

Many companies use Novell network products within LANs to operate file and print servers. Therefore, it is important to understand how Novell's NetWare products can be used in an Intranet implementation strategy. If a company has an existing NetWare network, it might choose not to provide TCP/IP software to its network clients. Instead, the local-area network features provided by the NetWare software can be used.

A true Intranet uses Internet technology. This implies that an internet protocol (IP) address is assigned to each network computer (actually to each network-interface card), and that the TCP/IP protocol is used in the network. However, it is possible to run an Intranet on top of a NetWare LAN using various software products that translate IPX to IP. Many of these software packages provide IPX to IP translation while leaving the existing LAN infrastructure unchanged.

For example, the Novell product, IntranetWare, does not require assignment of an IP address to each client on a network. Instead, an IP address is assigned only to the NetWare Web server. The software performs IPX to IP translation on the client side, translating the TCP/IP protocols used by a Web browser to IPX protocols. After the protocol translation on the client side, the messages travel across the network until they reach the NetWare Web server. At this point, the IntranetWare server running on the NetWare Web server translates the IPX messages back into TCP/IP so they can be sent on to other servers on the network.

Another way to create an Intranet using a NetWare-based LAN is to use the Internet gateway product Inetix from Micro Computer Systems, Inc. Inetix runs on NetWare, Windows NT, or UNIX servers. Inetix does not require that the NetWare client machines support TCP/IP, only that a single IP address be assigned to the Intranet Web server. The Inetix client software allows a Web browser to execute on the client side, even though the client machine does not use TCP/IP-based software.

In a Mac-based network, NetWare for Macs, or AppleTalk[2] can be used to support the Intranet.

---

[2] AppleTalk is Apple's proprietary network operating system for supporting a LAN. Unlike AppleTalk, TCP/IP supports both LANs and WANs.

6.2.2.3 Server hardware

Server machines run the network operating system and control how network computers share server resources. Large businesses with thousands of users typically use high-speed UNIX-based machines for their servers. Small and medium-sized companies normally use less expensive Intel-based machines. The load (i.e., the number of users and the amount of network traffic) on the Intranet server machine will influence the selection of a specific processor type.

There is considerable debate in the industry as to which machine makes a better Intranet server: a UNIX workstation, an Intel-based machine, or a PowerPC-based system. In general, the server choice depends on the plans for the Intranet and the level of familiarity the company has with each of these platforms. The server hardware selection also depends on the network operating system in use.

If a UNIX-based server is chosen, a company will pay more for an equivalent amount of computing power provided by an Intel-based machine. UNIX machines still carry a price premium over PCs, because they are made from custom parts, whereas Intel-based machines are made from commodity components available from many hardware vendors and suppliers. For example, a high-end Pentium machine with the capacity to serve over 1000 client machines can be bought for about $3,000. A comparable UNIX server would cost in the range of $15,000 to $30,000. Macintosh-based systems are more expensive than comparable Intel-based machines, but they are still much less expensive than UNIX-based machines.

The decision to use a UNIX-based machine vs. an Intel-based machine as the Intranet server is also influenced by maintenance costs. Maintaining a UNIX-based machine requires more resources than maintaining an Intel-based machine. Hardware upgrades for Intel-based machines are also cheaper than hardware upgrades for UNIX workstations. A Macintosh server will cost more to upgrade than an Intel-based machine. However, these costs are still lower than a comparable upgrade on a UNIX based machine.

The debate over UNIX-based and Intel-based machines focuses primarily on the performance of these machines in supporting business application servers. For example, companies that use large accounting and financial software packages often use UNIX servers. On the other hand, companies that do not want to pay the price premium for UNIX machines and/or are not familiar with UNIX machines often select Intel-based machines as their business application servers.

If an Intel-based machine is selected as the server, it is recommended that it be minimally configured with at least 32Mb of RAM (ideally 64Mb), a 2GB hard disk, and a Pentium 256 MHz microprocessor. These systems can be purchased from many personal computer vendors for approximately $2,000 to $2,500. A Pentium-class machine supports a vast array of software applications and server software. Many industry experts believe that Pentium-class machines will take a significant amount of the market share away from UNIX workstations. This, in turn, means that more and more Intranet-based applications will use Pentium-class machines in the future.

Principles of queuing analysis, presented in Chapter 2, can be applied to estimate the capacity needed in the server to support the anticipated traffic at the required throughput levels.

6.2.2.4 Web server software

A working Intranet requires server software that can handle requests from browsers. In addition, server software is needed to retrieve files and to run application programs (which might, for instance, be used to search a database or to process a form containing user-supplied information).

For the most part, selecting a Web server for an Intranet is similar to selecting a Web server for an Internet site. However, Internet servers must generally handle larger numbers of requests and must deal with more difficult security issues. The performance of the Web server has a major impact on the overall performance of the Intranet. Fortunately, it is fairly easy to migrate from a small Web server to a larger, high-performance Web server as the system usage increases over time.

Web servers are available for both Windows NT and UNIX from Microsoft, Netscape, and a number of other companies. The Web server selection is limited by the server operating system. In the discussion that follows, we describe Web servers for UNIX, Windows NT, Windows 95, and Macintosh operating systems.

It is expected that in the next two to five years, stand-alone Web servers will be replaced by servers that are an integral part of the operating system. In addition, these Web servers will handle many tasks that require custom programming today, such as seamless connection to databases, video and audio processing, and document management.

UNIX Web servers

One of the best and oldest Web servers for UNIX-based machines is the National Center for Supercomputing Application's HTTP (NCSA HTTP) Web server. Much of the Internet's growth is primarily due to the popularity of this server, which is free. NSCA is committed to the continued development of its Web server. Its Web server provides both common gateway scripting (CGI) capabilities and server side includes (SSI) software. SSI software is used by Web servers to display and/or capture dynamic (changing) information on an HTML page. For example, SSI can be used to display a counter showing the number of visitors to a Web site. The NCSA Web server also allows the creation of virtual servers on the same machine. The virtual servers can have their own unique universal resource identifier (URL). This is useful, for example, for assigning a different IP address to different departments using the same machine.

Netscape Communications Corporation offers the most popular commercial UNIX-based Web servers: Enterprise server and FastTrack server. The cost of Enterprise server software is about $1,000. It is designed for building large Intranets. FastTrack is easier to install than Enterprise server, but it offers less functionality. FastTrack sells for about $300. FastTrack is well suited for companies that plan to build small- to medium-sized Intranets. These Web servers are also available for Windows NT.

Windows NT Web servers

As discussed in the previous section, both Netscape's Enterprise server and FastTrack server are available for Windows NT. These servers should be considered when both Windows NT- and UNIX-based machines are used as servers.

If cross-platform Web server software is not needed, Microsoft's Internet Information Server (IIS) should be considered. This server is used to drive Microsoft's Internet site, and it works well for a large organization. Currently, Microsoft offers this server free of charge. It also comes bundled with Microsoft's Windows NT server software. IIS is easy to install and allows new users to be added to the Intranet with minimal effort. IIS comes with an FTP server.

WebSite Professional, from O'Reilly & Associates, Inc., is another popular Windows NT Web server. WebSite Professional sells for about $500 and is very easy to install. This server has built-in search capabilities, a Web-site management tool, and the popular HTML editing tool, HotDog.

Macintosh Web server

Not many Web servers are available for Macintosh computers. If your organization does not have a Macintosh network already in place, it is advisable to avoid installing a Mac-based Intranet server.

The largest market share of the Macintosh Web server market belongs to the WebStar server from Quarterdeck Corporation. This server is a mature product and is a good choice for a Mac-based Web server. WebStar is very easy to install and maintain. It sells for about $800.

NetWare Web server

The NetWare Web server from Novell is an excellent choice for companies that have a NetWare network already in place. This server sells for about $1,000. When using a NetWare Web server, IPX to IP translation software must be installed on each client machine. When this is done, an IP address does not need to be assigned to each client machine.

6.2.2.5   Desktop clients running TCP/IP

TCP/IP must be installed on each client machine running on the Internet.[3]
To use an Internet-based application (such as a Web browser) on a Windows-
based machine, a TCP/IP stack must be present. Windows 95, Windows NT,
and IBM's OS/2 Warp operating systems include the TCP/IP protocol suite.
Most UNIX-based systems use TCP/IP as their main network communication protocol.

If a company has Windows 3.1 clients, they should consider upgrading
the clients to Windows 95 or Windows NT. Many of the advanced Internet
and Intranet applications are only available for UNIX, Windows 95, and
Windows NT operating systems. If the Windows 3.1 clients cannot be
upgraded, then TCP/IP software must be installed on each client. One of
the more popular TCP/IP software applications for Windows 3.1 is Trumpet
Winsock, which can be downloaded from the Internet for free.

6.2.2.6   Web browsers

The last component needed to make a functional Intranet is a Web browser.
There are two major choices for a browser: Microsoft's Internet Explorer or
Netscape's Navigator. Netscape's Navigator product dominates the market;
however, Microsoft is pressing hard to make inroads.

It is expected that Netscape will retain it dominance in the UNIX market,
since Microsoft has chosen not to support a UNIX platform. Therefore, if the
Intranet must support UNIX, Macintosh, and Windows clients, and the company requires standardization on a single browser, Netscape's Navigator
product is the only viable option.

Browsers, as we know them today, will probably not exist in a few years.
For example, it is expected that eventually Microsoft will integrate the
browser's functionality into its business application software (such as Word,
Excel, etc.) and operating system.

6.2.2.7   Intranet component summary

In this section, the basic components of an Intranet are examined. We reca-
pitulate below some of the key concepts covered in this section.

- Intranets are based on a client-server network computing model. By
  definition, the client side of a network requests information or services
  and the server side responds to the client's requests.
- The physical components of an Intranet include network interface
  cards, cables, and computers.
- Suites of protocols, such as TCP/IP and IPX/SPX, manage data communication for various network technologies, network operating systems, and client operating systems.

---

[3] If the network does not support TCP/IP, a gateway application that translates TCP/IP for
the network operating system protocol must be used.

- IPX to IP translation programs provide NetWare users with the ability to build an Intranet without running the TCP/IP suite of protocols on their network.
- Windows-based Intranets are easier and less expensive to deploy than UNIX-based Intranets.
- Netscape and Microsoft provide both low- and high-end server software products designed to meet the needs of large, medium, and small organizations.
- Netscape's Navigator and Microsoft's Internet Explorer provide advanced browser features for Intranet applications.

6.2.3   Intranet implementation

6.2.3.1   Information organization

After the physical components of the Intranet are in place, the next step is to design the information content of the Intranet and/or Internet Web pages. This task involves identifying the major categories and topics of information that will be made available on the Intranet. Information can be organized by department, function, project, content, or any other useful categorization scheme. It is advisable to use cross-functional design teams to help define the appropriate informational categories that should be included on the corporate Web site. The following types of information are commonly found on corporate Intranet homepages:

- What's new
- Corporate information (history and contacts)
- Help desk and technical support
- Software and tools library
- Business resources
- Sales and marketing information
- Product information
- Human resources related information (benefits information, etc.)
- Internal job postings
- Customer feedback
- Telephone and e-mail directory
- Quality and system maintenance records
- Plant and equipment records
- Finance and accounting information
- Keyword search/index capability

6.2.3.2   Content structure

After the main topics of information to be displayed on the corporate Web page(s) have been identified, the flow and manner of presentation on the Intranet must be developed. Four primary flow models are used to structure the flow of presentation at an Intranet Web site: linear, hierarchical, nonlinear (or Web), and combination information structures.

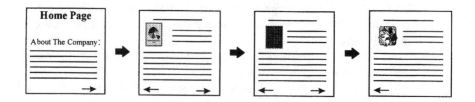

Figure 6.5   Linear hyperlink sequence.

A linear information structure is similar in layout to a book in that information is linked sequentially, page by page. When a linear layout is used, the Web pages are organized in a "slide show" format. This layout is good for presenting pages that should be read in a specific sequence or order. Since linear layouts are very structured, they limit the reader's ability to explore and browse the Web page contents in a nonsequential (or nonlinear) manner. This type of information structure is illustrated in Figure 6.5.

When a hierarchical layout is used to structure the information, all the Web pages branch off from the home page or main index. This layout is used when the material in the Web pages does not need to be read in any particular order. A hierarchical information structure creates linear paths that allow only up and down movements within the document structure. This type of information structure is illustrated in Figure 6.6.

A nonlinear, or Web, structure links information based on related content. It has no apparent structure. Nonlinear structures allow the reader to wander through information spontaneously by providing links that allow forward, backward, up and down, diagonal, and side to side movement within a document. A nonlinear structure can be confusing, and readers may get lost within the content, so this structure should be chosen with care. The World Wide Web uses a nonlinear structure. The advantage of a nonlinear structure is that is encourages the reader to browse freely. This type of information structure is illustrated in Figure 6.7.

The combination Web page layout combines, as the name implies, elements of the linear, Web, and hierarchical layouts. Regardless of the type of flow sequence employed, each Web page typically has links that allow the user to move back and forth between pages and back to the home page.

Over the lifetime of the Intranet, it is likely that the layout and organization of information on the corporate Web pages will change many times. It is often helpful to use flow charting tools to help manage and document the updated information flows. Visio for Windows by Visio Corp. and ABC Flowcharter by Micrografx are two excellent tools for developing flowcharts. In addition, some of the Web authoring tools offer flowcharting and organizational tools to help design and update the information structure on the Web pages.

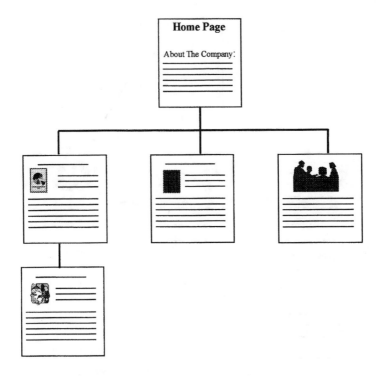

Figure 6.6 Hierarchical hyperlink sequence.

6.2.3.3 *Interface design*

After defining the Intranet's structure, the next step is to define the functionality and user interface. The Intranet design should be consistent with the organization's corporate image. For example, items such as corporate images, logos, trademarks, icons, and related design themes add a familiar look and feel to the content. Where possible, they should be included in the Web page design. It is also advisable to work with the marketing department when designing the web page layouts to ensure that a consistent theme is maintained in all the company communications that will be viewed by the outside world.

A technique called "storyboarding" is frequently used to design the web page layout. Storyboards are used by film producers, story writers, and comic strip artists to organize the content and sequence of their work. A storyboard depicts the content, images, and links between pages of the Intranet in the form of a rough outline.

Software — such as Microsoft PowerPoint or other similar presentation programs — can be used to develop a storyboard and sample Web pages. It is a good idea to test the interface design to ensure that the icons, buttons, and navigational tools are logical and intuitive. An Intranet without intuitive

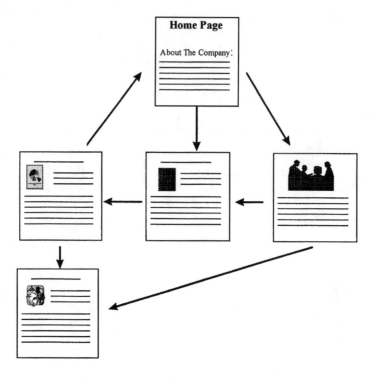

Figure 6.7   Web hyperlink layout.

navigational tools is like a road without signs. Just as it would be difficult for drivers to find their way from one city to another without the aid of signs, street names, and directional information, Intranet users will find it difficult to retrieve information without easy-to-follow categories, buttons, and links. It is often helpful to employ graphic designers and marketing-communications staff to create effective graphics and images for the web site.

6.2.3.4   Free clip art and images

Many icons and navigational signs are available as clip art that comes with word processing, page layout, and presentation software programs. In addition, many Web sites offer images and clip art, which can be downloaded for free. However, the licensing agreements for downloading free images may have restrictions and requirements that should be observed.

6.2.3.5   Intranet functionality

The required functionality of an Intranet dictates many of the design and user interface features. One of the goals in designing the Intranet should be to improve existing systems and infrastructures. After examining the current information structure, it may become clear which aspects of the structure work well and which ones need improvement.

Workflow processes, document management, and work collaboration are areas that the organization should strive to improve through the use of an Intranet. A work flow analysis should consider ways in which the Intranet can automate various organizational tasks and processes. For example, if a company has a geographically dispersed project team, the Intranet might be used to post and update project information as various tasks are completed. Other team members could then visit the Intranet page at any time to check the project status.

The following checklist is helpful when developing a list of the functions that need to be supported by the Intranet:

Functionality checklist

- The user interface must be intuitive and tested.
- The Intranet's design should support continuous updates.
- The Intranet may need to be integrated with database management systems to allow users to access information (such as customer and product data).
- The Intranet should support existing (legacy) applications, as needed.
- The Intranet should have built-in directories, such as corporate telephone numbers and e-mail addresses.
- The Intranet should incorporate groupware applications.
- Support (or future expansion) for on-line conferencing should be considered.
- The Intranet should provide division-specific and corporate-wide bulletin boards for electronic postings.
- The Intranet should be designed with a document sharing and management process in mind.
- The Intranet should foster teamwork and collaboration by enhancing channels of information distribution.
- Search engines — which simplify a user's ability to locate and access information — should be made available.
- The Intranet should support electronic mail.
- The Intranet should support (for future expansion) multimedia applications that use text, images, audio, and video.
- Automated real-time Web-page generation should be encouraged.
- The Intranet should be designed so it can interface, at least potentially, with factory equipment, other manufacturing devices, or other critical legacy systems.
- The Intranet should support the automation of organization work flows.

6.2.3.6 Content management

Many organizations struggle with the tasks of information creation, management, and dissemination. They are time consuming and difficult to control. The Intranet alone cannot solve information management problems unless specific Intranet solutions are implemented that directly address the need for document management. The following list identifies content-management tasks that should be considered in the Intranet plan:

- Users must have the ability to easily add or update content on a regular basis.
- Users must have the ability to protect their content from changes by other users.
- A content-approval process should be defined and in place. This process should encompass ways to manage and control document revisions, especially changes to shared documents.

As policies and procedures relating to content management are formulated, it is important to designate responsibilities to specific individuals to ensure that they are put into place and followed. An Intranet style guide should be developed that provides page layout, design elements, and HTML code guidelines. The style guide will help the organization to maintain a consistent look and feel throughout the Intranet's Web pages. The style guide should contain information on where to obtain standard icons, buttons, and graphics, as well as guidelines on page dimensions and how to link to other pages. As part of the style guide, it is helpful to create Web page templates. These templates consist of HTML files and are used to provide a starting point for anyone interested in developing Web pages or content for the Intranet. Although it is very easy to create a working Web page and to publish it for mass viewing, the real challenge is in producing a well-conceived Web page.

6.2.3.7 *Training and support*

After the Intranet is up and running, efforts should be focused on how to maintain the information content and on employee training. Part of the document-management strategy should encompass the selection of content stakeholders. Content stakeholders are individuals in different departments or work groups who are responsible for the creation and maintenance of specific content. Stakeholders can be department managers, team leaders, or content authors and publishers.

Some organizations create a position called a Webmaster. This position is responsible for maintaining and supporting the content published on the Intranet. A good Webmaster should have the following skills:

- Basic Internet skills including an understanding of e-mail, FTP, and Telnet
- A thorough understanding of HTML document creation
- Experience with CGI programming
- Programming experience with languages such as Perl, C/C++, and Java
- Experience with content creation and the conversion of text and images

- Knowledge of client-server processing
- Experience with server setup and maintenance
- Knowledge of your organization's structure and inner workings
- Organizational and training skills

It is possible that the organization may choose to decentralize the maintenance of the information content. In this case, individuals from various departments might be selected to maintain the content relating to their respective department. These individuals should be trained to handle a variety of maintenance issues. A decentralized approach depends on having more than one individual with the necessary skills available to maintain the web pages. A decentralized support structure gives authors and content owners direct control and responsibility for publishing and maintaining information. This can help prevent bottlenecks in making information available in a timely fashion.

Training for stakeholders, Webmasters, and Intranet users is an important part of an Intranet strategy. Intranet customers and content stakeholders should be trained to understand the Intranet and how it will improve the organization and the way the company does business. They should also be given training on how to create, utilize, and maintain content on the web page(s). Companies that invest in the education and training of their employees will have a better chance of creating and maintaining a successful Intranet.

6.2.4 Intranet deployment

Since Intranets are easy to set up, many companies do not realize what the true resource requirements are to maintain the Intranet with up-to-date information. The goal of this section is to provide a realistic perspective on how organizations are most likely to achieve long-lasting benefits from the Intranet.

Some companies invest much more than their competitors in information technology, such as an Intranet, but still fail to compete effectively in the marketplace. Computers alone do not, and cannot, create successful companies. A good start is to empower all employees to contribute to the Intranet. As is true for any collaborative effort, every team member is responsible for the overall success of the team.

6.2.4.1 Technological considerations

The major technological challenges facing the organization after the initial implementation of an Intranet include:

- Converting existing paper documents into electronic documents that employees can access electronically via the Intranet
- Connecting existing databases to the Intranet so they are accessible by a wide range of computing platforms (such as Windows-based and Mac-based systems)

Table 6.1   Intranet Document Tracking Information

Data for tracking Intranet documents
Name of document
Document description
Page owner
Type of document (i.e., official, unofficial, personal)
Confidentiality status (i.e., confidential, nonconfidential, etc.)
Original publish date
Date document last modified
Frequency of update (i.e., daily, weekly, monthly, etc.)

- Coordinating the use of multiple servers used across departmental lines
- Continuously enhancing the Intranet's features and capabilities to keep employees motivated to use the Intranet
- Installing security features within the Intranet to prevent unauthorized access to confidential or sensitive information

Intranet technology, and information technology in general, is changing so fast that keeping up with the latest software and hardware solutions requires a substantial ongoing organizational commitment.

Conversion of paper documents into electronic form

The first issue facing companies after the initial Intranet release is how to convert large numbers of existing paper documents into electronic format ready for distribution on an Intranet. There are many tools, such as HTML Transit, that can be used to convert documents from most electronic formats to HTML format. Microsoft's Internet Assistant for Microsoft Word can also be used to easily convert existing Word documents into HTML documents. After paper documents have been converted to HTML and placed on the Intranet, the next challenge is to keep the documents up to date.

Obsolete information can frustrate Intranet users and may encourage them to revert to old ways of information gathering (i.e., calling people, walking to various offices, and writing memos). One way to minimize this problem is to create a database containing the document title, date of last change, and frequency of update in a database. Other useful information that can be used to track the status and nature of documents on the Intranet is shown in Table 6.1. A program can then be written to search the Intranet for documents that have not been updated recently. The program can then issue e-mail to the document owner to request an update.

Interface to Legacy database(s)

Connecting databases to the Intranet is not an easy task, and may require additional staff or reassignment of current programming staff. Legacy database vendors are currently working on various Intranet solutions to facilitate the implementation of this requirement.

Companies may need to connect the Intranet to legacy databases in order to access:

- Financial reports (regarding project costs, product costs, the overall financial health of the enterprise, etc.)
- Document-management systems
- Human Resources information (e.g., so employees can review information on health care and benefits)

Use of multiple servers

As the Intranet becomes more complex, multiple servers will be needed. This is especially true for companies that have a large number of divisions and business units using the Intranet. For example, a product-development group may need to provide team members with the ability to search project-specific databases, submit forms to various databases, and use a private on-line discussion group. The Webmaster may find it impossible to support these service needs in a timely manner. When this happens, companies frequently relegate the task of server maintenance to each respective department.

Over the next five years, installing and using a Web server will become as easy as installing and using word processor software. Web servers will probably become part of the Windows NT server operating system. When each department is responsible for maintaining their own web server, it is particularly important to choose server software that is easy to install and maintain. A Pentium-class machine running Windows NT server software and Microsoft's Internet information server is a good choice for small departments. Another way to provide departments with their own domain name and disk space is to use a virtual domain name. Companies use virtual servers to reduce hardware costs. In the case of the Web, an HTTP-based server runs on a server computer. For example, a company may need two types of Web servers, one that allows easy access and one that requires usernames and passwords. In the past, the company would have purchased two different computers to run the Web server software. Today, however, the company can run both servers on the same system — as virtual servers.

Standardizing hardware and software

To avoid supporting multiple hardware and software components, it is important to standardize the server software, hardware, HTML editing tools, and browser software. This will help to minimize the potential for unexpected network errors and incompatibilities.

6.2.4.2   *Maintaining the information content on the Intranet*

One of the major challenges organizations must face is how to make the transition from paper-based systems to computer-based systems while keeping information up to date.

Automating HTML authoring

After establishing a policy for the distribution of Intranet documents, it is advisable to develop a set of guidelines that clearly specify who is responsible for keeping them current. Inaccurate information greatly reduces the effectiveness of the Intranet. If employees lose confidence in the accuracy of the on-line information, they will revert to calling people to find information. Unfortunately, many people tend to ignore the need to update information, irrespective of its form (i.e., electronic or print).

In some cases, the Intranet will contain information that employees must update daily, weekly, or monthly. Spreadsheets can be used to capture highly time-sensitive data. Macros can be written (typically, by a staff programmer) that automatically convert the spreadsheet data into HTML format.

Managing document links

In a traditional document-management system, documents often reference one another. In most cases, authors list the applicable references at the top of each new document. Intranets, unfortunately, create a situation where organizations cannot easily control the accuracy of links in documents.

HTML document developers can use links freely and, in many cases, without checking the accuracy of those links. Even if employees test the initial accuracy of their document links, it is difficult to maintain and check the accuracy of those links after the document is released. If you have ever encountered a "broken" link when surfing the Web, you know that it can be frustrating. People depend on links within a Web document to find information. Today, however, there are a few mechanisms available to ensure the accuracy of document links. Employees must understand that other people may link to their pages and that they should not freely move the location of their documents. Employees must view their needs in the total organizational context.

6.2.4.3   *Centralized vs. distributed control*

The implementation of an Intranet is a major change for any organization. Although change is not easy, people are more inclined to modify their behavior when leaders have a clear sense of direction, involve employees in developing that direction, and are able to demonstrate how the Intranet will positively affect the employees' well being. Managers should work with their employees to show that Intranets can free them from the routine aspects of their job. This, in turn, will allow employees to spend more time learning and developing new ideas for the corporation.

Some of the benefits that can be obtained using a distributed model of Intranet control are listed below:

- Employees can tap into the knowledge of everyone in the organization, making everyone a part of a solution.
- The power of any one Webmaster to dictate the Intranet's form and function is limited.
- It empowers departments to create their own information databases and to work with outside customers and vendors.

6.2.5 Intranet security issues

By their very nature, Intranets encourage a free flow of information. This means that it is also very easy for information to flow directly from the Intranet to the desktops of those who might seek to gain access to information they should not have. To guard against this situation, adequate security measures should be in place when the Intranet is deployed. In the discussion that follows, we review various security techniques to protect an Intranet from unauthorized external and internal use.

6.2.5.1 Firewalls

The Internet was designed to be resistant to network attacks in the form of equipment breakdowns, broken cabling, and power outages. Unfortunately, the Internet today needs additional technology to prevent attacks against user privacy and company security. Luckily, a variety of hardware and software solutions exist to help protect an Intranet. The term *firewall* is a basic component of network security. A firewall is a collection of hardware and software that interconnects two or more networks and, at the same time, provides a central location for managing security. It is essentially a computer specifically fortified to withstand various network attacks. Network designers place firewalls on a network as a first line of network defense. It becomes a "choke point" for all communications that lead in and out of an Intranet. By centralizing access through one computer (which is also known as a *firewall-bastion host*), it is easier to manage the network security and to configure appropriate software on one machine. The bastion host is also sometimes referred to as a server.

The firewall is a system that controls access between two networks. Normally, installing a firewall between an Intranet and the Internet is a way to prevent the rest of the world from accessing a private Intranet. Many companies provide their employees with access to the Internet long before they give them access to an Intranet. Thus, by the time the Intranet is deployed, the company has typically already installed a connection through a firewall. Besides protecting an Intranet from Internet users, the company may also need to protect or isolate various departments within the Intranet from one another, particularly when sensitive information is being accessed via the Intranet. A firewall can protect the organization from both internal and external security threats.

Most firewalls support some level of encryption, which means data can be sent from the Intranet, through the firewall, encrypted, and sent to the Internet. Likewise, encrypted data can come in from the Internet, and the firewall can decrypt the data before it reaches the Intranet. By using encryption, geographically dispersed Intranets can be connected through the Internet without worrying about someone intercepting and reading the data. Also, a company's mobile employees can also use encryption when they dial into your system (perhaps via the Internet) to access private Intranet files.

In addition to firewalls, a router can be used to filter out data packets based specific selection criteria. Thus, the router can allow certain packets into the network while rejecting others.

One way to prevent outsiders from gaining access to an Intranet is to physically isolate it from the Internet. The simplest way to isolate an Intranet is to not physically connect it to the Internet. Another method is to connect two sets of cables, one for the Intranet and the other for the Internet.

Even without a connection to the Internet, an organization is susceptible to unauthorized access. To reduce the opportunity for intrusions, a policy should be implemented that requires frequent password changes and keeping that information confidential. For example, disgruntled employees, including those who have been recently laid off, can be a serious security threat. Such employees might want to leak anything from source code to company strategies to the outside. In addition, casual business conversations, overheard in a restaurant or other public place, may lead to a compromise in security. Unfortunately, a firewall cannot solve all these specific security risks.

It should be noted that a firewall cannot keep viruses out of a network. Viruses are a growing and very serious security threat. Prevention of viruses from entering an Intranet from the Internet by users who upload files is necessary. To protect the network, everyone should run antivirus software on a regular basis.

The need for a firewall implies a connection to the outside world. By assessing the types of communications expected to cross between an Intranet and the Internet, one can formulate a specific firewall design. Some of the questions that should be asked when designing a firewall strategy include:

- Will Internet-based users be allowed to upload or download files to or from the company server?
- Are there particular users (such as competitors) that should be denied all access?
- Will the company publish a Web page?
- Will the site provide telnet support to Internet users?
- Should the company's Intranet users have unrestricted Web access?
- Are statistics needed on who is trying to access the system through the firewall?

- Will a dedicated staff be implemented to monitor firewall security?
- What is the worst case scenario if an attacker were to break into the Intranet? What can be done to limit the scope and impact of this type of scenario?
- Do users need to connect to geographically dispersed Intranets?

There are three main types of firewalls: network level, application level, and circuit level firewalls. Each type of firewall provides a somewhat different method of protecting the Intranet. Firewall selection should be based on the organization's security needs.

Network, application, and circuit-level firewalls

Network-level firewall. A network-level firewall is typically a router or special computer that examines packet addresses and then decides whether to pass the packet through or to block it from entering the Intranet. The packets contain the sender and recipient IP address and other packet information. The network-level router recognizes and performs specific actions for various predefined requests. Normally, the router (firewall) will examine the following information when deciding whether to allow a packet on the network:

- Source address from which the data is coming
- Destination address to which the data is going
- Session protocol such as TCP, UDP, or ICMP
- Source and destination application port for the desired service
- Whether the packet is the start of a connection request

If properly installed and configured, a network-level firewall will be fast and transparent to users.

Application-level firewall

An application-level firewall is normally a host computer running software known as a proxy server. A proxy server is an application that controls the traffic between two networks. When using an application-level firewall, the Intranet and the Internet are not physically connected. Thus, the traffic that flows on one network never mixes with the traffic of the other because the two network cables are not connected. The proxy server transfers copies of packets from one network to the other. This type of firewall effectively masks the origin of the initiating connection and protects the Intranet from Internet users.

Because proxy servers understand network protocols, they can be configured to control the services performed on the network. For example, a proxy server might allow ftp file downloads, while disallowing ftp file uploads. When implementing an application-level proxy server, users must use client programs that support proxy operations.

Application-level firewalls also provide the ability to audit the type and amount of traffic to and from a particular site. Because application-level firewalls make a distinct physical separation between an Intranet and the Internet, they are a good choice for networks with high-security require-ments. However, due to the software needed to analyze the packets and to make decisions about access control, application-level firewalls tend to reduce the network performance.

Circuit-level firewalls

A circuit-level firewall is similar to an application-level firewall in that it, too, is a proxy server. The difference between them is that a circuit-level firewall does not require special proxy-client applications. As discussed in the previous section, application-level firewalls require special proxy soft-ware for each service, such as ftp, telnet, and HTTP.

In contrast, a circuit-level firewall creates a circuit between a client and server without needing to know anything about the service required. The advantage of a circuit-level firewall is that it provides service for a wide variety of protocols, whereas an application-level firewall requires an appli-cation-level proxy for each and every service. For example, if a circuit-level firewall is used for HTTP, ftp, or telnet, the applications do not need to be changed. You simply run existing software. Another benefit of circuit-level firewalls is that they work with only a single proxy server. It easier to manage, log, and control a single server than multiple servers.

Firewall architectures.   Combining the use of both a router and a proxy server into the firewall can maximize the Intranet's security. The three most popular firewall architectures are the dual-homed host firewall, the screened host firewall, and the screened subnet firewall. The screened-host and screened-subnet firewalls use a combination of routers and proxy servers.

Dual-homed host firewalls

A dual-homed host firewall is a simple, yet very secure configuration in which one host computer is dedicated as the dividing line between the Intranet and the Internet. The host computer uses two separate network cards to connect to each network. When using a dual-home host firewall, the computer's routing capabilities should be disabled, so the two networks do not accidentally become connected. One of the drawbacks of this config-uration is that it is easy to inadvertently enable internal routing.

Dual-homed host firewalls use either an application-level or a circuit-level proxy. Proxy software controls the packet flow from one network to another. Because the host computer is dual-homed (i.e., it is connected to both networks), the host firewall can examine packets on both networks. It then uses proxy software to control the traffic between the networks.

Screened-host firewalls

Many network designers consider screened-host firewalls more secure than a dual-homed host firewall. This approach involves adding a router and placing the host computer away from the Internet. This is a very effective and easy-to-maintain firewall. A router connects the Internet to your Intranet and, at the same time, filters packets allowed on the network. The router can be configured so that it sees only one host computer on the Intranet network. Users on the network who want to connect to the Internet must do so through this host computer. Thus, internal users appear to have direct access to the Internet, but the host computer restricts access by external users.

Screened-subnet firewalls

A screened-subnet firewall architecture further isolates the Intranet from the Internet by incorporating an intermediate perimeter network. In a screened-subnet firewall, a host computer is placed on a perimeter network, which users can access through two separate routers. One router controls Intranet traffic and the second controls the Internet traffic. A screened-subnet firewall provides a formidable defense against attack. The firewall isolates the host computer on a separate network, thereby reducing the impact of an attack to the host computer. This minimizes the scope and chance of a network attack.

6.2.5.2   CGI scripting

Web sites that provide two-way communications use CGI (common gateway scripting). For example, if you fill in a form and click your mouse on the form's "Submit" button, your browser requests the server computer to run a special program, typically a CGI script, to process the form's content. The CGI script runs on the server computer, which processes the form. The server then returns the output to the browser for display.

From a security perspective, the danger of CGI scripts is that they give users the power to make a server perform a task. Normally, the CGI process works well, providing an easy way for users to access information. Unfortunately, it is also possible to use CGI scripts in ways they were never intended. In some cases, attackers can shut down a server by sending potentially damaging data through the use of CGI scripts. From a security perspective, it is important to make sure that users cannot use CGI scripts to execute potentially damaging commands on a server.

6.2.5.3   Encryption

Encryption prevents others from reading your documents by "jumbling" the contents of your file in such a way that it becomes unintelligible to anyone who views it. You must have a special key to decrypt the file so its contents can be read. A key is a special number, much like the combination of a padlock, which the encryption hardware or software uses to encrypt and decrypt files. Just as padlock numbers have a certain number of digits, so do encryption keys. When people talk about 40-bit or 128-bit keys, they are simply referring to the number of binary digits in the encryption key. The more bits in the key, the more secure the encryption and less likely an attacker can guess your key and unlock the file. However, attackers have already found ways to crack 40-bit keys.

Several forms of encryption can be used to secure the network, including link encryption, document encryption, secure-sockets layer (SSL), and secure HTTP (S-HTTP). The following sections describe these encryption methods in more detail. The reader is also referred to Section 2.3.6 in Chapter 2 for additional information on encryption schemes.

6.2.5.3.1 Public-key encryption. Public-key encryption uses two separate keys: a public key and a private key. Users give their public key to other users so anyone may send them encrypted files. The user uses his or her private key to decrypt the files (which were encrypted with a public key).

A public key only allows people to encrypt files, not to decrypt them. The private user key (designed to work in conjunction with a particular public key) is the only key that can decrypt the file. Therefore, the only person that can decrypt a message is the person holding the private key.

6.2.5.3.2 Digital signatures. A digital signature is used to validate the identity of the file sender. A digital signature prevents clever programmers from forging e-mail messages. For example, a programmer who is familiar with e-mail protocols can build and send an e-mail using anyone's e-mail address, such as BillGates@microsoft.com.

When using public-key encryption, a sender encrypts a document using a public key, and the recipient decodes the document using a private key. When a digital signature is used, the reverse occurs. The sender uses a private key to encrypt a signature, and the recipient decodes the signature using a public key. Because the sender is the only person who can encrypt his or her signature, only the sender can authenticate messages. To obtain a personal digital signature, you must register a private key with a certificate authority (CA), which can attest that you are on record as the only person with that key.

6.2.5.3.3 Link encryption. Link encryption is used to encrypt transmissions between two distant sites. It requires that both sites agree on the encryption keys that will be used. It is commonly used by parties that need to communicate with each other frequently. Link encryption requires a dedicated line and special encryption software. It is an expensive way to encrypt data. As an alternative to this, many routers have convenient built-in encryption options. The most common protocols used for link encryption are PAP (password authentication protocol) and CHAP (challenge handshake authentication protocol). Authentication occurs at the data link layer and is transparent to end users.

6.2.5.3.4 Document encryption. Document encryption is a process by which a sender encrypts documents that the recipient(s) must later decrypt. Document encryption places the burden of security directly on those involved in the communication. The major weakness of document encryption is that it adds an additional step to the process by which a sender and receiver exchange and receive documents. Because of this extra step, many users prefer to save time by skipping the encryption. The primary advantage of document encryption is that anyone with an e-mail account can use document encryption. Many document encryption systems are available free or for little cost on the Internet.

6.2.5.3.5 Pretty good privacy (PGP). Pretty good privacy (PGP) is a free (for personal use) e-mail security program developed in 1991 to support public-key encryption, digital signatures, and data compression. PGP is based on a 128-bit key. Before sending an e-mail message, PGP is used to encrypt the document. The recipient also uses PGP to decrypt the document. PGP also offers a document compression option. Besides making a document smaller, compression enhances the file security because compressed files are more difficult to decode without the appropriate key. According to the PGP documentation, it would take 300 billion years for someone to use brute force methods to decode a PGP-encrypted compressed message. A commercial version of PGP, ViaCrypt PGP, is available for around $150.

6.2.5.3.6 Secure socket layer (SSL). The Secure socket layer (SSL) was developed by Netscape Communications to encrypt TCP/IP communications between two host computers. SSL can be used to encrypt any TCP/IP protocol, such as HTTP, telnet, and ftp. SSL works at the system level. Therefore, any user can take advantage of SSL because the SSL software automatically encrypts messages before they are put onto the network. At the recipient's end, SSL software automatically converts messages into a readable document.

SSL is based on public-key encryption and works in two steps. First, the two computers wishing to communicate must obtain a special session key (the key is valid only for the duration of the current communication session). One computer encrypts the session key and transmits the key to the other computer. Second, after both sides know the session key, the transmitting computer uses the session key to encrypt messages. After the document transfer is complete, the recipient uses the same session key to decrypt the document.

6.2.5.3.7 Secure HTTP (S-HTTP). Secure HTTP is a protocol developed by the CommerceNet coalition. It operates at the level of the HTTP protocol. S-HTTP is less widely supported than Netscape's Secure Socket Layer. Because S-HTTP works only with HTTP, it does not address security concerns for other popular protocols, such as ftp and telnet.

S-HTTP works like SSL in that it requires both the sender and receiver to negotiate and use a secure key. Both SSL and S-HTTP require special server and browser software to perform the encryption.

6.3   Intranet security threats

This section examines additional network threats that should be considered when implementing Intranet security policies.

6.3.1   Source-routed traffic

As discussed earlier, packet address information is contained in the packet header. When source routing is used, an explicit routing path for the communication can be chosen. For example, a sender could map a route that sends packets from one specific computer to another through a specific set of network nodes. The road map information contained in the packet header is called "source routing," and is used mainly to debug network problems. It is also used in some specialized applications. Unfortunately, clever programmers can also use source routing to gain (unauthorized) access into a network. If a source-routed packet is modified so that it appears to be from a computer within your network, a router will obediently perform the packet routing instructions, permitting the packet to enter the network, *unless special precautions are taken*. One way to combat such attacks is simply to direct your firewall to block all source-routed packets. Most commercial routers provide an option to ignore source-routed packets.

6.3.2   Protecting against ICMP redirects (spoofing)

ICMP stands for Internet control message protocol. ICMP defines the rules routers use to exchange routing information. After a router sends a packet to another router, it waits to verify that the packet actually arrived at the specified router. Occasionally, a router may become overloaded or may malfunction. In such cases, the sending router might receive an ICMP-redirect message that indicates which new path the sending router should use for transmission.

It is fairly easy for knowledgeable "hackers" to forge ICMP-redirect messages to reroute communication traffic to some other destination. The term *spoofing* is used to describe the process of tricking a router into rerouting messages in this way. To prevent this type of unauthorized access, it may be necessary to implement a firewall that will screen ICMP traffic.

6.4   Summary

Intranets are being used to improve the overall productivity of the organization. Important Intranet concepts covered in this chapter are summarized below:

- TCP/IP was created because of the need for reliable networks that could span the globe. Because of its reliability and ease of implementation, TCP/IP has become the standard language (or protocol) of the Internet. TCP/IP defines how programs exchange information over the Internet.
- An Intranet is based on Internet technology. It consists of two types of computers: a client and a server. A client asks for and uses information that the server stores and manages.
- Telnet, ftp, and gopher are widely used network programs that help users connect to specific computers and to transfer and exchange files.
- The World Wide Web (or Web) is a collection of interlinked documents that users can access and view as "pages" using a special software program called a browser. The two most popular browser programs are Netscape Navigator and Microsoft Internet Explorer.
- HTML (hypertext markup language) is used to describe the layout and contents of pages on the Web.
- Java is a new computer programming language that allows users to execute special programs (called applets) while accessing and viewing a Web page.
- A network computer is a low-cost, specialized computer designed to work in conjunction with the Internet and Java application programs.

To be effective, the Intranet must deliver quality information content. To ensure this, management must play a proactive role in assigning staff who will keep the corporate information reservoirs on the Intranet current and relevant. The following is a checklist of some of the ways to encourage the development of a high-quality Intranet:

- Give users access to document management systems and various corporate databases.
- Distribute the responsibility of maintaining the Intranet to increase the number of staff involved in developing and enhancing Intranet content.
- Create a corporate culture based on information sharing.
- An Intranet deployment strategy must place employee training at its center.
- Design and implement appropriate security measures as soon as possible.
- Use firewalls to control access to the network.
- Use antivirus software.
- Implement a security plan that controls the access that employees and outsiders have to the network.
- Design and implement CGI scripts with security in mind.
- Encourage users to encrypt files before sending confidential data across the Internet/Intranet.

chapter seven

Outsourcing network design, planning and implementation

Contents

7.1 Process overview

Outsourcing has been a major consideration for many companies for many years. Outsourcing means the assignment of certain management or planning functions to external companies. The number of external companies offering professional services is increasing and can be categorized as follows:

- Device vendors and manufacturers, e.g., Lucent Technologies
- Telecommunication providers, e.g., AT&T
- Consulting companies, e.g., Anderson Consulting
- Special service providers, e.g., International Network Services, Net-Solve, EDS

The following steps describe the process of outsourcing the network design, planning and implementation (Figure 7.1):

1. Conducting diligency audits to identify status
2. Decision about functions to be outsourced
3. Decision about outsourcers to be considered

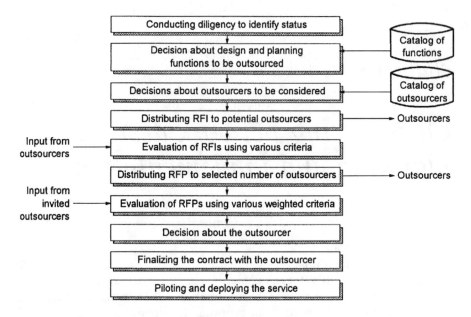

Figure 7.1 Process of selecting the outsourcer for design and planning.

4. Distributing the request for information (RFI) document
5. Evaluation of RFIs based on predetermined criteria
6. Distributing the request for proposal (RFP) document
7. Evaluating RFPs based on predetermined criteria
8. Making the final selection
9. Implementing the solution

7.2   Major considerations

The decision about outsourcing should be very thoroughly considered. In particular, in the area of network design and planning, the company is giving away strategic responsibility to an external company. Contract management and continuous control over the outsourcer is absolutely necessary. There are few cases where the network design and planning are outsourced completely. If support is required, consulting companies are contracted for preparing and/or validating network design alternatives. But the ownership of the network planning design data remains with the company. The ownership of planning and design instruments may vary; in some cases the consulting company uses its own tools. This is easy to understand because the use of tools needs knowledge, education, and experience and cannot be easily transferred from Tool A to Tool B. In special cases, where the networks are very complex or are changing very rapidly, consulting companies may be hired to use, and/or modify existing tools and models maintained by the user.

Before preparing RFIs and RFPs, a number of questions should be answered internally. The most important considerations are:

1. Present costs of equipment, communications, people, and network design and planning.
2. Full visibility of existing processes, instruments, and human resources. This is needed in order to decide which functions may be considered for outsourcing. Outsourcing is a good excuse to audit the present design and planning functions and to address areas that need improvement. After the preparations and planning for outsourcing are completed, the decision may be made to complete the work in-house instead. This type of analysis by internal or external analysts may result in substantial savings in the network design and planning expenses (30%–40%), and in staff reduction (25%–50%). It is also helpful in stabilizing network budgets.
3. Dependence on network availability. This involves assessing the level of risk that can be tolerated. These considerations should be reflected in the service contract; many times certain vendors will fall short at the very beginning, and are not able to guarantee the targeted availability. The outsourcer should be requested to guarantee future service levels.
4. Grade of service required by users and applications may dictate that the outsourcing company not share the network design and planning resources with other clients.
5. Security standards and tolerable risks may prohibit third-party vendors from gaining access to the network and its carried traffic.
6. Assessment of business needs and the appropriate emphasis that should be placed on technological simplicity or sophistication.
7. Availability of planning and design instruments. If the company has to invest substantial amounts on instrumentation, outsourcing should be favored; if not, outsourcing may still be considered, but with lower priority.
8. Availability of skilled design and planning personnel is one of the most critical issues; most frequently, it is the only driving factor for outsourcing. Not only present status, but also future needs and how they will be satisfied should be quantified prior to the outsourcing decision.
9. The stability of the environment and its growth rate have serious impacts on the contract with the vendor. Acquisitions, mergers, business unit sales, application portfolio changes need special and careful treatment in the contracts.

After answering all of these questions, the company must decide which design, planning and modeling, and related functions should be outsourced. In a broad sense, the following functions may be considered for outsourcing:

- Baselining network performance: This function concentrates on measuring key service and utilization indicators to determine optimal operational thresholds.
- Conducting performance measurements: In various segments of local area networks, measurements can be conducted using standard information sources, such as SNMP-MIBs and RMON-MIBs.
- Performance reporting: Based on the performance information collected and maintained, periodic and ad-hoc reports are generated and distributed.
- Design and planning: Based on present status on service and resource utilization, and load estimates, modeling tools help to evaluate "what-if" scenarios.
- Implementation and maintenance: It includes the physical implementation of the network, service activation, service assurance, and maintaining components of the network.
- Operations: This means turning over the day-to-day responsibilities, including fault management, trouble ticketing and tracking, tests, service restoration and repair to the outsourcer.

Table 7.1 identifies which outsourcing functions are best outsourced to external companies. Outsourcing is often recommended for baselining, implementation, and operations. Performance measurements and reporting, when outsourced, typically require close internal coordination and supervision. Network design and planning, in contrast, are usually performed in-house, due to the sensitive and proprietary nature of these functions. When these functions are outsourced, it is to obtain the benefit of multiple expert opinions. In general, it is best to outsource functions furthest away from the company's core competencies. In addition, if the company's dependence on information is high, and the corporate environment is stable, then outsourcing may make sense. In highly innovative corporate settings, the risk of outsourcing is greater, due to the potential for revealing proprietary and sensitive information to outsiders.

Table 7.1 Outsourcing Network Design, Planning,
and Modeling Functions in Broad Sense

Functions	Answers		
	High	Medium	Low
Baselining	x		
Performance measurements		x	
Performance reporting		x	
Design and planning			x
Implementation and maintenance	x		
Operations	x		

7.3   Outsourcing industry examples

Recently, outsourcing has been penetrating the performance measurement and reporting area. Not everybody wants to equip the whole network with monitoring instruments, because processing measurement data requires significant human resources. Instead, they turn to companies who offer measurement, reporting, and planning services.

International network services with its NETracker

NETracker is exactly the solution for which many users are looking. The company provides performance monitoring and provides information for proactive fault management.

INS NETracker works within the Internet Network Management Framework and relies on the SNMP protocol and the manager/agent model. Currently, the system uses SNMP to obtain performance information from wiring hubs, routers, and RMON agents.

Many vendors have developed private MIBs that support their equipment in ways that go beyond the capabilities supported by the standard MIB-II. Currently, Cisco MIBs are used to monitor the processing load and memory utilization on Cisco routers. Other private MIB monitoring implementations are being developed for NETracker.

In addition to SNMP-gathered information, the system also uses two non-SNMP mechanisms to gather performance data. Resource uptime is measured using ICMP echoes (PING) and ICMP echo replies. Any IP device — usually IP hosts — that is able to respond to a PING can be monitored. An intelligent platform — referred to as an INS meter — placed on one of the client's LAN segments will periodically PING specified client hosts and will record the results. This mechanism gives an indication of the reachability of the host, which is a function of the host's condition and the network path to the host.

Response times to IP hosts are determined using ICMP echoes (PINGs) and echo replies. PINGs and their replies are timed and these times are used to determine the round trip transit time from the INS meter to the host. Because most IP hosts implement the ICMP echoes and replies at low levels in the operating system (UNIX kernel), the response time is normally not affected by application load on the host. Therefore, this is a good measure of true network delay.

Figure 7.2 shows the NETracker system in operation. The system consists of one or more INS meters that are connected to one or more client LAN segments. The meters have INS software that communicates with the SNMP agents and IP hosts, collects performance data, analyzes and consolidates the data into hourly statistics, and uploads this data daily via a dial-out modem connection to an INS server at an INS location that maintains current and archived data.

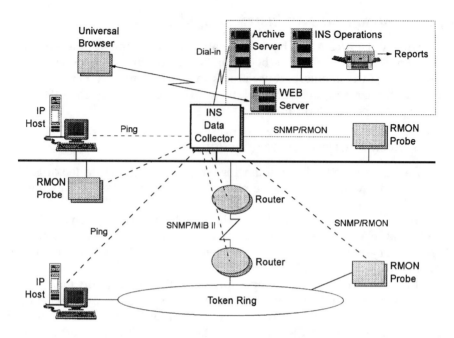

Figure 7.2   NETracker system in operation.

Database applications create weekly and monthly reports from the information on the INS server. These reports are printed and delivered on the following day to the client. INS considers each resource monitored as an element. Typically, an element is either an IP host being monitored for uptime and response time, a router interface, or a Cisco router CPU or memory. NETracker uses both MIB and RMON entries for the performance evaluation. RMON probes see all traffic on the LAN segment. Utilization and error reports, based on data gathered by RMON, reflect the total traffic and total errors on the segment.

When data are collected via MIB-II, the utilization and error information applies to a particular interface on the device, which is typically a router, but could be a bridge, hub, or switch. Router interface statistics provide a good indication of traffic entering and exiting the LAN and can be useful in sizing routers and WAN links, and in determining the effects of reconfiguring the local LAN. Error information collected from a LAN interface in this manner can provide a cost-effective way to monitor LAN health.

The reports are organized into three categories depending on the time frame for reporting. Current detail reports show the hour-by-hour history of each indicator for the most current reporting period. Current summary reports show how metrics varied by hour of the day, on average, for a definite number of days. The historical reports show daily statistics for the most recent 91 days. Included in the set of historical reports is one unique report, the device count report, which shows a history of how many devices were on the network during each day of the reporting period.

Availability (%)

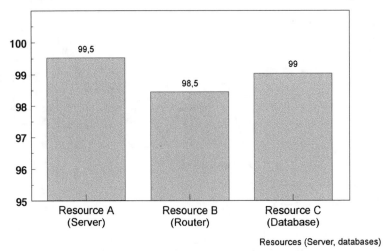

Figure 7.3 Uptime of resources.

The general types of performance metrics monitored and reported are:

- Uptime of resources, such as file servers, database servers, interconnecting devices, measured by responding to PINGs (Figure 7.3)
- Response time for each monitored device in relation to the INS meter (Figure 7.4)
- Use of Ethernet and token ring segments
- Use of WAN-links from the perspective of in- and outgoing traffic of routers (Figure 7.5)
- Distribution of protocols to determine bandwidth occupancy by applications (Figure 7.6)
- Errors and events
- Collisions
- Distribution of frame sizes
- Communigrams for the highest traffic transmitters and receivers
- CPU utilization of routers to highlight saturation levels; proactive view of load peaks and their impacts on routers helps to avoid crashes and discarded traffic (Figure 7.7)

This way of outsourcing performance monitoring and report generation offloads network management staff from these activities for a reasonable fee. But performance-related data are managed outside the company, making ad-hoc reporting, trending, and capacity planning very difficult.

Advanced network support system from Network Defenders, Inc.

ANSS is a set of Windows NT/95 applications designed to enable Work-Group based proactive network management of LAN/WAN networks. Its capabilities include:

Response time in ms

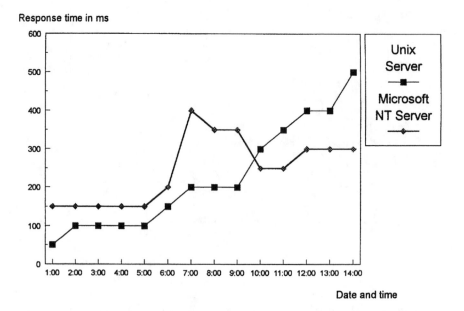

Figure 7.4 Response time measurements.

Utilization of WAN-links (%)

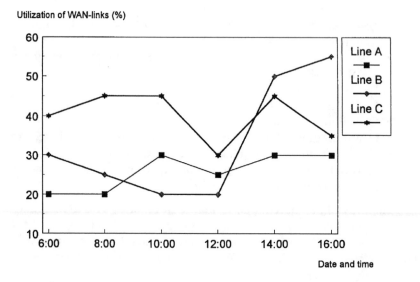

Figure 7.5 Utilization of WAN links.

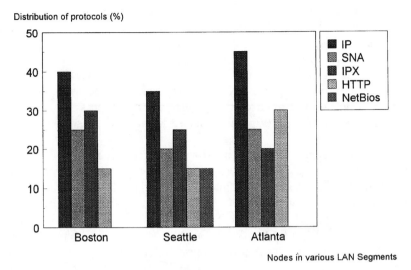

Figure 7.6 Distribution of load by protocols in use.

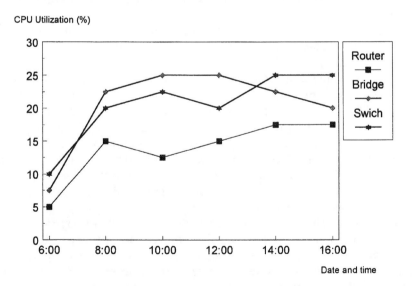

Figure 7.7 CPU utilization of routers.

- Network performance measurement and improvement
- Automated data collection and processing
- Automated reporting
- Online custom report generation
- Trending and baselining
- Capacity planning of LAN and WAN links, nodes, and servers.
- Troubleshooting of chronic or intermittent problems
- Interactive work-group based network analysis tools

ANSS is designed to complement existing real-time Network Management Systems and services. It does not require on-premises operator intervention. It was designed for easy installation and operation while providing immediate results. NDI's ANSS tools operate independent of, but complementary to, the popular real-time Network Management platforms such as IBM's NetView AIX, HP OpenView, SunNet Manager, etc. Since the ANSS tools are designed for workgroup reporting, analyzing, and trending of historical data, the tools are provided in a 32-bit Windows NT/95–compatible format so that network operations and planning personnel can conduct their work from any networked or remote PC workstation. The major user benefits of ANSS are gained through dramatic improvements in:

- Productivity of skilled network resources
- Reduced cycle times in planning, analysis, and report generation
- Network up-time
- Service level measurement reporting and improvement
- Network cost optimization

ANSS's on-net data collection engine (the ProVision Server) systematically collects comprehensive statistics, events, and traps from all network nodal devices and servers — using standards-based TCP/IP and SNMP protocols and standard MIB2 definitions. Vendor specific extensions or Enterprise MIB support is included for the major routing and switching vendors such as Cisco Systems, Bay Networks, 3Com, and others. Windows NT and Novell servers are also supported by the system integrating applications and file server performance management capabilities with the network to provide a complete enterprise view: total network quality management.

ANSS's reports and analysis capabilities enable network operators, planners, and designers to perform detailed analyses, including baseline performance and trend analysis. The system produces ready-to-use management and operational reports, providing a consistent and concise presentation of network reliability, availability, utilization, response time, error rates and more. Some of the unique capabilities of the system include:

- Automated data collection with flexible user definable scripts
- Automatic device configuration protection and configuration change audit trails

- Real-time display of network node availability and response time with Web linkage
- Multilevel report generation, scheduled or on-demand
- Workgroup-based interactive graphical analysis
- Customized reporting and trending
- IntraWeb report publishing, which aids troubleshooting personnel dispatching

At a detailed operational level, ANSS and the ProVision services provide the information and tools needed to systematically troubleshoot chronic and intermittent network performance problems with a sophisticated set of data manipulation and graphical presentation technology.

The ANSS applications suite and associated ProVision services are particularly useful for monitoring and reporting on delivered levels of service and bring a cohesive view of the network during periods of network change, upgrade, or migration. It can also bring significant value and automation to the operational impact and evaluation of new device adds, new sites, new applications, firmware upgrades or new carrier services, easing the evaluation and acceptance testing process. It is also an excellent source of information in monitoring the delivered level of contracted services from outsourced network operations and/or carrier service providers.

Typical report contents include:

- Device/link reliability, availability, and utilization
- Device configuration and rate of changes
- Round trip delay/response time
- Loading and error rates of: frame relay links, network device serial ports, LAN interfaces, Novell servers, and more...

Proactive network management is used to detect and eliminate problems before they cause serious impacts. It provides the basis for proper planning of network growth and technology change. A well-designed proactive system will ensure that the information needs of management and professionals are met by providing periodic summary reports, trend information, and detailed supporting data. Due to escalating workload and limited technical resources proactive management usually takes second place to the reactive activities triggered by day-to-day problems and ongoing network changes. The *ProVision* service offering is designed to provide the benefits of proactive management without placing high demands on limited technical resources.

A baseline study is an essential first step in a proactive network management program. A baseline provides a view of the level of service the network is delivering to end users. It also provides essential information relating to the loading and error rates of critical circuits and network devices.

NDI offers an initial baseline study of an enterprise network. The baseline provides measurements of traffic and error rates of wide area network circuits and the loading and critical resource utilization of your networking devices.

The baseline is performed for a flat fee for WAN/LAN networks of up to 100 network devices (routers, hubs, and servers). The cost of a baseline analysis for a larger enterprise must be negotiated.

The highlights of the baseline program include:

- Installation and configuration of a ProVision Server
- Data collection and analysis for a two-week period
- Data reduction and production of NDI standard reports
- First-pass analysis of data
- Review of results with NDI engineers
- Delivery of final summary report and CD-ROM with detailed graphs and reports

NDI's ProVision services focus on the systematic delivery of network performance, measurement, and improvement. The service is based on a core set of technology developed by NDI that takes advantage of the statistical information available from the installed network devices: routers, hubs, switches, servers, RMON agents, etc. NDI service personnel perform all on-site installation, configuration, and maintenance of the ProVision server, a PC-based data-collection engine. Critical quality of service metrics of key LAN/WAN network elements are continuously monitored by the on-net ProVision server. Extensive network statistics are systematically gathered and stored to disc on the server in a highly compressed format. There is no need for continuous remote connectivity to the ProVision server. Simple dial-up (or other switched or FTP enabled) connectivity is sufficient to provide the service.

Once installed, the ProVision server requires no operator intervention. All required software and configuration maintenance is performed remotely by NDI service technicians. In addition any required hardware maintenance and service is provided by NDI.

The ProVision server focuses on the statistical data needed to enable enterprise-wide reporting and analysis of:

- Availability
- Reliability
- Response time
- Traffic load
- Resource utilization
- Error rates
- Configuration

ProVision basic services include:

- Continuous non-disruptive data collection
- Weekly data extraction and processing

- Weekly report generation
- Graphical object database creation
- ProVision client (interactive graphical analysis and report-viewing tool)
- Data trending for predictive analysis
- Reports, databases, and ProVision client delivery on CD-ROM
- Off-site database maintenance and archiving
- Device configuration protection and back-up
- Unlimited technical support

In addition, the ProVision basic service includes some start-up and periodic value added services, including:

- Two first-pass analysis services performed by NDI ProVision analysis engineers
- 8 hours of graphical report customization services
- Quarterly network analysis service and review
- Ongoing collection optimization and network device list maintenance

ProVision flexibility in service level choices:

- ProVision optional value added services
- Printing and binding
- First pass data analysis
- Network baseline service
- Report customization services
- ProVision scalability

7.4 Preparation of the request for information document

Prior to issuing an RFI, potential outsourcing partners should be evaluated. Table 7.2 lists the different types of outsourcers. The final decision depends on the networking environment and the budgets for external professional services. Outsourcers are expected to meet the following requirements:

- Financial strength and stability over a long period of time
- Proven experience in managing domestic and multinational networks
- Powerful pool of skilled personnel
- Tailored network-management instruments that may be used exclusively or in shared modes for the clients under consideration
- Proven ability of implementing the most advanced technology
- Outstanding reputation in conducting business
- Willingness for revenue sharing
- Fair employee transfers

Table 7.2 Service Provider Evaluation

Vendor type	Benefits	Disadvantages
Mainframe vendor	Knowledge of data networks	Dependency on proprietary architecture
	Reputation	Data orientation
	Knowledge of logical network design and management	Usually horizontal management integration
	Vertical integration	Little experience with physical network design
Telco suppliers	Knowledge of voice networks	Dependency on proprietary architectures
	Reputation	Little experience with logical network design
	Knowledge of physical networks and their management	Voice-data integration is slow
	Good knowledge about tariffs	
System integrators	Knowledge of both data and voice networks	High cost
	Knowledge of both logical and physical network design	Reputation
	Implementation experiences	Project driven
Consulting companies	Knowledge of both logical and physical network design	High cost
	Knowledge of both data and voice networks	Reputation
	Aware of leading edge technology	Implementation experiences
Device vendor	Knowledge of own equipment	Little knowledge of other equipment
	Knowledge about design tools	Reputation
	Implementation experiences	Biased toward own products

The weights of the criteria are set by the client during the evaluation process. Multiple functions should be considered in the service contract, including on-site, off-site, and virtual office functions. Table 7.3 compares these three choices.

The request for information (RFI) should include the following items:

Table 7.3   Service Alternatives

Type of service contract	Benefits	Disadvantages
On-site design work	Rapid prototyping	High cost
	Dedicated personnel	No shared instrumentation
	Continuous consultation	Space requirements
Off-site design work	Shared instrumentation	Slow communication
	No space requirements	Ad-hoc consultation
	Moderate cost	No dedicated personnel
	More concentrated knowhow	Longer cycles for evaluating design alternatives
Virtual office	Moderate costs	No dedicated personnel
	No space requirements	Periodic consultation
	More concentrated know-how	Slower prototyping

- Company size
- Revenue and profit
- Submission of annual reports
- References with products and services
- Submission of product descriptions
- Hardware and software requirements
- Support for certain geographic area
- Memberships in standard bodies
- Ownership of the product
- Subcontractors
- Development and sales personnel

7.5   Preparation of the request for proposal document

The responses from bidders provide information about capabilities, company background, and financial strength. It is recommended that the evaluation criteria be agreed upon prior to receiving the responses. Usually, the evaluation team checks the so-called "k.o." criteria first. These criteria may include:

- Company size
- Revenue and profit
- Reputation by evaluating references with products and services
- Support for a certain geographic area.

In the second step, technical capabilities are also evaluated, but not in great depth. The team should be convinced that the outsourcer under consideration could do the job. In the third step, financial figures may be evaluated. But, again not yet in great depth. Contract negotiations are far away from this phase.

The expected result is a shorter list consisting of outsourcers who do not violate any k.o. criteria and who could do the job at a reasonable price. These companies are invited for the bid by sending out RFPs. There are no guidelines as to how many companies should be invited in this phase. Industry analysts recommend that more rather than less should be invited. If in doubt, it is better to invite too many as opposed to too few vendors to participate in the RFP process. If just a few companies are invited, the evaluation basis will not be representative. If too many are invited, the evaluation team may be stressed.

The request for proposal (RFP) should include the following items:

- Proposal format
- Cover letter
- Submission date
- Table of contents
- Proposed copyright releases
- Executive summary
- LAN architectures and topologies
- Vendors and equipment
- Technical requirements
- Implementation requirements
- Price details
- Product references
- Clarification
- Evaluation criteria without weights

7.6 Vendor evaluation and selection checklists

The evaluation scenario should be prepared prior to the receipt of proposals. The evaluation table or tables is (are) distributed to each member of the team. The table (Table 7.4) should contain:

- Weights for each criteria
- Weights for each subcriteria
- The range of scores
- Guidelines for scoring

After the scoring process, which may require several dialogs with the outsourcers, the evaluation team determines the two or three leading proposals. The outsourcers are invited to make presentations, demonstrations, tests, and to present their financial terms.

Prior to signing an outsourcing contract, the following items should be addressed and agreed upon:

Table 7.4 Evaluation Matrix

Criteria Proposals	Weight	A	B	C
Group 1 — Company background				
Subcriteria	100%			
Group 2 — Financial strength				
Subcriteria	100%			
Group 3 — Technical capabilities				
Subcriteria	100%			
Group 4 — Experiences in the market				
Subcriteria	100%			
Group 5 — Conformance to company standards				
Subcriteria	100%			
	100%	Weighted scores for A, B and C		

Who is authorized to approve modifications? Somewhere in the agreement, preferably in the beginning, the provisions for modifying the agreement should appear. Basically, the provisions should allow either the outsourcer or the customer to reopen negotiations and for management approval if the priority of the work is to be changed. Generally, changes should be made only after an analysis has demonstrated that the problem is not an aberration. Trial agreements for new outsourcing services can be very helpful.

Agreement duration. The agreement should be written as ending after a certain period of time, such as a certain number of years. Alternatively, all outsourcing agreements might be scheduled for revision after expected changes to hardware, networking nodes, facilities, or software are scheduled to occur. Above all, no one should be under the impression that the agreement is a commitment for eternity, despite changes in the business or networking environment.

Reviews. Reviews are necessary to consider the impacts of a dynamically changing environment. For mutual benefits, those impacts should be openly discussed and the necessary changes should be written in the existing contract.

Service-level indicators. Service may be agreed upon for various levels of detail, including on-premise monitoring and off-premise solutions. Service level indicators should be quantified in the design and planning models.

User commitments. These commitments include informing the outsourcers about lines of business, strategic goals, critical success factors, networking environments, organizational changes, application portfolios, service expectations and indicators, directions of technology in networks design, and early warning of the need to renegotiate the contract if necessary.

Reporting periods.   Performance reports on key service indicators should be regular (weekly, monthly, etc.), and copies should go to the organization entity in charge, and to information technology (IT) or information systems (IS) management. Reports have to have a format agreed upon by both parties. Performance reports provide an opportunity for identifying potential problems and solutions in advance. They are key tools for avoiding crisis management.

Costs and charge-back policy.   Contracting parties have to agree on the conditions of payments, charge-back report, alternative of bill verification, and expected inflation rates. The transfer of human resources also has to be negotiated and agreed upon.

Penalties for noncompliance.   Using an appropriate costing and charge-back policy, creating penalties for noncompliance is relatively simple. For noncompliance of service objectives, payments have to be reduced to outsourcers. But, in most cases, the damage caused by noncompliance is much more severe than the penalty reimbursed in monetary units.

Employee transition.   Somewhere in the agreement, usually in the closing section, questions of employee transition have to be addressed. This section has to include the names of employees transferred to the outsourcers, and the conditions of takeover, such as salaries, job security, title, position, and so on. Training and education to be provided also have to be identified.

Billing and currency issues.   It is beneficial to tie the contract, be it distributed or centralized, to a single currency. This is particularly important with multinational companies.

Periodic pricing reviews.   Subcontractors lower costs or raise them over time. The contract must find a way to accommodate these changes and to adjust the contract, if necessary.

7.7   Strategic partnerships and agreements

Five different parties are involved in the network design and planning business. These parties are:

- Manufacturers of monitoring tools
- Manufacturers of networking equipment
- Vendors of design and planning tools
- Outsourcers with or without own design and planning tools
- Users

These five groups represent very interesting partnerships. Some of them are described below:

- **Users and outsourcers:** The emphasis of this chapter has been on this relationship. The user may sign a consulting or outsourcing contract for all or some of network design and planning related functions.

- **Outsourcers and vendors of design and planning tools:** Outsourcers are interested to acquire or lease tools to complete their professional service portfolios. They may even consider multiple tools to cover many networking architectures and protocols.
- **Manufacturers of networking equipment and vendors of design and planning tools:** Manufacturers are motivated to provide tools to configure and size their equipment. In addition, change management could be a target for modeling tools by evaluating the impacts of changes.
- **Manufacturers of monitoring tools and vendors of design and planning tools:** Monitoring the networks can greatly facilitate model verification. Baseline models use actual measurement data on services and utilization.

7.8 *Summary*

Outsourcing is a well-known discipline. Businesses concentrate on their core functions and outsource support functions to external professionals. In the area of network design and capacity planning, outsourcing is limited. There are particular reasons for this. Very frequently, telecommunications suppliers include modeling and simulation results in their proposals. Users assume that the results are biased, and sometimes they are. Also equipment vendors offer optimization services within their offers. Their internal modeling and simulation packages are fine-tuned for their equipment only. When users implement other equipment or a combination of equipment, the optimum solution is no longer assured. Independent companies, such as consulting firms, explode budgets with their high rates, making the outsourcing decision even more difficult.

The most promising way today is to combine performance management, design, and capacity planning into one package, and to outsource it to an independent company that supports both. The mixture of instruments implemented remains the responsibility of the outsourcer.

Closing remarks

As telecommunications and information technology become more closely aligned strategically to a company's core competencies, network designers and managers need to match new technical innovations with their company's business objectives.

Considering that technology changes on a daily basis, one needs to keep current with both existing and new technical breakthroughs through the analysis of vendor products, design specifications, technical journals, and trade publications. Equally important are formal educational classes that explain the theory behind technologies and why they were created. In addition, training seminars can provide essential information on how various manufacturers and service providers have implemented these technologies.

The following areas of technological study serve as general guidelines when considering what is available in designing a network. They should not be considered a complete set; rather, they provide some of the core technologies that network designers need to examine throughout their technical careers:

- Network design tools and traffic simulation products;
- Technical standards including CCITT ITU, Internic RFCs, Bellcore, ANSI, and IEEE;
- Network Management technologies including SNMP, RMON, and CMIP;
- Internetworking protocols such as TCP/IP, OSPF, BGP, RIP, IGRP, EIGRP, PPP/MPPP, and IPX/SPX;
- LAN/WAN technology including Ethernet, Token Ring, FDDI, Frame Relay, ATM, SONET/SDH, ISDN/B-ISDN, SMDS, X.25/X.75, and Private Lines;
- ADSL/HDSL high-speed modems;
- Voice communications including PBX/PABX, voice compression, and fax demodulation techniques;
- Video compression for both point-to-point and multipoint sessions;
- Fiber optics technology including Single-mode and Multi-mode methodologies;

- Satellite for low-orbit and geostationary transmissions;
- Microwave for alternate local loop connectivity;
- Security techniques including PAP, CHAP, TACACS, RADIUS, and Kerboros;
- Public and private encryption technologies.

Besides having an in-depth knowledge of the above mentioned technologies, the telecom and information technology designer must understand how to integrate the various components into a seamless network that is cost-efficient, scaleable, easy to upgrade, and compatible across various vendors. One also needs to develop a balance between technical innovations, migration strategies, and systems integration solutions. This requires thorough analysis of internetworking topologies, utilizations, traffic loads and flows, networking and performance trends, disaster recovery plans, and network impact statements. These technical considerations also need to be coupled with financial cost analysis, RFI/RFP development, and implementation planning.

In summary, network design and planning is a challenging and formidable task. To do it effectively, requires in-depth knowledge, creativity and imagination in envisioning ways to piece together a multitude of often conflicting options, and the right organizational and technological tools to support the process.

appendix A

Return on investment calculation

Contents

In this Appendix, we demonstrate how the costs and benefits of a network are evaluated from a financial perspective using return on investment (ROI) analysis. This analysis helps the decision-maker compare and contrast the proposed network project with alternative investment options. It is also helpful in justifying the funding for the network project.

For the purposes of illustration, we present a case study involving the analysis and justification of a proposed Intranet project for a Fortune 500 company that has 1000 employees located at one primary site.

ROI analysis

Calculating the ROI involves a thorough investigation of the network costs and benefits. In general, the start-up costs for Intranets are not high. However, as the number and variety of applications supported by the network increases, so will the network costs. These costs relate to additional purchases of hardware and software, and the staffing and related support required to maintain the applications on an ongoing basis. The ROI analysis attempts to assess all the costs and benefits of the Intranet over its expected life.

The first step in analyzing the Intranet's ROI is to develop an inventory of all the activities that the Intranet's deployment will affect and to quantify the associated costs. This is not an easy task; nonetheless, it is an essential one for the ROI analysis.

The ROI equation

ROI is expressed as a ratio of the benefits that an investment will provide (i.e., the potential savings and income expressed in dollars) divided by the total investment needed to derive those benefits (expressed in dollars). The ROI formula is expressed as:

$$\text{ROI} = \frac{\text{Savings + Income (revenue - cost)}}{\text{Investment Cost}} \qquad (A.1)$$

where   Savings = how much money will be saved after the Intranet is in place
Income = how much money the network investment will generate (i.e., revenue − costs)
Investment costs = how much money is needed to build the Intranet

All components of the ROI formula should be computed over the total life expectancy of the Intranet, and should be expressed in the same dollar and time units.

Depreciation and capitalization

Since the Intranet will be used over a period of years, depreciation has to be taken into account in the ROI analysis. Several methods of depreciation are available for amortizing the investment costs over the expected life of the project.[1] To preserve simplicity in our presentation, we use straight-line depreciation. The annual straight-line depreciation expense is computed as:

$$\text{Annual Depreciation Expense} =$$

$$\frac{\text{Total Capitalization Costs Over Life Of Project}}{\text{Expected Life of Project In Years}} \qquad (A.2)$$

---

[1] For a comprehensive treatment of various depreciation methods, the interested reader is referred to *Economic Analysis for Engineering and Managerial Decision Making*, by N. Barish and S. Kaplan, McGraw-Hill, Inc., copyright 1978.

Throughout the useful life of the Intranet, additional investment costs will occur in the form of hardware upgrades (such as hard disks, memory) and various software expenses. These expenditures are referred to as capitalization (new investment) costs. They must be added to the denominator of the ROI formula when they are present.

Expanded ROI formula

Taking into account the depreciation and expenses incurred to build and maintain the Intranet, we amend the ROI formula accordingly. A more accurate expression of the ROI equation is stated as follows:

$$\text{ROI} = \frac{\text{Savings} + \text{Income} - \text{Depreciation}}{\text{Depreciated Investment Cost}} \qquad (A.3)$$

where   Savings = how much money is saved after the network is in place
Income = how much money the investment will generate =
    (Revenue – costs)
Depreciation = the devaluation of capital expenditures over time
Depreciated investment costs = how much money is invested
    (capital) to build, depreciated over time

All components of the ROI formula should be expressed in the same units of time and dollars.

ROI numerator

Net benefits: Additions.   To calculate the ROI numerator, the benefits of the Intranet must be defined and calculated. Benefits may take two forms: *savings from costs avoided and returns from efficiencies created.* For example, if a cost is no longer necessary due to the Intranet's implementation, this can be counted as a benefit of the Intranet. Similarly, if a job can be completed faster due to the Intranet's implementation, the company has saved money, which can also be counted as a benefit of the Intranet.

Costs avoided.   Many of the advantages of the Intranet will result from its use as a form of electronic publishing. The company can place virtually anything that it prints and distributes on paper onto its internal Web site, thereby reducing printing costs. Visitors to the company Web site can complete and return materials in an electronic format. A driving force in business over the last several decades has been the pursuit of the paperless office. The Intranet may provide a significant step toward that elusive vision.

In addition, the storage area for the boxes of paper and toner cartridges can be reduced. When hard copies are distributed to the employee offices, they must be filed and stored. In addition to reducing the distribution expenses (postage, shipping, and courier costs can be significant), the company plans to rely on the Intranet to reduce travel expenses associated with off-site meetings. The cost of airfare, cabs, hotels, and meals for each trip avoided saves hundred of dollars per person. In the analysis to follow, a per-employee expenditure figure is used to represent the potential cost savings of using the Intranet to avoid unnecessary travel expenses.

Efficiencies created.   The Intranet not only shortens the time between the request for information and its delivery, but it also reduces the number of people involved in delivering the information. While a major advantage of an Intranet lies in its ability to disseminate information across many platforms, the ease with which users can access the information is another crucial advantage. Potentially, the Intranet can also improve the productivity of company employees. In the example that follows, the time employees save performing their required assignments as a result of using the Intranet will be factored into the ROI analysis. If using the Intranet saves an hour, that one hour of salary should be factored in as a benefit.

Costs incurred: Minuses.   Costs fall into two major categories: "hard" costs and "soft" costs. Hard costs are measurable and concrete. For an Intranet, the hard costs include hardware and software purchases and the costs of installing the hardware or software. Soft costs relate to intangible factors, such as lost productivity or inefficiency during the period the users are learning to use the Intranet. Soft costs may be easy to identify, yet difficult to quantify. For example, it is difficult to estimate how long it will take employees to learn the new Intranet technology and to become as proficient as they are with the existing technology. However, both hard and soft expenses must be quantified in concrete numbers for the ROI analysis.

Common expenses associated with the deployment of an Intranet include:

- System personnel: A system administrator is needed to keep the Intranet running, install new equipment, upgrade desktop computers, and maintain system security. While current staff may be able to absorb these responsibilities, a new position may be necessary for larger installations. As the number of employees using the Intranet increases, there may also be an increase in the number of customized software applications needed. An application administrator may also be needed to manage existing software programs and to write custom applications.
- Training: The system and application administrators will need to attend classes and certification programs to stay on the cutting edge of the technology. Users should also have ongoing instruction on the use of the Intranet.

- Ongoing planning: The company should put in place a cross-functional team that continuously guides the Intranet's development. While these team members are involved in planning activities, they are not accomplishing their regular assignments. These costs should be calculated based on their salaries.
- Authoring: The Intranet is only as useful as the information it provides. The company should encourage each employee (possibly in some controlled way) to put information on the internal Web. When employees are authoring such information, they are not performing their regular tasks. The cost of these activities should be based on employee salaries.
- Miscellaneous hardware and software: This should include budget allowances for anticipated and unexpected (i.e., this is a form of fudge factor) hardware and software expenses.

In summary, the ROI numerator (net benefits) can be expressed by the following equation:

Intranet Benefits =

Cost Avoided + Efficiencies Created − Cost Incurred   (A.4)

In the equation above, the cost incurred should be discounted, taking depreciation into account.

ROI denominator

Intranet investment.   The Intranet investment consists primarily of hardware and software expenditures. In the discussion that follows, we assume that the company has an existing network in place. Some companies may have enough surplus equipment to start a small Intranet with no hardware costs and can download free software from Internet sites, so they will have very minimal start-up costs. Even in this situation, the most balanced and accurate ROI analysis will use retail prices for the hardware and software used in the Intranet deployment. Investment costs for an Intranet include:

- Preinstallation planning: The company should formulate a cross-functional team to design the Intranet and its uses. While these team members are defining the Intranet plans, they are not accomplishing their regular assignments. This lost productivity cost should be calculated based on each employee's salary.
- Hardware: The Intranet will require at least one server computer with adequate hard disk space and speed to house the Web sites. The servers will require replacement or expansion during the expected life of the Intranet. In addition, additional servers will be needed to serve as security firewalls.

- Software: The Intranet will require server software, as well as client software (browsers), firewall software, HTML-authoring tools, collaborative groupware, a document-management system, a search engine, and possibly more, depending on the company's needs.
- Installation: The company must install the proper software on each server. The company may also need to install software (i.e., browser software, etc.) on each user's computer system. One or more skilled technicians may spend many hours completing this work.
- Depreciation: As discussed previously, annual depreciation should be subtracted.

To finalize the calculation of the ROI denominator, add all the costs incurred in the original investment and any subsequent years and then subtract all the annualized depreciation expenses. In summary, the ROI denominator is calculated as follows:

Depreciated Investment Costs = Investment Costs − Depreciation   (A.5)

All components of this equation should be expressed in the same time and dollar units.

Sample ROI analysis

In the following example, we provide the details of an ROI analysis for our case study company. Our analysis is based on an expected project life of five years.

ROI numerator.   Refer to Table A.1 for the specific items and costs that are relevant for the calculation of the ROI numerator for this sample problem. A detailed explanation of each benefit and cost line item listed in Table A.1 is included in the discussion that follows. The ten (10) items listed in Table A.1 represent the costs avoided, efficiencies created, and expenses incurred that are factored into the ROI numerator.

Savings and costs related to Intranet project

- Salary savings: This item represents the money saved by making employees more efficient through the use of the Intranet. The company has six basic employee categories: executives, engineering, managers, technicians, clerical, and consultants. Salaries are based on the job category. For this example, we make an informed assumption on the amount of time the Intranet will save each worker during a typical week. We conservatively estimate this amount at one hour per week. The annual savings associated with this is calculated below:

Productivity Savings By Job Category =
(Number of Emplyess × Hourly Salary) × Number of Hours Saved
Per Week × 50 Weeks

For example, for the Executives category, the first year's productivity savings is:

$$\text{Year 1} = (50) \times (\$100 \times 1) \times 50 = \$250,000$$

In each successive year, the Intranet will become increasingly useful, and therefore the company will accrue more benefits over time. We conservatively estimate that the time savings associated with increased employee productivity due to Intranet usage will increase 5% a year.

- Communications savings: This number represents the amount of money the company will save on printed material, travel, telephone, and courier costs per employee per year. For the purposes of this analysis, our estimate is $50 savings per person per year. We also expect that these savings will increase 25% per year in each subsequent year of the Intranet's use.
- System administrator costs: The estimated annual salary expense for one full-time administrator is $75,000 per year. Salary increases in subsequent years two through five are estimated at 10% per year.
- Application developer: The estimated annual salary expense for one full-time developer is $50,000 per year for years one and two. In years two through five, an additional full-time developer will be needed, so that a total of two application developers will be available to support the Intranet project. The salary increases are estimated at 10% per year.
- User training costs: Instructor costs are estimated at $150 per class. All users are scheduled to receive two hours of classroom experience each year. Thus, the total training costs are estimated at $15,000 ($15 X 1,000 users).
- Administrator training costs: Estimates are based on a team of five, working 40 hours each at an estimated hourly salary cost of $50 per person (50*40*5).
- Miscellaneous hardware and software: Unplanned equipment and software needs are estimated at $10,000 per year.
- Salary authoring costs: These costs are calculated based on the assumption that the average time saved by each employee will be one hour per week, and that 20% of the employees will use the time saved in this manner to author Intranet material.
- Depreciation: These costs are estimated by multiplying each previous year's total investment by 0.5 (50%). This high rate reflects the rapid rate of replacement needed to keep the high-tech computer and networking equipment up to current standards.

Table A.1   Total Benefits Calculated for ROI Numerator

Item	Year 1	Year 2	Year 3	Year 4	Year 5	Cumulative
Annual savings						
Salary	$1,587,500	$1,666,875	$1,750,219	$1,837,730	$1,929,616	$8,771,940
Hard costs	$50,000	$60,000	$72,000	$86,400	$103,680	$372,080
Total annual svgs.	**$1,637,500**	**$1,726,875**	**$1,822,219**	**$1,924,130**	**$2,033,296**	**$9,144,020**
Annual expenses						
Systems admin.	$75,000	$150,000	$165,000	$181,500	$199,650	
Apps. developer	$50,000	$55,000	$121,000	$133,100	$146,410	
Training - users	$15,000	$15,000	$15,000	$15,000	$15,000	
Admin costs	$10,000	$10,000	$10,000	$10,000	$10,000	
Hardware/software	$10,000	$10,000	$10,000	$10,000	$10,000	
Content creation	$317,500	$333,375	$350,044	$367,546	$385,923	
Future planning	$0	$15,000	$15,000	$15,000	$15,000	
Depreciation	$80,750	$66,775	$78,900	$78,900	$78,900	
Total annual exps.	**$558,250**	**$655,150**	**$764,944**	**$811,046**	**$860,883**	**$3,650,273**
Total net benefits	**$1,079,250**	**$1,071,725**	**$1,057,275**	**$1,113,084**	**$1,172,413**	**$5,493,747**

Table A.2   Annual Savings Due to Intranet Deployment

Item	# of	Avg. salary/yr.	Total salary	Salary/hr	Savings annual (1 hr/wk.
Annual savings					
Salary - Executive	50	$200,000	$10,000,000	$100	$250,000
Salary - Engineers	400	$50,000	$20,000,000	$25	$500,000
Salary - Managers	100	$75,000	$7,500,000	$38	$187,500
Salary - Technicians	200	$40,000	$8,000,000	$20	$200,000
Salary - Clerical	100	$30,000	$3,000,000	$15	$75,000
Salary - Consultants	150	$100,000	$15,000,000	$50	$375,000
Total	**1000**		**$63,500,000**		**$1,587,500**

ROI denominator.   The ROI denominator represents the total net investment made in the Intranet. Table A.3 lists the 14 items relating to the investment costs and the depreciation costs associated with the Intranet deployment. We describe these items below.

ROI denominator costs related to Intranet project

- Preinstallation planning: These estimates are based on the assumption that a team of 20 working 40 hours each, at an average hourly salary of $50 per person, will be required for this task.

Table A.3 ROI Denominator Items for Intranet Project

Item	Year 0	Year 1	Year 2	Year 3	Year 4	Year 5
			Invstmnt. Exps.			
Initial Planning	$40,000					
Servers	$10,000					
Server software	$1,000					
Client software	$50,000					
Installation costs	$50,000					
Addl. servers		$20,000	$20,000	$20,000	$20,000	$20,000
HTML software		$15,000	$15,000	$15,000	$15,000	$15,000
Software updates			$60,000		$60,000	
Collaborative sftwre.		$1,800		$1,800		$1,800
Dbase connect tool	$500					
Document mngmt.		$6,500		$6,500		$6,500
Search tools		$4,500		$4,500		$4,500
Firewall hardare	$3,000		$3,000		$3,000	
Firewall software	$2,000		$2,000		$2,000	
Misc.	$5,000	$5,000	$5,000	$5,000	$5,000	$5,000
Total investment	$161,500	$52,800	$105,000	$52,800	$131,400	$105,300
Dep. invest cost		$133,550	$131,400	$105,300	$131,400	$105,300
Total net benefits		$1,079,250	$1,071,725	$1,057,275	$1,113,084	$1,172,413
Yearly ROI		808%	816%	1004%	847%	1113%

- Server: These estimates are based on the assumption that a Pentium Pro NT 200Hz, 128Mb DRAM, 8Gb SCSI hard drive with RAID technology will be used as the system server, at an estimated cost of $10,000.
- Server software: The estimate for the required Web Server software is $1,000.
- Client software: The Netscape Navigator browser costs are estimated at $49 per desktop.
- Installation: The installation cost estimate is based on the assumption that 500 hours at $100 per hour (average three client installations per hour plus the server installation) will be needed.
- Additional/replacement servers: Each year the original server must be replaced or modified to support increasing workloads. It is estimated that one new server will be needed each year to provide the necessary capacity to process the system workloads. We will use the original purchase price as a basis for these estimates.
- HTML authoring software: The company plans to purchase software for 25% of desktops each year beginning in year two. The costs for this are estimated at $60 per copy.
- Intranet software upgrades: Software upgrades will be needed for each server and each client in years 2 and 4. The original prices of this software are estimated at $1,000 (10 servers) and $49, respectively.

- Collaborative software: This expense is estimated at $1,800, and reflects purchases planned for years 1, 3, and 5.
- Database connectivity tools: This expense is estimated at $500.
- Document management system: This expense is estimated at $6,500, to be incurred at the time of purchase in years 2 and 4.
- Search engine: The costs for the search engine are estimated at $4,500, to be purchased in years 1, 3, and 5.
- Firewall server and software: Firewall servers, estimated to cost $3,000, will be purchased in years 1, 3, and 5. The associated firewall software is estimated to cost $2,000, and will be incurred over the same three-year period.
- Depreciation: Given the rapid rate of replacement and technical innovation, the total investment is depreciated 50% per year. The denominator of the ROI formula is computed by multiplying 0.5 times the previous year's total investment plus the current year's total investment.

Results

The results of the ROI analysis indicate that the Intranet project is an exceptional investment opportunity. From the first year of the project, to the last year of the project, the ROI is estimated at a low of around 800% to a ROI of over 1,100%. Thus, the Intranet promises to be an important tool to augment the company productivity, and a strong business case can be made for proceeding with the project. The final ROI calculations for years one through five of the project are summarized in Table A.4 below.

Table A.4 Final ROI Calculations for Intranet Project

	Year 1	Year 2	Year 3	Year 4	Year 5
Annualized ROI	808%	816%	1,004%	847%	1,113%

The results for our hypothetical case study are not inconsistent with the results achieved, in general, by companies making use of the Internet, since it is not uncommon for companies to realize huge returns on their Intranet investments. In fact, most companies find that within 10 to 12 weeks they have fully recovered the start-up costs associated with Intranet deployment. The Netscape web site cites several examples of the returns obtained by companies from their Intranet investments. A few of these examples include: Cadence Designs Systems, Inc. with a ROI of 1,766%, Booz, Allen & Hamilton with a ROI of 1,389%, Amdahl Corporation with a ROI of 2,063%, and Silicon Graphics, Inc. with a ROI of 1,427%.

Index